BIOLOGICAL ACTIVITIES AND APPLICATION OF MARINE POLYSACCHARIDES

Edited by **Emad A. Shalaby**

Biological Activities and Application of Marine Polysaccharides
http://dx.doi.org/10.5772/62752
Edited by Emad A. Shalaby

Contributors

Noemi Zaritzky, Jimena Bernadette Dima, Cynthia Sequeiros, Rajendra Sukhadeorao Dongre, Melissa Gurgel Adeodato Vieira, Meuris Gurgel Carlos Da Silva, Thiago Lopes Da Silva, Jacyara Moreira Martins Vidart, Marcelino Luiz Gimenes, Fabio Zuluaga, Gustavo Muñoz, Nechita Petronela, Claudia Casalongué, David Lucio, María Cristina Martínez-Ohárriz, Alberto Tecante, G. Thirumurugan, M.D. Dhanaraju, Sherif Amr, Basma Ekram, Fabien Salaün, Ada Ferri, Stéphane Giraud, Jagadish Chandra Roy, Jinping Guan, Laura Orzali, Luca Riccioni, Beatrice Corsi, Cinzia Forni, Waldo Argüelles-Monal, Maricarmen Recillas Mota, Daniel Fernández-Quiroz, Zurina Osman, Abdul Kariem Arof

Published by InTech

Janeza Trdine 9, 51000 Rijeka, Croatia

Publishing Process Manager Dajana Pemac

Technical Editor SPi Global

Cover InTech Design team

First published January, 2017

Printed in Croatia

Additional hard copies can be obtained from orders@intechopen.com

Biological Activities and Application of Marine Polysaccharides, Edited by Emad A. Shalaby

p. cm.

Print ISBN 978-953-51-2859-5
Online ISBN 978-953-51-2860-1

For EU product safety concerns :
IN TECH d . o . o . , Prolaz Marije Krucifikse Kozulić 3 , 51000 Rijeka , Croatia ,
info@intechopen . com or visit our website at intechopen . com .

Meet the editor

Dr. Emad A. Shalaby is an associate professor (specialty: Biochemistry) in the Biochemistry Department Faculty of Agriculture, Cairo University. He received a short-term scholarship to carry out his postdoctoral studies abroad, from JICA, Japan to Egyptian government. He fluently speaks two languages: English (TOEFL, 520 points) and Arabic (the native language). Furthermore, he has 73 international published research papers, and he has attended 13 international conferences. Moreover, he is a member of seven international specialized scientific societies besides the local one. He has 13 international books and chapters. He has six prizes. He works as a reviewer in more than 100 international journals, and he joined the editorial board for more than 15 international journals.

Contents

Preface

Marine organisms have been under research for the last decades as a source for different active compounds with various biological activities and application in agriculture, pharmacy, medicine, environment, and industries. Marine polysaccharides from these active compounds are used as antibacterial, antiviral, antioxidant, anti-inflammation, bioremediations, etc. During the last three decades, several important factors that control the production of phytoplankton polysaccharides have been identified such as chemical concentrations, temperature and light.

This book includes 14 chapters contributed by experts around the world; the chapters are categorized into three sections: Marine Polysaccharides and Agriculture, Marine Polysaccharides and Biological Activities, and Marine Polysaccharides and Industries.

Section 1 (Chapters 1–2) describes the bioactivity and application of marine polysaccharides in agriculture. Chapter 1 presents chitosan as one form of marine polysaccharides used for pesticide formulations; Chapter 2 discusses chitosan as an active molecule that finds many possibilities for application in agriculture, including plant disease control.

Section 2 (Chapters 3–9) focuses on biological activities and application of marine polysaccharides in medicine and pharmacy.

Section 3 (Chapters 10–14) describes the properties and application of marine polysaccharides in industries. Chapter 10 focuses on chitosan application in wastewater treatments as well as the preliminary results on its chemical modification to obtain and utilize zeolite—chitosan composites in adsorption of organic pollutants from industrial wastewaters. Chapter 11 discusses the flow behavior and viscoelastic properties of l-carrageenan in aqueous solution, without or with the addition of sodium to characterize the polyelectrolyte behavior of the polysaccharide, and evaluates the effect of its concentration and that of the counterion on these behaviors. Chapter 12 emphasizes on chitosan-based formulations of fibers, fabrics, coatings, and functional textiles. Chapter 13 focuses on the production of chitosan-based thermosensitive materials, as well as their most relevant physicochemical properties and applications. Chapter 14 discusses chitosan and only the phthaloyl chitosan derivative and their applications in several electrochemical devices.

I would like to thank all the contributing authors for their time and great efforts in the careful construction of the chapters and for making this project realizable.

I am grateful to Ms. Dajana Pemac (Publishing Process Manager) for her great efforts and encouragement and guidance during my preparation of the current book.

Finally, I would like to express my deepest gratitude to my wife (Dr. Ghada M. Azzam) and my daughters for their kind cooperation and encouragement, which helped me in the completion of this book.

Dr. Emad A. Shalaby
Biochemistry Department
Faculty of Agriculture
Cairo University, Cairo, Egypt

Section 1

Marine Polysachharides and Agriculture

Chapter 1

Chitosan as Source for Pesticide Formulations

Sebastián D. Ippólito, Julieta R. Mendieta,
María C. Terrile, Claudia V. Tonón,
Andrea Y. Mansilla, Silvana Colman,
Liliana Albertengo, María S. Rodríguez and
Claudia A. Casalongué

Additional information is available at the end of the chapter

http://dx.doi.org/10.5772/65588

Abstract

Late blight and wilt caused by the oomycete, *Phytophthora infestans*, and the fungus, *Fusarium solani* f. sp. *eumartii*, respectively, are severe diseases in Solanaceae crops worldwide. Although traditional approaches to control plant diseases have mainly relied on toxic chemical compounds, current studies are focused to identify more sustainable options. Finding alternatives, a low molecular weight chitosan (LMWCh) obtained from biomass of Argentine Sea's crustaceans was assayed. In an attempt to characterize the action of LMWCh alone or in combination with the synthetic fungicide Mancozeb, the antimicrobial properties of LMWCh were assayed. In a side-by-side comparison with the SYTOX Green nucleic acid stain and the nitric oxide–specific probe, diaminofluorescein-FM diacetate (DAF-FM DA), yielded a similar tendency, revealing LMWCh-mediated cell death. The efficacy of LMWCh, Mancozeb, and the mixture LMWCh–Mancozeb was in turn tested. A synergistic effect in the reduction of *F. eumartii* spore germination was measured in the presence of subinhibitory dosis of 0.025 mg ml^{-1} LMWCh and 0.008 mg ml^{-1} Mancozeb. This mixture was efficient to increase the effectiveness of the single treatments in protecting against biotic stress judged by a drastic reduction of lesion area in *P. infestans*–inoculated tissues and activation of the potato defense responses.

Keywords: chitosan, *Fusarium* f. sp. *Eumartii*, Mancozeb, *Phythophtora infestans*, potato, tomato

1. Introduction

The Solanaceae species potato (*Solanum tuberosum* L.) and tomato (*Solanum lycopersicum*) are important horticulture crops. The oomycete, *Phytophthora infestans* (Mont.) de Bary and the fungus, *Fusarium solani* (Mart.) Sacc. f. sp. *eumartii* (Carp.) Snyder & Hansen, isolate 3122 (*F. eumartii*) are potentially pathogens in both Solanaceae species [1]. Late blight and fusariosis caused by *P. infestans* and *F. eumartii*, respectively, are frequently controlled with toxic and chemical fungicides. Mancozeb is a broad-spectrum contact fungicide commonly used to control early and late blights, rusts, downy mildews, and black spots, including *Fusarium* wilt [2]. In potato, Mancozeb treatments have been used by application rates of 2 and 4 kg ha^{-1} [3]. In the US, approximately 3.4 million kg of Mancozeb is applied annually in agriculture. The Mancozeb breakdown metabolite, ethylene thiourea (ETU) is an industrial contaminant [4]. ETU has been shown to produce tumors, birth defects, cell mutations, and thyroid effects in human and animals.

Despite the valuable contribution of Mancozeb to control plant diseases, alternatives and more sustainable options are still hot topics in phytopathology. In this sense, chitosan has been proved as a nontoxic and environmental-friendly compound for agricultural uses [5]. Chitosan is a linear polysaccharide composed of randomly distributed β-(1-4)-linked D-glucosamine and *N*-acetyl-D-glucosamine. This polymer is obtained from chitin which is commonly isolated from the crustacean exoskeletons by enzymatic or nonenzymatic procedures. Chitosan of high and low molecular weights has been found to be differentially effective against fungal diseases [5, 6]. The mode of action of chitosan is rather variable, depending on its biological target and interacting lipids in the cell plasma membrane [7, 8]. Chitosan in combination with different chemicals (e.g., antioxidants, saccharine, essential oil) has been successful for the control of foliar diseases in cucumber, cantaloupe, pepper, and tomato [9]. Low-level copper and chitosan have been provided to confer protection against late blight in potato [10].

This work provides data from *in vitro* and *in vivo* studies revealing insights into the water-soluble chitosan, low molecular weight chitosan (LMWCh) action in phytopathogens and plants. An effective action of LMWCh in combination with suboptimal doses of Mancozeb for the control of late blight was demonstrated. Additional field trials could also provide knowledge on its efficiency and environmental implications.

2. Materials and methods

2.1. Isolation and characterization of chitosan

Chitin was isolated from shells waste of the shrimp *Pleoticus mülleri* from Argentine seacoast. Chitosan was prepared by heterogeneous deacetylation of chitin with 19 M NaOH. The water-soluble LMWCh (2.764 kDa) with a degree of deacetylation (DD) of 68% was prepared from chitosan by oxidative degradation with 1 M H_2O_2 under microwave irradiation (700 W) for 4 min. The molecular mass was estimated by using Mark-Houwink-Kuhn-Sakurada's equation [11]. The sample contained 13.3% moisture and 2.60% ash content [12].

2.2. Biological materials

Solanum tuberosum L. var. Spunta plants were grown to the seven-leaf stage under greenhouse conditions at 18°C with 16:8 h light/dark cycles in a growth chamber. It has been described as a susceptible genotype to *P. infestans* [13]. Expanded leaves from the third to the sixth node of potato plants were excised at the stem [14] and used for detached leaflet assays. *P. infestans* race R2 R3 R6 R7 R9, mating type A2 was cultivated as described by Andreu et al. [15], Andreu et al. [16], and Lobato et al. [17]. *F. eumartii* was obtained from Agricultural Experimental Station, INTA, Balcarce (Argentine). The fungus was grown as described by Terrile et al. [18].

2.3. Chemical treatments and inoculations

Commercial Mancozeb (800 g kg^{-1}) WP (Dow AgroSciences, Argentina) was used at 8 mg ml^{-1} (8000 ppm) or at 0.08 mg ml^{-1} and 0.008 mg ml^{-1} in aqueous solution. Each solution was sprayed with a hand-held spray separately, or in combination with 0.25 mg ml^{-1} LMWCh on detached leaflets placed in Petri dishes. Controls were sprayed with distilled water. After being sprayed, leaflets were incubated at 18°C with light intensity of 120 μmol photons m^{-2} s^{-1} with 16:8 h light/dark cycles in a growth chamber. Each treatment combination was tested in a series of three independent trials.

After chemical treatments on detached leaflets, 10 μl of spore suspension *containing* 2×10^4 spores ml^{-1} was infiltrated in the center of the abaxial side of each lateral leaflets using a 1 ml needleless syringe. Petri dishes with leaflets were sealed with parafilm to maintain 90–100% relative humidity and incubated at 18°C. After 5 days, diameter of lesion area was measured in each leaflet using the image-processing software ImageJ (NIH, Maryland, USA).

2.4. *P. infestans* sporangium germination assay

The sporangium germination assay was conducted as described by Mendieta et al. [19].

2.5. Measurements of *F. eumartii* spore germination

Determination of *F. eumartii* spore germination was performed as described by Mendieta et al. [19].

2.6. Measurements of endogenous NO production in *F. eumartii* spores

Determination of endogenous NO production was monitored by incubating 10^7 spores ml^{-1} in 20 mM of HEPES-NaOH, pH 7.5, with different concentrations of LMWCh or LMWCh in combination with NO-specific scavenger, 2-(4-carboxyphenyl)-4,4,5,5,-tetramethylimidazoline-1-oxyl-3-oxide (cPTIO, 1 mM) as described in Terrile et al. [18].

2.7. Cell death stain and fluorescence microscopy

F. eumartii spore suspension (2×10^5 spores ml^{-1}) was incubated with different concentrations of LMWCh for 16 h and 100% relative humidity at 25°C in the dark. SYTOX Green probe

(Molecular Probes, Thermo Fisher Scientific, USA) was added at a final concentration of 1 mM, and detection of SYTOX Green uptake was done after 30 min of incubation with a Nikon Eclipse E200 fluorescence microscope equipped with a B-2A Fluorescein filter set. SYTOX® Green nucleic acid stain is a green-fluorescent nuclear and chromosome counterstain that penetrates the compromised membranes characteristic of dead cells.

2.8. Protein extraction and western blot assay

Total proteins from potato leaflets were extracted as described by Terrile et al. [18]. Protein samples were boiled for 5 min and running on SDS-PAGE 12% polyacrylamide gels according to the method of Laemmli [20]. Proteins were transferred onto nitrocellulose using a semi-dry blotter (Invitrogen, USA) [21]. Immunodetection was performed using polyclonal antibodies raised against chitinase [22].

2.9. Statistical analysis

Treatments were established in a randomized complete block design, typically with four to seven treatments in each trial. The values shown in each figure are mean values ± SD of at least three experiments. Data were subjected to analysis of variance (one-way ANOVA), and *post hoc* comparisons were done with Tukey's multiple range test at $p < 0.05$ level. SigmaStat 3.1 was used as the statistical software program. Limpel's formula as described by Richer [23] was used to determine synergistic interactions between LMWCh and Mancozeb. This formula corresponds to $E_e = X + Y - (XY. 10^{-2})$, in which E_e is the expected effect from additive response of two treatments, and X and Y are the percentages of inhibition of germination relative to each agent used alone. Thus, if the combination of the two agents produces any value of inhibition of germination greater than E_e, the synergism does exist [24].

3. Results

3.1. LMWCh exerts antimicrobial action on *P. infestans* and *F. eumartii*.

The high water solubility and easy handling for agricultural application make LMWCh an attractive compound to deepen on its properties in the control of plant diseases. In this work, we hypothesized that LMWCh exerts protection against biotic stress in Solanaceae species, and in turn, it can be used in combination with reduced doses of Mancozeb (hundred times less than recommended dosage) for utilization in putative safer formulations. With the intention to move in that direction, we characterized the antimicrobial properties of LMWCh on *P. infestans* and *F. eumartii* as phytopathogen models of potato and tomato, respectively. LMWCh exerted inhibitory activity on germination of both *P. infestans* and *F. eumartii* in a dose-dependent mode (**Figure 1**). The estimated IC_{50} value (the concentration needed to inhibit half of the maximum spore germination) was IC_{50} = 0.1 μg ml^{-1} LMWCh and IC_{50} = 130 μg ml^{-1} LMWCh for *P. infestans* and *F. eumartii*, respectively.

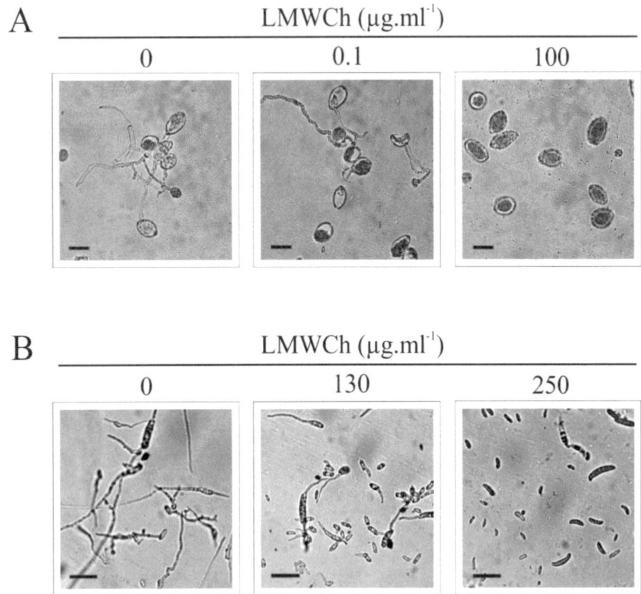

Figure 1. LMWCh affects germination of reproductive structures in phytopathogenic microorganisms. *P. infestans* sporangia (A) and *F. eumartii* spores (B) were incubated with different concentrations of LMWCh for 16 h and 100% relative humidity at 18°C in the dark. Scale bar: 30 μm (A) and 25 μm (B). Values are the mean (±SD) of three independent experiments.

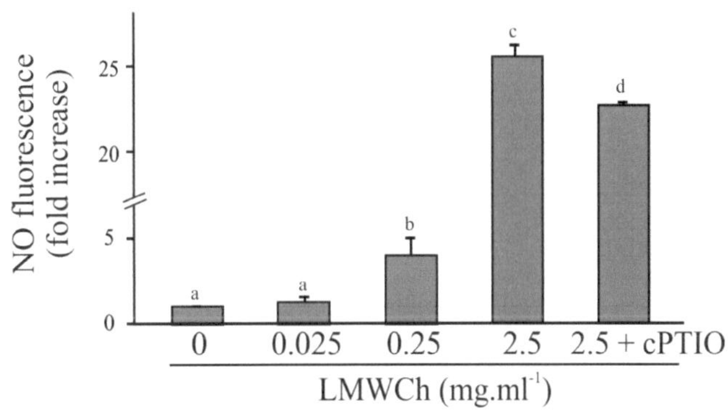

Figure 2. LMWCh induces NO production in *F. eumartii* spores. Spore suspension was incubated with different concentrations of LMWCh in combination with the NO scavenger, cPTIO for 16 h. Then, the suspension was loaded with the NO-specific fluorescent probe DAF-FM DA, and the fluorescence was determined in a fluorometer. Data are expressed as the fold increase with respect to the control. Values are the mean (±SD) of at least three independent experiments. Different letters indicate statistically significant differences (Tukey's test, $p < 0.05$).

In turn, due to the easy handle and reproducibility, *F. eumartii* was used to characterize LMWCh-mediated biological action. NO-specific fluorophore diaminofluorescein-FM diacetate (DAF-FM DA) was assayed to compare endogenous NO production in LMWCh- and non-treated spores as a biomarker of cytotoxicity (**Figure 2**) [18]. Exogenous LMWCh application ranging from 0.25 to 2.5 mg ml^{-1} induced NO production indicating a dose-dependent LMWCh-mediated effect on *F. eumartii* spores. Since DAF-FM DA fluorescence was significantly reduced in the presence of the NO scavenger, cPTIO, we propose that endogenous NO production is a downstream event upon LMWCh treatment in fungal spores.

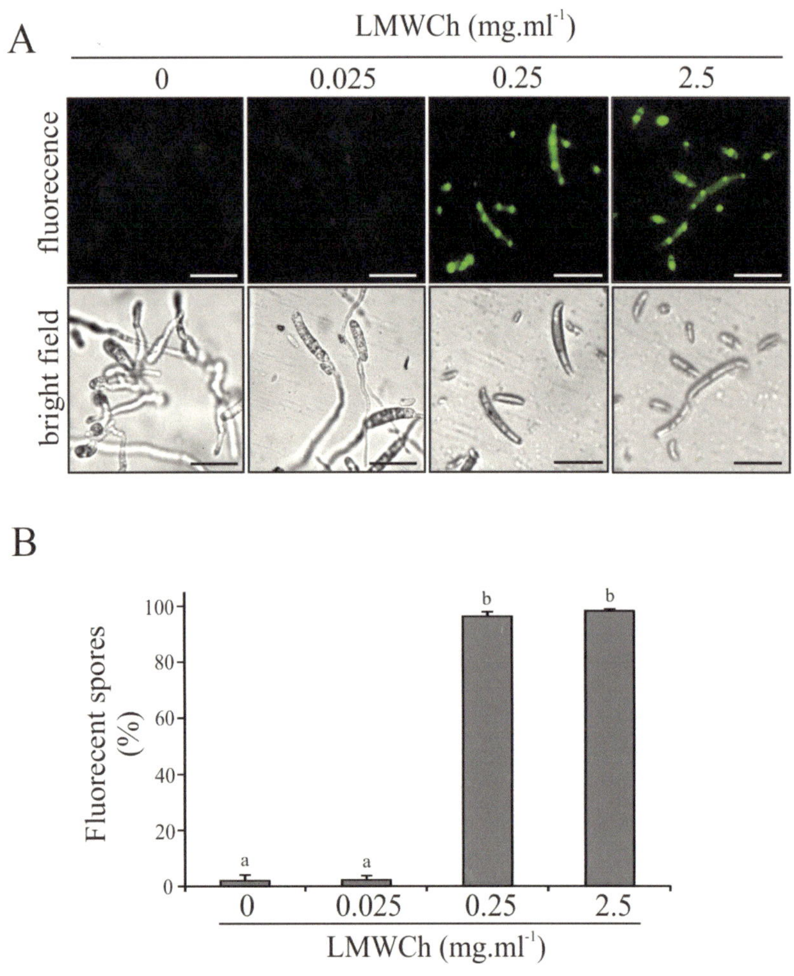

Figure 3. LMWCh treatment induces cell death in *F. eumartii* spores. Fungal spores were exposed to different concentrations of LMWCh for 16 h and then incubated with the fluorescent probe SYTOX Green. (A) Dead cell was visualized as green fluorescence. Pictures show general phenomena representative of at least three individual experiments. A bright field image for each treatment is shown below fluorescent images. Scale bar: 25 μm. (B) Quantification of fluorescent spores. Values are the mean (±SD) of at least three independent experiments. Different letters indicate statistically significant differences (Tukey's test, $p < 0.05$).

SYTOX Green dye is often used to distinguish between live and dead cells. The fluorescence emission of SYTOX Green stain was measured in the presence of increasing concentrations of LMWCh. A substantial enhancement was measured in 0.25 and 2.5 mg ml^{-1} LMWCh-treated spores (**Figure 3**).

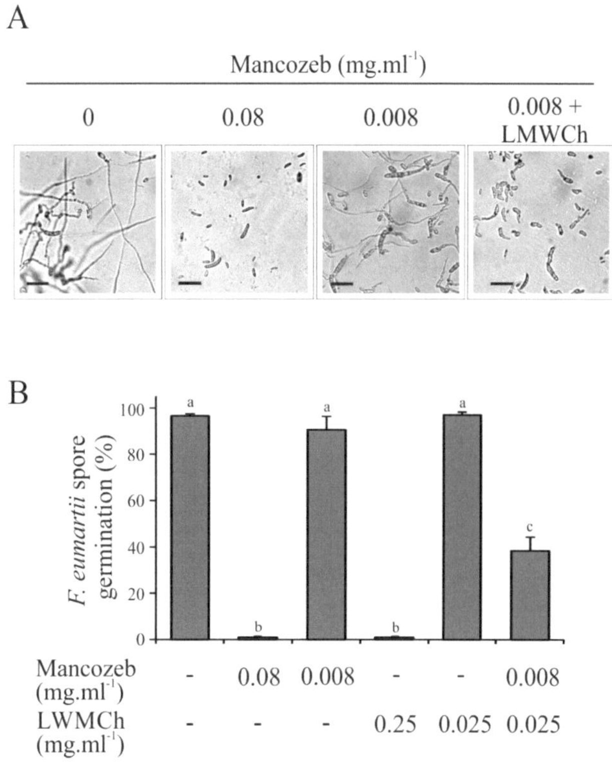

Figure 4. LMWCh plus Mancozeb have a synergistic action on *F. eumartii* spore germination. *F. eumartii* spores were incubated with different concentrations of LMWCh, Mancozeb, or a combination of both, for 16 h and 100% relative humidity in the dark. (A) Pictures show general phenomena representative of at least three individual experiments. Scale bar: 25 μm. (B) Quantification of spore germination. Values are the mean (±SD) of at least three independent experiments. Different letters indicate statistically significant differences (Tukey's test, $p < 0.05$).

Then, LMWCh at 0.025 and 0.25 mg ml^{-1} and Mancozeb at 10^{-1} and 10^{-2} dilutions from the recommended field dosage were tested on fungal spore germination (**Figure 4**). Compared with control, the combination of suboptimal doses of 0.025 mg ml^{-1} LMWCh and 0.008 mg ml^{-1} Mancozeb had a remarkable inhibitory activity on spore germination (**Figure 4**). The treatment with the combined solution resulted in a synergistic effect in the reduction of spore germination. According to Limpel's formula, the $E_e$value calculated for percent reduction of germinated spores using LMWCh and Mancozeb alone and their combination was 61.51%. Thus, we proposed that the mixture of LMWCh and Mancozeb exerted a synergistic effect on germination of *F. eumartii* spores.

3.2. LMWCh potentiates Mancozeb effect and protects against late blight in potato

The effect of LMWCh and Mancozeb treatments, alone or in combination, to control late blight in potato was tested. Meanwhile, single treatments revealed no significant differences, and the combined treatment of 0.25 mg ml^{-1} LMWCh and 0.08 mg ml^{-1} Mancozeb evidenced a reduction of at least sixfold in the late blight symptoms on potato leaflets (**Figure 5**).

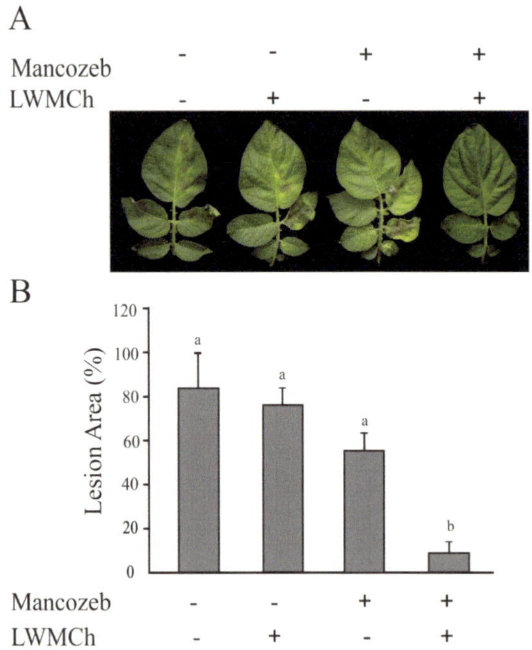

Figure 5. LMWCh and Mancozeb protect plant from microbial attack. Potato leaves were pretreated with 0.25 mg ml^{-1} LMWCh, 0.08 mg ml^{-1} Mancozeb or a combined solution of LMWCh and Mancozeb and then inoculated with *P. infestans*. (A) Representative images from leaflets at 5 days upon *P. infestans* inoculation. (B) Quantification of lesion area; 100% represents the total area in each leaflet. Values are the mean (±SD) of at least three independent experiments. Different letters indicate statistically significant differences (Tukey's test, $p < 0.001$).

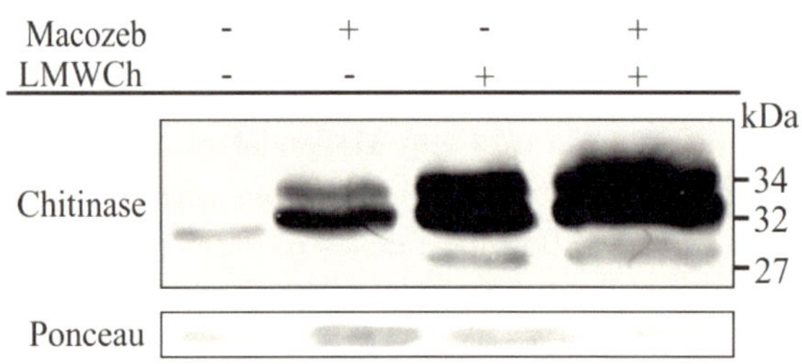

Figure 6. LMWCh and Mancozeb activate plant defense responses. Potato leaves were pretreated with 0.25 mg ml^{-1} LMWCh, 0.08 mg ml^{-1} Mancozeb, or the combined solution of LMWCh and Mancozeb. Western blot analysis was assayed with antichitinase antibody (upper panel). Ponceau staining was used as loading control (lower panel). Picture is representative of two independent experiments.

With the aim to provide broader evidence on LMWCh-mediated mechanism, we assayed its elicitor properties in potato plants (**Figure 6**). Chitinases, the well-known pathogenesis-related proteins used as defense markers, are constitutively expressed at low levels, but highly induced by biotic stresses in potato leaves [22, 25]. Meanwhile, the 32 kDa chitinase was only detected at very low level in control; it moderately increased in Mancozeb-treated leaflets and was remarkably high in LMWCh-treated leaflets. Otherwise, the highest levels of chitinase isoforms were detected in potato leaflets upon LMWCh and Mancozeb treatment. Particularly, in the presence of LMWCh, the level of a third 27 kDa isoform was increased, suggesting that specific defense proteins are elicited by LMWCh in potato tissues (**Figure 6**).

4. Discussion

LMWCh was effective as an antimicrobial compound on spores from both *P. infestans* and *F. eumartii* under assayed conditions. As *spores* were treated with LMWCh, the SYTOX Green signal went up. The fact that NO production has been postulated as stress signals and is detected in LMWCh-treated spores allowed us to suggest that it probably downstream modulates cell toxicity in *F. eumartii* [18]. The combination of LMWCh and Mancozeb had a synergistic effect in reducing spore germination and also remarkably induced chitinase accumulation in potato leaflets. The sensitivity to LMWCh as judged by its estimated IC_{50} values on the reduction of germination was higher in the oomycete than in fungal spores. The inhibition of *P. infestans* [26] and fungal growth by chitosan has been extensively demonstrated [7, 18, 27–29]. Thus, the positive effect of LMWCh on the control of late blight might derive from the combination of its antimicrobial and eliciting properties.

In this work, we demonstrated that the addition of LMWCh to an ineffective Mancozeb dose became a highly effective treatment to control late blight. The synergistic effect of LMWCh and commercially synthetic fungicides has been showed on *Botrytis cinerea, Alternaria brassicicola*, and *Muocor piriformis* [30]. The combination of chitosan and ethanol in reduced doses showed an effective control of gray mold [31]; however, the mechanism of action has not still been elucidated. Due to the intrinsic polycationic chemistry of chitosan, we speculated that LMWCh might modify cell membrane permeability [32] and in turn facilitate the uptake of Mancozeb when both compounds are applied simultaneously, on plant tissues. In our work, consistent with the less damage, a potentiated accumulation of three chitinase isoforms was detected on potato leaflets, allowing us to suggest that LMWCh and Mancozeb might activate natural defense mechanisms and maximize the known action of chitosan as plant elicitor [33]. Similar accumulation patterns were also obtained in tomato seedlings subjected to identical treatments (data not shown). In addition to chitinases, the host defense responses elicited by chitosan might include glucanase activation, cell wall lignifications, phytoalexin biosynthesis, and generation of reactive oxygen species, among others [33]. Majid et al. [34] demonstrated that Mancozeb induces antioxidant compounds in *Cassia angustifolia* Vahl. Considering that the antioxidant metabolism is also activated during the plant defense response and adding the fact that chitosan is a suitable candidate to enhance antioxidative enzymes activities [35], we also speculated that summative antioxidant activity might be exerted by Mancozeb and LMWCh on the plant tissue. Since at least two degraded metabolites, ETU and ethylene urea from Mancozeb, have been found on potato tissue [36], we cannot rule out that Mancozeb derivatives also had action on potato defense mechanisms. It has also been accepted that the risk of *P. infestans* resistance against Mancozeb is rather low [37]. Thus, chitosan applied in combination with Mancozeb could represent a useful strategy with additional advantages within environmental-friendly plant protection strategies.

5. Conclusion

In summary, water-soluble LMWCh is an effective antimicrobial compound on spores from both *P. infestans* and *F. eumartii*. NO is a downstream stress signal mediating LMWCh cytotoxicity in spores. In addition, the combination of LMWCh and the synthetic fungicide Mancozeb showed an effective control late blight in potato leaflets. However, major studies are needed on new agro-inputs and environmental science to provide the conditions for the use of more eco-friendly pesticide formulations in plant crops.

Author details

Sebastián D. Ippólito†[1], Julieta R. Mendieta†[1], María C. Terrile†[1], Claudia V. Tonón[1], Andrea Y. Mansilla[1], Silvana Colman[1], Liliana Albertengo[2], María S. Rodríguez[2] and Claudia A. Casalongué[1*]

*Address all correspondence to: casalong@mdp.edu.ar

1 Institute for Biological Research, Faculty of Natural Sciences, UE CONICET-UNMDP, National University of Mar del Plata, Mar del Plata, Argentina

2 Department of Chemistry, Basic Research and Chitin Application Laboratory, INQUISUR UE CONICET-UNS, National University of SouthBahía Blanca, Argentina

† Equal contribution to this work.

References

[1] Nowicki M, Foolad MR, Nowakowska M, Kozik EU. Potato and tomato late blight caused by *Phytophthora infestans*: an overview of pathology and resistance breeding. Plant Disease. 2011;96: 4–17.

[2] Hibar K, El Mahjoub M. Evaluation of fungicides for control of *Fusarium* wilt of potato. Plant Pathology Journal. 2006;5: 239–243.

[3] Caldiz D, Rolon D, Di Rico J, Andreu A. Performance of dimethomorph + Mancozeb applied to seed potatoes in early management of late blight (*Phytophthora infestans*). Potato Research. 2007;50: 59–70.

[4] Burwell SM. Report on Carcinogens, 13th ed., Department of Health and Human Services, Public Health Service, Research Triangle Park, NC, USA; 2014.

[5] Badawy ME, Rabea EI. A biopolymer chitosan and its derivatives as promising antimicrobial agents against plant pathogens and their applications in crop protection. International Journal of Carbohydrate Chemistry. 2011;1–29.

[6] Tikhonov VE, Stepnova EA, Babak VG, Yamskov IA, Palma-Guerrero J, Jansson H-B, Lopez-Llorca LV, Salinas J, Gerasimenko DV, Avdienko ID. Bactericidal and antifungal activities of a low molecular weight chitosan and its N-/2 (3)-(dodec-2-enyl) succinoyl/-derivatives. Carbohydrate Polymers. 2006;64: 66–72.

[7] Palma-Guerrero J, Jansson HB, Salinas J, Lopez-Llorca LV. Effect of chitosan on hyphal growth and spore germination of plant pathogenic and biocontrol fungi. Journal of Applied Microbiology. 2008;104: 541–553.

[8] Palma-Guerrero J, Lopez-Jimenez JA, Pérez-Berná AJ, Huang IC, Jansson HB, Salinas J, Villalaín J, Read ND, Lopez-Llorca LV. Membrane fluidity determines sensitivity of filamentous fungi to chitosan. Molecular Microbiology. 2010;75: 1021–1032.

[9] Abdel-Kader M, El-Mougy NS, Aly M, Lashin S. Integration of biological and fungicidal alternatives for controlling foliar diseases of vegetables under greenhouse conditions. International Journal of Agriculture and Forestry. 2012;2: 38–48.

[10] Hadwiger LA, McBride PO. Low-level copper plus chitosan applications provide protection against late blight of potato. Plant Health Progress. 2006;22: 150–168.

[11] Wagner HL. The Mark–Houwink–Sakurada equation for the viscosity of atactic polystyrene. Journal of Physical and Chemical Reference Data. 1985;14: 1101–1106.

[12] Vela Gurovic M, Staffolo MD, Montero M, Debbaudt A, Albertengo L, Rodríguez M. Chitooligosaccharides as novel ingredients of fermented foods. Food and Function. 2015;6: 3437–3443.

[13] Fernández MB, Pagano MR, Daleo GR, Guevara MG. Hydrophobic proteins secreted into the apoplast may contribute to resistance against *Phytophthora infestans* in potato. Plant Physiology and Biochemistry. 2012;60: 59–66.

[14] Goth R, Keane J. A detached-leaf method to evaluate late blight resistance in potato and tomato. American Potato Journal. 1997;74: 347–352.

[15] Andreu AB, Guevara MG, Wolski EA, Daleo GR, Caldiz DO. Enhancement of natural disease resistance in potatoes by chemicals. Pest Management Science. 2006;62: 162–170.

[16] Andreu AB, Caldiz DO, Forbes GA. Phenotypic expression of resistance to *Phytophthora infestans* in processing potatoes in Argentina. American Journal of Potato Research. 2010;87: 177–187.

[17] Lobato MC, Machinandiarena MF, Tambascio C, Dosio GA, Caldiz DO, Daleo GR, Andreu AB, Olivieri FP. Effect of foliar applications of phosphite on post-harvest potato tubers. European Journal of Plant Pathology. 2011;130: 155–163.

[18] Terrile MC, Mansilla AY, Albertengo L, Rodríguez MS, Casalongué CA. Nitric-oxide-mediated cell death is triggered by chitosan in *Fusarium eumartii* spores. Pest Management Science. 2014;71: 668–674.

[19] Mendieta JR, Pagano MR, Munoz FF, Daleo GR, Guevara MG. Antimicrobial activity of potato aspartic proteases (StAPs) involves membrane permeabilization. Microbiology. 2006;152: 2039–2047.

[20] Laemmli UK. Cleavage of structural proteins during the assembly of the head of bacteriophage T4. Nature. 1970;227: 680–685.

[21] Mahmood T, Yang P-C. Western blot: technique, theory, and trouble shooting. North American Journal of Medical Sciences. 2012;4: 429–434.

[22] Kombrink E, Schröder M, Hahlbrock K. Several "pathogenesis-related" proteins in potato are 1, 3-β-glucanases and chitinases. Proceedings of the National Academy of Sciences. 1988;85: 782–786.

[23] Richer DL. Synergism—a patent view. Pesticide Science. 1987;19: 309–315.

[24] Romanazzi G, Nigro F, Ippolito A. Short hypobaric treatments potentiate the effect of chitosan in reducing storage decay of sweet cherries. Postharvest Biology and Technology. 2003;29: 73–80.

[25] Wu C-T, Bradford KJ. Class I chitinase and β-1, 3-glucanase are differentially regulated by wounding, methyl jasmonate, ethylene, and gibberellin in tomato seeds and leaves. Plant Physiology. 2003;133: 263–273.

[26] Atia M, Buchenauer H, Aly A, Abou-Zaid M. Antifungal activity of chitosan against *Phytophthora infestans* and activation of defence mechanisms in tomato to late blight. Biological Agriculture and Horticulture. 2005;23: 175–197.

[27] Palma-Guerrero J, Huang IC, Jansson HB, Salinas J, Lopez-Llorca LV, Read ND. Chitosan permeabilizes the plasma membrane and kills cells of *Neurospora crassa* in an energy dependent manner. Fungal Genetics and Biology. 2009;46: 585–594.

[28] Meng X, Yang L, Kennedy JF, Tian S. Effects of chitosan and oligochitosan on growth of two fungal pathogens and physiological properties in pear fruit. Carbohydrate Polymers. 2010;81: 70–75.

[29] Mandal S, Kar I, Mukherjee AK, Acharya P. Elicitor-induced defense responses in *Solanum lycopersicum* against *Ralstonia solanacearum*. The Scientific World Journal. 2013;1–9.

[30] Rahman MH, Shovan LR, Hjeljord LG, Aam BB, Eijsink VG, Sørlie M, Tronsmo A. Inhibition of fungal plant pathogens by synergistic action of chito-oligosaccharides and commercially available fungicides. PLoS One. 2014;9: e93192.

[31] Romanazzi G, Karabulut OA, Smilanick JL. Combination of chitosan and ethanol to control postharvest gray mold of table grapes. Postharvest Biology and Technology. 2007;45: 134–140.

[32] Falcón-Rodríguez AB, Wégria G, Cabrera J-C. Exploiting plant innate immunity to protect crops against biotic stress: chitosaccharides as natural and suitable candidates for this purpose. InTech Open Access Publisher, Croatia; 2012.

[33] El Hadrami A, Adam LR, El Hadrami I, Daayf F. Chitosan in plant protection. Marine Drugs. 2010;8: 968–987.

[34] Majid U, Siddiqi TO, Iqbal M. Antioxidant response of *Cassia angustifolia* Vahl. to oxidative stress caused by Mancozeb, a pyrethroid fungicide. Acta physiologiae plantarum. 2014;36: 307–314.

[35] Yang F, Hu J, Li J, Wu X, Qian Y. Chitosan enhances leaf membrane stability and antioxidant enzyme activities in apple seedlings under drought stress. Plant Growth Regulation. 2009;58: 131–136.

[36] Environmental fate of Mancozeb. 2000. Available from: http://www.cdpr.ca.gov/docs/emon/pubs/fatememo/mancozeb.pdf [Accessed: 2016-03-01].

[37] Kato M, Mizubuti ES, Goodwin SB, Fry WE. Sensitivity to protectant fungicides and pathogenic fitness of clonal lineages of *Phytophthora infestans* in the United States. Phytopathology. 1997;87: 973–978.

Chapter 2

Chitosan in Agriculture: A New Challenge for Managing Plant Disease

Laura Orzali, Beatrice Corsi, Cinzia Forni and
Luca Riccioni

Additional information is available at the end of the chapter

http://dx.doi.org/10.5772/66840

Abstract

In recent years, environmental-friendly measures have been developed for managing crop diseases as alternative to chemical pesticides, including the use of natural compounds such as chitosan. In this chapter, the common uses of this natural product in agriculture and its potential uses in plant disease control are reviewed. The last advanced researches as seed coating, plant resistance elicitation and soil amendment applications are also described. Chitosan is a deacetylated derivative of chitin that is naturally present in the fungal cell wall and in crustacean shells from which it can be easily extracted. Chitosan has been reported to possess antifungal and antibacterial activity and it showed to be effective against seedborne pathogens when applied as seed treatment. It can form physical barriers (film) around the seed surface, and it can vehicular other antimicrobial compounds that could be added to the seed treatments. Chitosan behaves as a resistance elicitor inducing both local and systemic plant defence responses even when applied to the seeds. The chitosan used as soil amendment was shown to give many benefits to different plant species by reducing the pathogen attack and infection. Concluding, the chitosan is an active molecule that finds many possibilities for application in agriculture, including plant disease control.

Keywords: chitosan, seed treatment, soil amendment, plant resistance elicitation, induced defence, elicitor

1. Introduction

The control and management of crop plants diseases has always been considered a subject of great interest for the huge economic losses associated with them. For many years, the control of pathogens has been performed mainly through the application of chemical pesticides, due to their easy application, the relatively low cost and the broad spectrum of action. Pesticides application for crop defence has been widely used since post-war years and led to a large yield growth in agriculture, contributing to economical development, reducing endemic diseases and protecting/restoring plantations, forests, harvested wood products [1]. In fact, plant diseases represent a critical problem to successful production. Agricultural productivity has demonstrated to get benefits from the utilization of pesticides both at quantitative and healthy level, e.g. when pesticides are properly used, they contribute to the higher production and quality characteristics of crops. However, the advantages in their use comprise several drawbacks problems related to two main aspects: the human health and the environmental impact. In fact, the chemical plant protection products, including even copper that is allowed in organic agriculture, are mostly toxic, persistent, bio-accumulative and extremely harmful not only for human health, but also for many living organisms [2]. Pesticides can contaminate environmental matrices coming up the aquifer [3], causing direct and permanent damage to the ecosystem. In addition, there is the real possibility that their residues can get into the food chain of consumers [4]. The massive use of these chemical substances has also favoured the emergence of resistance phenomena in the major crop pests [5] and the contemporary disappearance of many pests' natural enemies, such as bumblebees, butterflies and bees [6].

The pesticides application and their related effects constitute a topic of major concern, so that, according to the new Europe directive in favour of a sustainable agriculture, many plant protection products currently in use will be replaced with lower environmental impact substances. For this reason, many scientific works and researches have been focused in developing alternative approaches to the use of pesticides for managing crop diseases, through experimentations that have followed different paths including physical methods [7] integrated pest management and biological control [8]. A promising approach consists in the use of natural compounds such as plant extracts and their active principles (alkaloids, phenols, monoterpenes and sesquiterpenes, isoprenoids), which have been studied for their various antifungal, antibacterial and antioxidant properties [9–11] and animal derived compounds like chitosan. It has proved to be very interesting for controlling plant diseases [12]. In fact, it has been shown both to possess a broad-spectrum antimicrobial activity against several phytopathogenic organisms and to induce numerous biological responses in plants [13].

2. Chitosan

Chitosan is a linear polysaccharide that can be obtained from the deacetylation of chitin, a long-chain polymer of N-acetyl-glucosamine present and easily extracted from fungal cell wall and crustacean shells (**Figure 1**). From a practical viewpoint, the shells of marine crustaceans such as crabs and shrimps are very affordable for a commercial production of chitin. They represent a practical challenge because they are available as waste from the seafood processing industry. Recent advances in fermentation technologies suggest that the cultivation of selected fungi can provide an alternative source of chitin and chitosan [14].

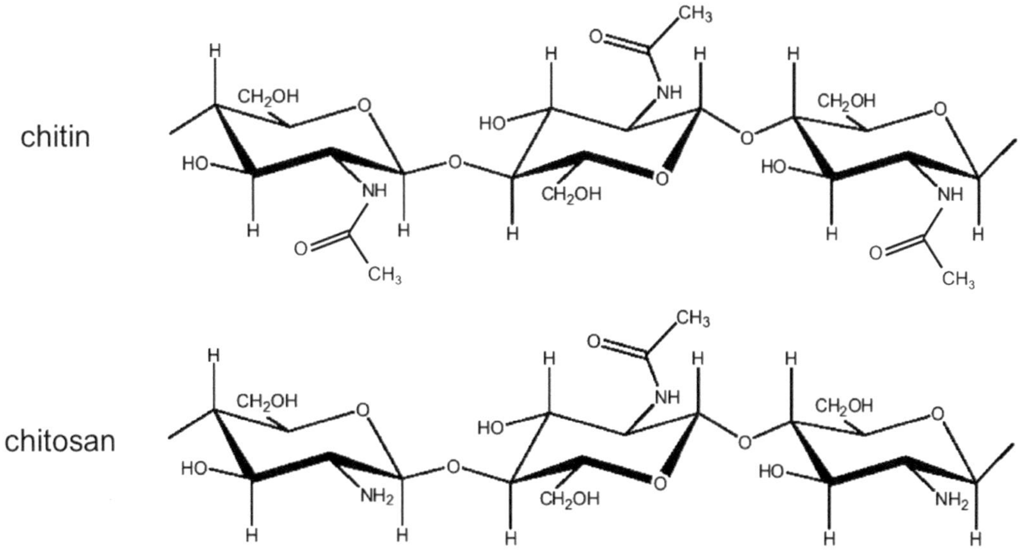

Figure 1. The structure of the molecules of chitin (β-(1-4)-N-acetyl-D-glucosamine) and chitosan (β-(1-4)-D-glucosamine) that results from the partial deacetylation of chitin.

In addition to its low cost production, chitosan also possesses other biological properties such as non-toxicity, biocompatibility and biodegradability, which make chitosan a sustainable and eco-friendly molecule.

Chitin is the second renewable carbon source after lignocelluloses biomass and, in fact, about 1600 tons of chitin are annually produced [15]. For industrial production, solid chitin is soaked in 40–50% (w/v) NaOH. This process removes more than 80% of the acetyl residues and converts N-acetyl-D-glucosamine into β-1,4-D-glucosamine. Complete deacetylation is possible by repeating the alkaline treatment. Therefore, the term "chitosan" is not uniquely related to a defined compound, but to a group of commercially available copolymers that are heterogeneous for deacetylation degree, molecular mass, polymerization degree and acid dissociation constant (pKa value) [13]. These different characteristics, in particular the degree of deacetylation and the molecular weight, influence the physic-chemical properties (including viscosity and solubility), and they have a direct influence on the biological properties of the substance and the effects in plants and pathogens.

All these characteristics make chitosan very useful for several industrial applications, namely cosmetology, food, biotechnology, pharmacology, medicine and, more recently, agriculture [16].

Based on the current state of research and progress in agriculture, this chapter will consider the potential uses of this natural compound in plant disease control, based mainly on a dual mode of action involving a direct antimicrobial activity and an indirect resistance induction that elicit several defence responses in the plants. The upgrading of the plant defence mechanisms against pathogen represents a very innovative approach for crop protection [15], and it will be described separately. Moreover, the last advanced researches such as seed treatment and soil amendment applications will be described. Postharvest fruit application will not be included because extensively discussed in the literature [17].

2.1. Antimicrobial activity

One of the most studied proprieties of chitosan is its high antimicrobial activity against a wide variety of microorganisms such as fungi, bacteria and viruses (**Table 1**). An antimicrobial substance is defined as a substance that kills or inhibits the growth of microorganisms [18].

A broad spectrum fungicidal activity of chitosan has been described: it inhibits *in vitro* fungal growth of many pathogenic fungi, for example, *Botrytis cinerea, Alternaria alternata, Colletotrichum gleosporoides* and *Rhizopus stolonifer*. The inhibition was observed at different pathogen development stages such as mycelial growth, sporulation, spore viability and germination, and on the production of fungal virulence factors [14]. Moreover, the antifungal activity was also demonstrated *in vivo* in many different plant-pathogen systems, such as in pear against *A. kikuchiana* and *Physalophora piricola* [19], in grapevine and in strawberry against *B. cinerea* [20, 21], and in dragon fruit against *C. gleosporoides* [22]. In rice, the antifungal activity against *R. solani* was further demonstrated by transmission electron microscope observations and pathogenicity testing [23].

Chitosan also prevents the growth of several pathogenic bacteria including *Xanthomonas* [24], *Pseudomonas syringae* [25] *Agrobacterium tumefaciens* and *Erwinia carotovora* [26]. However, the antimicrobial effectiveness of chitosan seems to be higher against fungi than bacteria [27], and among bacteria often been higher against Gram-positive than Gram-negative ones, even if this efficiency is somewhat controversial. This could be explained by the different structure of the bacteria surface and cell wall composition [28].

Besides these activities, chitosan is able to inactivate the replication of viruses and viroids thus limiting their spread [29], even though relatively few research studies on its antiviral activity have been reported [30].

The exact mechanism of the direct antimicrobial action of chitosan is still ambiguous, and different mechanisms have been proposed and described [14, 28], but none of them are mutually exclusive. The main mode of action proposed is related to its cationic properties [31] hypothesis supported by the lack of antifungal activity of uncharged chitin oligomers [32]. In fact, unlike chitosan, the polymeric form of chitin is naturally uncharged, and it does not show substantial antimicrobial activity. Basing on this model, the positive charges on the chitosan molecules interact with negatively charged pathogen surfaces (electrostatic interactions); this leads to the cell structure destruction, causing an extensive cell surface alterations and increasing membrane permeability [33–35]. Another proposed mechanism involved the alteration of cell permeability by chitosan that includes its deposition onto the pathogen cell surface, and consequent creation of an impermeable polymeric layer that prevents the uptake of nutrients in the cell, and in the meantime changes of the metabolite excretion in the extracellular matrix [28].

Chitosan is also able to chelate some essential nutrients, metal ions and trace elements necessary for bacterial and fungal growth [12, 28], inhibiting thereby toxin production and microbial growth [36].

2.2. Plant resistance elicitation

All plants, whether they are resistant or susceptible, respond to pathogen attack by the induction of a coordinate signalling system, which results in the accumulation of different gene products. The responses to pathogen attack are effective at different levels: at first, the pathogen recognition leads to the development in the plant of a rapid localized cell death, also known as the hypersensitive response (HR), that causes necrosis at the site of infection (local response). Then, even in uninfected parts of the plants, a systemic expression of a broad spectrum and long-lasting resistance against further pathogen infection is triggered. This leads to the production of reactive oxygen species (ROS), the activation of defence-related genes as well as an enhanced expression of genes related to the production of molecules, such as phytoalexins, terpens, pathogenesis-related (PR) proteins and many enzymes involved in defence mechanisms (phenylalanine ammonia-lyase [PAL], polyphenol-oxidases [PPOs] and peroxidases such as guaiacol-peroxidase (G-POD) and ascorbate peroxidase [APX]) [37–39].

The signals able to trigger the defence mechanisms in plants are called elicitors, and they can be produced in the site of infection both by infected plant cells (endogenous elicitors, released by the plants upon contact with the pathogen) and by the pathogen itself (exogenous pathogenic elicitors). They consist of several compounds including oligosaccharides, lipids, peptides and proteins, and they are capable, even at low concentrations, to act as signal molecules inducing the plant to trigger the defensive responses [40]. The elicitors, once recognized by transmembrane receptors of plant cells, induce an immune response, both locally (around the infection site/application) and systemic, through the translocation of signalling molecules in distal tissues [41]. The increasing knowledge of the mechanisms underlying the plant response to pathogen attacks has supported the idea that it is possible to achieve a broad-spectrum disease control and an increased protection against virulent pathogens by artificially inducing the plant's own resistance mechanisms using substances with elicitor function. It is now well-documented that treatment of plants with various eliciting agents leads to an induced resistance against subsequent pathogen attacks, both locally and systemically. Therefore in order to enhance plant resistance in agriculture the use of elicitors is becoming a very attractive and eco-friendly tool that could provide efficient alternatives to the usage of chemical pesticides for managing plant diseases, thus reducing their environmental negative impacts.

Moreover, it is known that chitosan at low molecular weight acts as a potent biotic elicitor, able to induce plant defence responses and to activate different pathways that increase the crop resistance to diseases [13, 15, 28, 42, 43]. The most studied plant responses to chitosan treatment are the formation of chemical and mechanical barriers and the synthesis of new molecules and enzymes involved in the defence response [15, 37]. In some cases, chitosan causes the induction of the hypersensitive response, mainly around the infection site, that leads to the programmed cell death [44]. This hypersensitive response can be followed by systemic response of the plant defence mechanisms. These latter mainly include the synthesis and accumulation of secondary metabolites with active roles in defence: phenolic compounds such as lignin, callose, phytoalexins, PR proteins (pathogenesis-related proteins) and the modulation of the activity of key enzymes of metabolic pathways involved in the defensive response, such as the PAL, peroxidases and chitinase (**Figure 2**) [45–48].

The mechanism of action of chitosan and the responses induced by the latter in plant-pathogen interaction is not yet entirely clear. As written above, the plant through transmembrane cell receptors recognizes the elicitors, but the specific receptor for the chitosan has not yet been identified [37]. Protein kinase cascades that may relay the signal to transcription factors (TFs) have not been identified as well. Various models have been proposed to explain the role of chitosan in the activation of plant defence genes. A proposed direct elicitation of gene activity in plant defence implicates chitosan/DNA interactions (**Figure 2**). The proposed predictive models assume that chitosan induces the activation of defence genes by modifying the structure of DNA (chromatin remodelling) along with reductions in the architectural transcription factor high mobility group HMG A [43, 49] or by the interaction with the DNA polymerase complex [50].

The defensive responses that are induced by chitosan treatment may depend on the pathosystem and, even in the same crop, on numerous factors, including the type of treatment application (**Table 1**).

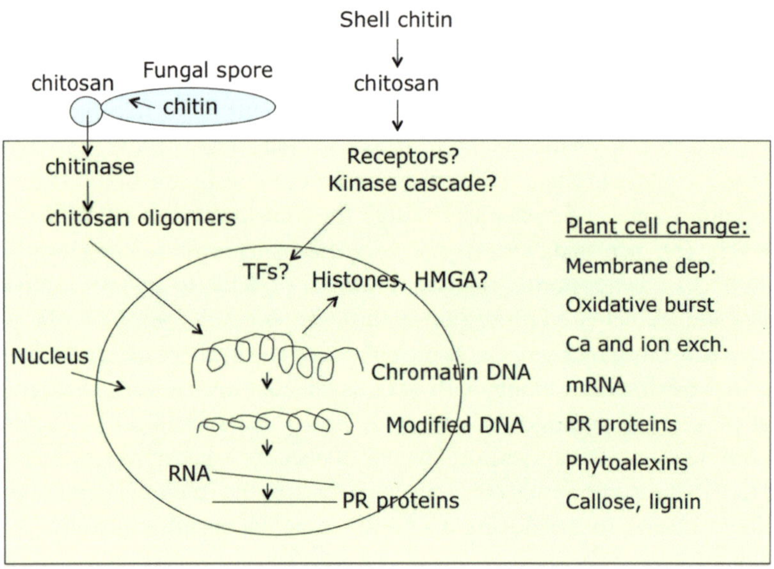

Figure 2. Some proposed effects of chitosan as elicitor of plant defence responses [43]. The cellular and molecular changes elicitated by chitosan can be summarized in: membrane depolarization, oxidative burst, influx and exit of ions such as Ca^{2+}, activation of MAP-kinases, chromatin and DNA alteration, increase in PR gene mRNA, PR proteins synthesis, phytoalexins accumulation, lignification and callose deposition. TFs: transcriptional factors. HMG A: architectural transcription factor high mobility group. PR proteins: pathogenesis-related proteins.

3. Chitosan seed treatments

In agriculture, "Seed treatments are the biological, physical and chemical agents and techniques applied to seed to provide protection and improve the establishment of healthy crops" [International Seed Federation (ISF)]. It represents the first line of defence for seeds and seedlings against pests infecting the seed teguments or living in the soil (seedborne and soilborne pathogens). It provides protection during the critical stage of germination and the very first seedling development, when seeds and seedlings are unable to protect themselves from invasive pathogens [51]. The substance applied to the seeds can be of various nature such as chemical pesticides, biochemical substance, natural compounds [8, 52], and there are many different techniques that can be used for this purpose. Among them, seed coating and dressing represent a common procedure of seed treatment applied for preventing diseases and pests [53, 54] other than improving the seedling performances, i.e. the seedling emergence time, synchronized emergence, improve germination percentage, emergence rate and yield in many field crops [52, 55–57].

The technique of seed dressing involves the application on the seed surface of thin layer of the active product, such as pesticides, fertilizers or growth promoters [57], often in combination with other additives. These other components may include colour (effective pigment), filming agents, surfactants or tackifiers. These products come in dry formulations as powders or liquid formulations are also available for sprays, dips, fluid drilling gels and solid matrix priming.

Chitosan represents an interesting prospective in this field because it could cover different aspects thanks to its variety of properties mainly related to the molecular weight [15]. The high molecular weight confers to chitosan biopolymer characteristics, so that it can be used as film, forming physical barriers (film) around the seeds preventing the pathogen infection [58, 59].

The low molecular weight chitosan posses high antimicrobial activity, which increases with weight decreasing, demonstrated against a wide variety of microorganisms such as bacteria, yeast and fungi [60, 61] even if some controversial evidences for a correlation between bactericidal activity and chitosan molecular weight have been found [62]. Thanks also to its ability to induce plant resistance, low molecular weigh chitosan has great potential as protector against diseases.

An interesting application on seed is the use of chitosan as film coating as a delivery system for fertilizers, plant protection products and micronutrients for crop growth promotion [13]. Chitosan in fact can vehicular and protect other antimicrobial compounds such as essential oils. Essential oils have demonstrated antimicrobic activity [10, 11, 63] but are very volatile and their incorporation into coating can ensure a better persistence of the active ingredient on the surface and maintain high concentration of active molecules [64].

For example, different kinds of chitosan seed coatings with or without essential oils like thyme (*Thymus vulgaris*) and tea tree (*Melaleuca alternifolia*) essential oils, incorporated at different concentrations and applied with different thickness, have been studied for controlling disease and reduce the risk of pathogen attack [65]. The treatment effectiveness in reducing fungal development was evaluated both on *Fusarium* spp. naturally infected wheat seeds and on seeds artificially infected with *F. graminearum*, one of the causal agents of root and foot rot in cereals (**Figure 3**). Results showed that coating treatment with essential oils reduced fungal growth on seeds without affecting germinability and lowered severity on seedlings at the first developing stages. Scanning electron microscope studies allowed to monitor the superficial structure stability of the coating treatment on seeds after the imbibition processes [65].

Another main application of chitosan as seed treatment concerns the elicitation of the systemic resistance in plants. Basing on recent evidences, chitosan, when applied as a seed treatment, behaves as a resistance elicitor, inducing a physiologically enhanced defensive ability in seedlings and plants, whereby the plant's innate defences are potentiated [45, 66]. Chitosan is able to cross the seed teguments, probably by diffusing through microscopic ruptures caused by the imbibition [67] and to interact with embryo cells, influencing the plant cellular metabolism in the following stages of development. Using radio labelled chitosan, it has been showed that the chitosan applied to seed is transferred to the emerging seedling during their development [68]. The major effects produced by the seed/chitosan interaction can be summarize as follows: (a) the seeds germination index is enhanced, (b) the mean germination time and flowering time are reduced; (c) plant growth (e.g. shoot height, root length, and seedling, vegetative growth vigour) and biomass are increased [52]. In maize [69], rice [70] and wheat [71], the chitosan seed treatment increased the germination percentage and stress tolerance, and improved the vigour of the seedlings. In artichoke [53], chitosan seeds treatment resulted in a better growth of the seedlings (e.g. longer and better developed radicle and greener hypocotyls) and lower chance of being infected by fungi in comparison with the untreated seeds. The observed growth improvement by chitosan could be also related to the incorporation of nutrients (nitrogen) from chitosan. Chitosan seed treatment is also able to increase the content of important resistance markers, like phenols and the activities of defence-related enzyme, thus improving the plant resistance. Biochemical analyses on durum wheat and sunflower confirmed the ability of chitosan to induce plant defence increasing PAL, PPO, peroxidases, and chitinase activities and phenolic content in seedling. The enhancement of these plant defence mechanisms suggests the activation of systemic resistance processes. Laboratory results on the chitosan-induced resistance were also confirmed under field and greenhouse condition, where an enhancement of the number of emerged plants (**Figure 4**) and a reduc-

Figure 3. Blotter test for seed health analysis after 7 days of incubation at 25°C of durum wheat seeds artificially infected with *F. graminearum*, one of the causal agents of root and foot rot in cereals. Seeds were then coated with a solution of chitosan and tea tree oil. The chitosan/tea tree oil treatment reduced significantly the fungal infection on seeds (right) compared with the inoculated and not treated seeds (left).

Figure 4. Parcels of field trial (Italy) sowed with durum wheat seeds cv. Simeto artificially infected with *F. graminearum* and then treated with chitosan (CH) or with a copper sulphate-based treatment (CU), commonly used as chemical treatment in organic farming. Chitosan treatment enhanced the number of emerged plants, if compared with no treatment (NT), as well as copper treatment.

tion in the disease severity (root and foot rot and downy mildew, respectively) were observed [45, 66]. Chitosan seed treatment also induced lignifications process, considered one of the first line of defences in plant-pathogen interaction: in chilli, the lignin content of seedlings obtained from chitosan-treated seeds was higher than that of untreated ones [72], thus giving to the plant a major protection against potentially penetration of invasive pathogen.

Chitosan seed treatment can also be effective in insect control because it stimulates plants to produce systemic antibodies with repellent effects on insect pests, as reported in soybean against *Agrotis ypsilon*, soybean pod borer and soybean aphid. The chitosan treatment developed an antifeedant rate of more than 80% against all these insects, together with increases in seed germination, plant growth and soybean yield [73].

4. Chitosan as soil amendment

As previously described, chitosan can be used in several ways to reduce plant disease levels and prevent the development and spread of diseases, thus preserving crop yield and quality. Chitosan as soil amendment was found to successfully decrease *Fusarium* wilt in several plant species [33, 74, 75]. Similar results were reported against *Cylindrocladium floridanum* [76], *Alternaria solani* [77] and *Aspergillus flavus* infections [78] after soil treatment with chitosan. Part of the observed effect of chitosan on the reduction in these pathogens comes from the fact that it enhances plant defence responses. It has been demonstrated that chitosan acts as an elicitor of plant systemic immunity by the accumulation of defence-related antimicrobial compounds, and it plays an important role in the activation of induced resistance [43]. For example, the innate immunity induction was observed in micropropagated kiwi plants after the addition of chitosan to the growth medium [46]. This includes the modulation of several enzyme activities, involved in detoxification processes as well as the increasing of the activity of enzymes involved in plant defence barriers (G-POD, APX, PAL and PPO) [46]. In addition, chitosan amendments were reported to induce callose formation, proteinase inhibitors and phytoalexin biosynthesis in many dicot species [79–81].

The amendment of soil with chitosan is eco-friendly, since in the soil chitosan can be degraded at a substantial rate, due to the enormous abundance and diversity of bacteria in most soils and the presumed presence of chitinases in a considerable fraction of the bacterial populations. Chitin degradation is mainly a bacterial process [82–85]. However, it still remains unknown the wideness of chitinolytic process due to soil bacteria population with different chitin degradation and whether fungi can also play a major role in this process. Works on microbial community members hypothesized a role of chitin in stimulating bacterial communities to a greater extent than the fungal ones [85, 86]. Among the bacteria genera isolated from chitin-treated soils, there were *Streptomyces* [86], *Stenotrophomonas* and *Bacillus* [87, 88]. However, due to the complexity of bacterial response in soil treated with chitin, the mechanisms behind the observed effects, in particular what are the active bacterial community, the timing of the chitinolytic activity and bacterial succession is still poorly understood. Probably, members of the actinobacteria have a key role in the degradation of complex organic molecules like chitinosan in field [88, 89].

In field conditions, chitosan alters the equilibrium of the rhizosphere, disadvantaging microbial pathogens and promoting the activity of beneficial microorganisms, such as *Bacillus*, *Pseudomonas fluorescens*, actinomycetes, mycorrhiza and rhizobacteria [90–93]. For instance, soil treatment with chitin and/or chitosan from shrimp waste has been shown to decrease the rate of infection of plant roots by nematodes [82, 84] and to enhance the suppressiveness against soilborne diseases [91]. Although not definitely proven in all cases, the mechanisms behind chitosan effectiveness are most often related to a change in the structure and/or activity of soil microbiota [94, 95]. Two hypotheses have been formulated regarding the response of the soil microbial communities to chitosan addition: (a) chitinolytic microorganisms, which are capable of hydrolyzing the chitinous hyphae of pathogenic fungi, increased their numbers and/or activities (b) secondary responders to the added chitosan have a detrimental activity against pathogens [96].

The beneficial effect of the chitosan seems to be linked not only to its impact on soil microbiota, but also on plant itself. Recently, an innovative bioremediation strategy uses the ability of chitosan to chelate minerals and other nutrients, making them more available for the uptake by the plant [97, 98]. This is important, since crop production is many times limited by low availability of essential mineral elements [99]. In agreement with this strategy, in Ref. [100], the effect of using chitosan oligosaccharides as a soil conditioner was demonstrated on the flowering and fruit growth of purple passion fruit. It was found that this soil conditioner increased significantly the numbers of flowers, fruit weight and juice production. The inclusion of soluble chitosan to hydroponic fertigation streams also promoted the growth and final yield of hydroponically cultivated potato microtubers [101].

Thus, if chitosan can increase absorption of essential minerals, enhancing the plant's nutritional value (biofortification), it is possible that it can also help plants to take up higher concentrations of toxic elements. In fact, the ability of chitosan to chelate certain ions also makes it an interesting compound to be used in phytoremediation. In Refs. [102, 103], the remediation of metal contaminated soil using chitosan as soil amendment was shown to be possible (**Table 1**).

Plant	Disease/pathogen	Activity/defence response	Application	References
Pear	*A. kikuchiana* and *P. piricola*	Antifungal activity	Growth medium addition and postharvest treatment on fruits	[19]
Grapevine Strawberry	*B. cinerea*	Antifungal activity	Growth medium addition and preharvest spray treatment	[20–21]
Dragon fruit	*C. gleosporoides*	Antifungal activity	Growth medium addition and spray treatment on plants	[22]
Rise	*R. solani*	Antifungal activity	Growth medium addition, seed treatment and treatment on plants	[23]
	Xanthomonas, P. syringae, A. tumefaciens and *E. carotovora*	Bacterial growth inhibition	Growth medium addition and treatment on plants	[24–26]
Wheat	Root and foot rot/*F. graminearum*	Enhancing of phenol content and G-POD, APX, PAL and PPO activities; decreased disease incidence	Seed treatment	[45]
Kiwi	Healthy plants	Modulation of G-POD, APX, PAL and PPO activities	Growth medium addition	[46]
Maize	Healthy plants	Increased plant growth and biomass	Seed coating	[52]
Artichoke	Healthy plants	Enhanced seedlings growth	Seed coating	[53]
Soybean	*Agrotis ypsilon,* soybean pod borer and soybean aphid	Stimulation of systemic antibodies production with repellent effects	Seed treatment	[73]
Sunflower	Downy mildew/*P. halstedii*	PAL, PPO, peroxidases, and chitinase activities and phenolic content in seedling	Seed treatment	[66]
Chilli	*Colletotrichum* sp.	Increased lignin content	Seed treatment	[72]
Tomato	*Fusarium oxysporum* f. sp. *radicis-lycopersici*	Decreased disease severity	Soil addition	[74]
Celery	*Fusarium* wilt	Decreased disease severity	Soil addition	[75]
Purple passion fruit	Healthy plants	Increased flowers number, fruit weight and juice production	Soil addition	[100]
Potato	Healthy plants	Improved growth and final yield	Hydroponic fertigation	[101]

Table 1. Listing of some possible applications of chitosan in agriculture and the related effects (activity and plant defence responces).

5. Conclusion

The chitosan is an active molecule that finds many possible applications in agriculture with the aim of reducing or replacing more environmentally damaging chemical pesticides. Although it is a good alternative even in conventional farming, chitosan applications would find interesting opportunities particularly in organic farming, disadvantaged by the lack of effective tools for managing biotic diseases. The plant disease control in organic farming, especially those caused by fungal and bacterial pathogen, is currently based on copper treatments. However, the research of an ecological alternative is mandatory because of the environmental impact problems related to the use of this heavy metal. Thus, chitosan could represent an innovative eco-friendly strategy for managing plant diseases and replacing copper or reducing its use, thanks to its several properties such as those previously described. In fact, several studies have demonstrated the effectiveness of chitosan in protecting plants from biotic stresses by direct and/or indirect actions, but its interaction with pathogens and plants are still not fully understood. Chitosan application in the field, including formulation aspects, is one of the least studied issues and it needs further testing and validation.

Acknowledgements

The authors wish to thank the Office PQAI I "Organic farming and National food quality Systems and General Affairs" of the Italian Ministry of Agriculture, which funded the project "Copper reduction Strategies and possible alternatives to its use in organic farming - ALT. RAMEINBIO", within which the reported study was realized.

Author details

Laura Orzali[1], Beatrice Corsi[1], Cinzia Forni[2], and Luca Riccioni[1,*]

*Address all correspondence to: luca.riccioni@crea.gov.it

1 Council for Agricultural Research and Economics, Plant Pathology Research Center (CREA-PAV), Rome, Italy

2 Department of Biology, University of Rome Tor Vergata, Rome, Italy

References

[1] Ecobichon D.J. Our changing perspectives on benefits and risks of pesticides: a historical overview. Neurotoxicology. 2000;**21**(1–2), 211–218.

[2] Aktar W., Sengupta D. and Chouwdhury A. Impact of pesticides use in agriculture: benefits and hazards. Interdiscip. Toxicol. 2009;**2**(1):1–12.

[3] Haria A.H., Hodnett M.G. and Johnson A.C. Mechanisms of groundwater recharge and pesticides penetration to a chalk aquifer in Southern England. J. Hydrol. 2003;**275**(1):122–137.

[4] Boobis A.R., Ossendorp B.C., Banasiak U., Hamey P.Y., Sebestyen I. and Moretto A. Cumulative risk assessment of pesticides residues in food. 180. Toxicol. Lett. 2008;**180**:137–150.

[5] Hollingworth R.M. and Dong K. The biochemical and molecular genetic basis of resistance to pesticides in arthropods. In: Whalon, M.D., Mota-Sanchez, D., Hollingworth, R.M., editors. Global Pesticide Resistance in Arthropods. Michigan State University USA. 2008. p. 40–89.

[6] Van Lenteren J.C. Need for quality control of mass-produced biological control agents. In: van Lenteren J.C., editor. Quality Control and Production of Biological Control Agents: Theory and Testing Procedures. Wageningen University, The Netherlands. 2003. p. 1–18.

[7] Schmitt A., Koch E., Stephan D., Kromphardt C., Jahn M., Krauthausen H.J., Forsberg G., Werner S., Amein T., Wright S.A.I., Tinivella F., van der Wolf J. and Groot S.P.C. Evaluation of non-chemical seed treatment methods for the control of *Phoma valerianellae* on lamb`s lettuce seeds. J. Plant Dis. Prot. 2009;**116**(5):200–207.

[8] Tinivella F., Hirata L.M., Celan M.A., Wright S.A.I., Amein T., Schmitt A., Koch E., Van Der Wolf J.M., Groot S.P.C., Stephan D., Garibaldi A. and Gullino M.L. Control of seed-borne pathogens on legume by microbial and other alternative seed treatments. Eur. J. Plant Pathol. 2009;**123**:139–151.

[9] Isman M.B. Plant essential oils for pest and disease management. Crop Prot. 2000;**19**:603–608.

[10] Riccioni L. and Orzali L. Activity of tea tree (*Melaleuca alternifolia,* Cheel) and thyme (*Thymus vulgaris,* Linnaeus.) essential oils against some pathogenic seed borne fungi. J. Essent. Oil Res. 2011;**23**(6):43–47.

[11] Marinelli E., Orzali L., Lotti E. and Riccioni L. Activity of some essential oils against pathogenic seed borne fungi on legumes. Asian J. Plant Pathol. 2012;**6**(3):66–74.

[12] El Hadrami A., Adam L.R., El Hadrami I. and Daayf F. Chitosan in plant protection. Mar. Drugs. 2010;**8**:968–987.

[13] Malerba M. and Cerana R. Chitosan effects on plant systems. Int. J. Mol. Sci. 2016;**17**:996.

[14] Badawy M.E.I. and Rabea E.I. A biopolymer chitosan and its derivatives as promising antimicrobial agents against plant pathogens and their applications in crop protection. Int. J. Carbohydr. Chem. 2011;**2011**:29.

[15] Falcón-Rodríguez A.B., Wégria G. and Cabrera J.C. Exploiting plant innate immunity to protect crops against biotic stress: Chitosaccharides as natural and suitable

candidates for this purpose. In: Ali R. Bandani, editors. New Perspectives in Plant Protection. InTech. Rijeka, Croatia. 2012;**7**:139–166.

[16] Hamed I., Ozogul F. and Regenstein J.M. Industrial applications of crustacean by-products (chitin, chitosan, and chitooligosaccharides): A review. Trends Food Sci. Technol. 2016;**48**:40–50.

[17] Zhang H., Li R., and Liu W. Effects of chitin and its derivative chitosan on postharvest decay of fruits: A review. Int. J. Mol. Sci. 2011;**12**(2):917–934.

[18] Andrews J.M. Determination of minimum inhibitory concentrations. J. Antimicr. Chemoth. 2001;**48**:5.

[19] Meng X.H., Yang L.Y., Kennedy J.F. and Tian S.P. Effects of chitosan and oligochitosan on growth of two fungal pathogens and physiological properties in pear fruit. Carbohydr. Polym. 2010;**81**:70–75.

[20] Feliziani E., Landi L., and Romanazzi G. Preharvest treatments with chitosan and other alternatives to conventional fungicides to control postharvest decay of strawberry. Carbohydr Polym. 2015;**132**:111–117.

[21] Reglinski T., Elmer P.A.G., Taylor J.T., Wood P.N. and Hoyte S.M. Inhibition of *Botrytis cinerea* growth and suppression of botrytis bunch rot in grapes using chitosan. Plant Pathol. 2010;**59**:882–890.

[22] Zahid N., Maqbool M., Siddiqui Y., Manickam S. and Ali A. Regulation of inducible enzymes and suppression of anthracnose using submicron chitosan dispersions. Sci. Hortic. 2015;**193**:381–388.

[23] Liu H., Tian W.X., Li B., Wu G.X., Ibrahim M., Tao Z.Y., Wang Y.L., Xie G.L., Li H.Y. and Sun G.C. Antifungal effect and mechanism of chitosan against the rice sheath blight pathogen, *Rhizoctonia solani*. Biotechnol. Lett. 2012;**34**:2291–2298.

[24] Li B., Wang X., Chen R.X., Huangfu W.G. and Xie G.L. Antibacterial activity of chitosan solution against *Xanthomonas* pathogenic bacteria isolated from *Euphorbia pulcherrima*. Carbohyd. Polym. 2008;**72**:287–292.

[25] Mansilla A.Y., Albertengo L., Rodríguez M.S., Debbaudt A., Zúñiga A. and Casalongué C.A. Evidence on antimicrobial properties and mode of action of a chitosan obtained from crustacean exoskeletons on *Pseudomonas syringae* pv. *tomato* DC3000. Appl. Microbiol. Biotechnol. 2013;**97**:6957–6966.

[26] Badawy M.E., Rabea E.I. and Taktak N.E. Antimicrobial and inhibitory enzyme activity of N-(benzyl) and quaternary N-(benzyl) chitosan derivatives on plant pathogens. Carbohydr. Polym. 2014;**111**:670–682.

[27] Kong M., Chen X.G., Xing K. and Park H.J. Antimicrobial properties of chitosan and mode of action: a state of the art review. Int. J. Food Microbiol. 2010;**144**(1):51–63.

[28] Xing K., Zhu X., Peng X. and Qin S. Chitosan antimicrobial and eliciting properties for pest control in agriculture: A review. Agron. Sustain. Dev. 2015;**35**:569–588.

[29] Kulikov S.N., Chirkov S.N., Il'ina A.V., Lopatin S.A. and Varlamov V.P. Effect of the molecular weight of chitosan on its antiviral activity in plants. Prik. Biokhim. Mikrobiol. 2006;**42**(2):224–228.

[30] Su X.W., Zivanovic S. and D'Souza D.H. Effect of chitosan on the infectivity of murine norovirus, feline calicivirus, and bacteriophage MS2. J. Food Protect. 2009;**72**:2623–2628.

[31] Sharp R.G. A review of the applications of chitin and its derivatives in agriculture to modify plant-microbial interactions and improve crop yields. Agronomy. 2013;**3**(4):757–793.

[32] Parra Y. and Ramírez M.A. Efecto de diferentes derivados de quitina sobre el crecimiento in vitro del hongo *Rhizoctonia solani* Kuhn. [Effect of different chitin derivatives on in vitro growth of the fungi *Rhizoctonia solani* Kuhn]. Cultivos Tropicales. 2002;**23**:73–75.

[33] Rabea E.I., Badawy M.E.-T., Stevens C.V., Smagghe G. and Steurbaut W. Chitosan as antimicrobial agent: applications and mode of action. Biomacromolecules. 2003;**4**:1457–1465.

[34] Chung Y.C., Su Y.P., Chen C.C., Jia G., Wang H.L., Wu J.C., Lin J.G. Relationship between antibacterial activity of chitosan and surface characteristics of cell wall. Acta Pharmacol. Sin. 2004;**25**:932–936.

[35] Liu H., Du Y.M., Wang X.H., Sun L.P. Chitosan kills bacteria through cell membrane damage. Int J Food Microbiol. 2004;**95**:147–155.

[36] Reddy M.V.B., Arul J., Ait-Barka E., Angers P., Richard C. and Castaigne F. Effect of chitosan on growth and toxin production by *Alternaria alternata* f. sp. lycopersici. Biocontrol. Sci. Technol. 1998;**8**:33–43.

[37] Iriti M. and Faoro F. Chitosan as a MAMP, searching for a PRR. Plant Signal. Behav. 2009;**4**(1):66–68.

[38] Heil M. and Boostock R.M. Induced Systemic Resistance (ISR) against pathogens in the context of induced plant defences. Ann. Botany. 2002;**89**(5):503–512.

[39] Pieterse C.M.J., Leon-Reyes A., Van der Ent S. and Van Wees S.C.M. Networking by small-molecules hormones in plant immunity. Nat. Chem. Biol. 2009;**5**:308–316.

[40] Montesano M., Brader G. and Palva E.T. Pathogen derived elicitors: searching for receptors in plants. Mol. Plant Pathol. 2003;**4**:73–79.

[41] Kumar D. and Klessig D.F. The search for the salicylic acid receptor led to discovery of the SAR signal receptor. Plant Signal. Behav. 2008;**3**:691–692.

[42] Hadwiger L.A. Multiple effects of chitosan on plant systems: solid science or hype. Plant Sci. 2013;**208**:42–49.

[43] Katiyar D., Hemantaranjan, A., Bharti, S., and Nishant Bhanu, A. A Future perspective in crop protection: chitosan and its oligosaccharides. Adv. Plants Agric. Res. 2014;**1**(1):00006.

[44] Vasil'ev L.A., Dzyubinskaya E.V., Zinovkin R.A., Kiselevsky D.B., Lobysheva N.V. and Samuilov V.D. Chitosan-induced programmed cell death in plants. Biochem. Moscow. 2009;**74**:1035–1043.

[45] Orzali L., Forni C. and Riccioni L. Effect of chitosan seed treatment as elicitor of resistance to Fusarium graminearum in wheat. Seed Sci. Technol. 2014;**42**:132–149.

[46] Corsi B., Riccioni L. and Forni C. In vitro cultures of Actinidia deliciosa (A. Chev) C.F. Liang & A.R. Ferguson: a tool to study the SAR induction of chitosan treatment. Org. Agric. 2015;**5**:189–198.

[47] Chatterjee S., Chatterjee B.P., Guha A.K. A study an antifungal activity of water-soluble chitosan against *Macrophomina phaseolina*. Int J Biol Macromol. 2014;**67**:452–457.

[48] Li S.J., Zhu T.H. Biochemical response and induced resistance against anthracnose (*Colletotrichum camelliae*) of camellia (*Camellia pitardii*) by chitosan oligosaccharide application. For. Path. 2013;**43**:67–76.

[49] Hadwiger L.A. and Polashock J. Fungal mitochondrial DNases: effectors with the potential to activate plant defences in non-host resistance. Phytopathology. 2013;**103**:81–90.

[50] Weake V.M. and Workman J.I. Histone ubiquitination triggering gene activity. Mol. Cell. 2008;**29**:653–663.

[51] Castañeda L.M., Genro C., Roggia I., Bender S.S., Bender R.J., and Pereira C.N. Innovative rice seed coating (*Oryza sativa*) with polymer nanofibres and microparticles using the electrospinning method. J. Res. Updates Polym. Sci. 2014;**3**(1):33–39.

[52] Lizárraga-Paulín E.-G., Miranda-Castro S.-P., Moreno-Martínez E., Lara-Sagahón A.-V. and Torres-Pacheco I. Maize seed coatings and seedling sprayings with chitosan and hydrogen peroxide: their influence on some phenological and biochemical behaviors. J. Zhejiang Univ. Sci. B, 2013;**14**(2):87–96.

[53] Ziani K., Ursúa B., and Maté J.I. Application of bioactive coatings based on chitosan for artichoke seed protection. Crop Prot. 2010;**29**(8):853–859.

[54] Khanzada K.A., Rajput M.A., Shah G.S., Lodhi A.M., and Mehboob F. Effect of seed dressing fungicides for the control of seedborne mycoflora of wheat. Asian J. Plant Sci. 2002;**1**(4):441–444.

[55] Jett L.W., Welbaum G.E., and Morse R.D. Effects of matric and osmotic priming treatments on broccoli seed germination. J. Am. Soc. Hort. Sci. 1996;**12**:423–429.

[56] Ahmed N.E., Kanan H.O., Inanaga S., Ma Y.Q. and Sugimoto Y. Impact of pesticide seed treatments on aphid control and yield of wheat in the Sudan. Crop Prot. 2001;**20**(10):929–934.

[57] Thobunluepop P., Pawelzik E., and Vearasilp S. Possibility of biological seed coating application on direct-seed rice production: Emphasis on plant productivity and environment awareness. Agric. Sci. J. 2008;**39**(3 Suppl.):449–452.

[58] Qin Q.X. and GUO S.Y. Filming of chitosan and its applications. Mod. Food Sci. Technol. 2007;**4**:029.

[59] Chen J.L., and Zhao Y. Effect of molecular weight, acid, and plasticizer on the physicochemical and antibacterial properties of β-chitosan based films. J. Food Sci. 2012;**77**(5):E127–E136.

[60] Tikhonov V.E., Stepnova E.A., Babak V.G., Yamskov I.A., Palma-Guerrero J., Jansson H.B., Lopez-Llorca L.V., Salinas J., Gerasimenko D.V., Avdienko I.D. and Varlamov, V.P. Bactericidal and antifungal activities of a low molecular weight chitosan and its N-/2 (3)-(dodec-2-enyl) succinoyl/-derivatives. Carbohydr. Polym. 2006;64(1):66–72.

[61] Alburquenque C., Bucarey S.A., Neira-Carrillo A., Urzúa B., Hermosilla G., and Tapia C.V. Antifungal activity of low molecular weight chitosan against clinical isolates of *Candida* spp. Med. Mycol. 2010;**48**(8):1018–1023.

[62] Zheng L.Y. and Zhu J.F. Study on antimicrobial activity of chitosan with different molecular weights. Carbohydr. Polym. 2003;**54(**4):527–530.

[63] Kalemba D. and Kunicka A. Antibacterial and antifungal properties of essential oils. Curr. Med. Chem. 2003;**10**(10):813–829.

[64] Aloui H., Khwaldia K., Licciardello F., Mazzaglia A., Muratore G., Hamdi M. and Restuc- cia C. Efficacy of the combined application of chitosan and Locust Bean Gum with different citrus essential oils to control postharvest spoilage caused by *Aspergillus flavus* in dates. Int. J. Food Microbiol. 2014;**170**:21–28.

[65] Riccioni L., Immirzi B., Orzali L., Santagata G. and Malinconico M. The use of natural film as seed-coating. *In*: Proceedings of the Conference "L'agricoltura biologica in risposta alle sfide del futuro: il sostegno della ricerca e dell'innovazione [The organic farming in response to the challenges of the future: the support of research and innovation] ". Catania, 7–8-nov 2011.

[66] Nandeeshkumar P., Sudisha J., Ramachandra K.K., Prakash H.S., Niranjana S.R. and Shekar S.H. Chitosan induced resistance to downy mildew in sunflower caused by *Plasmopara halstedii*. Physiological Mol. Plant Pathol. 2008;**72**:188–194.

[67] Benhamou N., Kloepper J.W., and Tuzun S. Induction of systemic resistance to Fusarium crown rot and root rot in tomato plants by seed treatment with chitosan. Phytopathology. 1994;**84**:1432–1444.

[68] Hadwiger L.A., Fristensky B., and Riggleman R.C. Chitosan, a natural regulator in plant-fungal pathogen interactions, increases crop yields. In Zikakis J.P. eds. Chitin, Chitosan and Related Enzymes. Academic, New York, NY. 1984;291–302.

[69] Guan Y.J., Hu J., Wang X.J. and Shao C.X. Seed priming with chitosan improves maize germination and seedling growth in relation to physiological changes under low temperature stress. J. Zhejiang Univ. Sci. B. 2009;**10**(6):427–433.

[70] Ruan S.L. and Xue Q.Z. Effects of chitosan coating on seed germination and salt-tolerance of seedlings in hybrid rice (*Oryza sativa* L.). Acta Agron. Sinica. 2002;**28**:803–808.

[71] Reddy M.V.B., Arul J., Angers P. and Couture L. Chitosan treatment of wheat seeds induces resistance to Fusarium graminearum and improves seeds quality. J. Agric. Food Chem. 1999;**47**(3):67–72.

[72] Photchanachai S., Singkaew J. and Thamthong J. Effects of chitosan seed treatment on *Colletotrichum* sp. and seedling growth of chili cv. ′jinda′. Acta Hortic. 2006;**712**:585–590.

[73] Zeng D., Luo X. and Tu R. Application of bioactive coatings based on chitosan for soybean seed protection. Int. J. Carbohydr. Chem. 2012(2012) 5 pp.

[74] Lafontaine J.P. and Benhamou N. Chitosan treatment: An emerging strategy for enhancing resistance of greenhouse tomato plants to infection by *Fusarium oxysporum* f. sp. *radicis lycopersici*. Biocontrol Sci. Tech. 1996;**6**:111–124.

[75] Bell A.A., Hubbard J.C., Liu L., Davis R.M. and Subbarao K.V. Effects of chitin and chitosan on the incidence and severity of *Fusarium* yellows in celery. Plant Dis. 1998;**82**:322–328.

[76] Laflamme P., Benhamou N., Bussieres G. and Dessureault M. Differential effect of chitosan on root rot fungal pathogens in forest nurseries. Can. J. Bot. 2000;**77**:1460–1468.

[77] Abd-El-Kareem F. and Hagga W.M. Chitosan and citral alone or in combination for controlling early blight disease of potato plants under field conditions. Res. J. Pharmaceut. Biol. Chem. Sci., 2014;**5**:941–949.

[78] El Ghaouth A., Arul J., Asselin A., Benhamou N. Antifungal activity of chitosan on post-harvest pathogens: induction of morphological and cytological alterations in *Rhizopus stolonifer*. Mycol. Res. 1992;**96**:769–779.

[79] Elwagia M.E.F. and Algam S. Evaluation of chitosan efficacy on tomato growth and control of early blight disease. Proceedings of the 3rd Conference of Pests Management in Sudan, February 3–4, 2014, Wad Medani, Sudan, pp: 10.

[80] Mishra S., Jagadeesh K.S., Krishnaraj P.U. and Prem S. Biocontrol of tomato leaf curl virus (ToLCV) in tomato with chitosan supplemented formulations of *Pseudomonas* sp. under field conditions. Aust. J. Crop Sci., 2014;**8**:347–355.

[81] Saied-Nehal M. New approaches for controlling *Fusarium* wilt disease of watermelon plants. [Ph.D. Thesis], Benha University, Egypt; 2015.

[82] Sarathchandra S.U., Watson R.N., Cox N.R., di Menna M.E., Brown J.A., Burch G., and Neville F.J. Effects of chitin amendment of soil on microorganisms, nematodes, and growth of white clover (*Trifolium repens* L.) and perennial ryegrass (*Lolium perenne* L.). Biol. Fertil. Soils. 1996;**22**:221–226.

[83] Green S.J., Inbar E., Michel F.C. Jr, Hadar Y. and Minz D. Succession of bacterial communities during early plant development: transition from seed to root and effect of compost amendment. Appl. Environ. Microbiol. 2006;**72**:3975–3983.

[84] Radwan M.A., Farrag S.A.A., Abu-Elamayem M.M. and Ahmed N.S. Extraction, characterization, and nematicidal activity of chitin and chitosan derived from shrimp shell waste. Biol. Fertil. Soils. 2012;**48**:463–468.

[85] Gooday G.W. Physiology of microbial degradation of chitin and chitosan. Biodegradation. 1990;**1**:177–190.

[86] Manucharova N.A., Yaroslavtsev A.M., Senchenko D.V., Stepanov A.L. and Zvyagintsev D.G. Microbial transformation of chitin in soil under anaerobic conditions. Biol. Bull. 2006;**33**:191–194.

[87] Whips J.M. Microbial interactions and biocontrol in the rhizosphere. J. Exp. Bot. 2001;**52**:487–511.

[88] Hjort K., Bergstrom M., Adesina M.F., Jansson J.K., Smalla K. and Sjoling S. Chitinase genes revealed and compared in bacterial isolates, DNA extracts and a metagenomic library from a phytopathogen-suppressive soil. FEMS Microbiol. Ecol. 2010;**71**:197–207.

[89] Kawase T., Yokokawa S., Saito A., Fuji T., Nikaidou N., Miyashita K. and Watanabe T. Comparison of enzymatic and antifungal properties between family 18 and 19 chitinases from *S. coelicolor* A3(2). Biosci. Biotechnol. Biochem. 2006;**70**:988–998.

[90] Bell A.A., Hubbard J.C., Liu L., Davis R.M. and Subbarao K.V. Effects of chitin and chitosan on the incidence and severity of *Fusarium* yellows in celery. Plant Dis. 1998;**82**:322–328.

[91] Weller D.M., Raaijmakers J.M., McSpadden Gardener B.B and Thomashow L.S. Microbial populations responsible for specific soil suppressiveness to plant pathogens. Annu. Rev. Phytopathol. 2002;**40**:309–348.

[92] Murphy J.G., Rafferty S.M. and Cassells A.C. Stimulation of wild strawberry (*Fragaria vesca*) arbuscular mycorrhizas by Addition of shellfish waste to the growth substrate: interaction between mycorrhization, substrate amendment and susceptibility to red core (Phytophthora fragariae). Appl. Soil Ecol. 2000;**15**:153–158.

[93] Mendes R., Kruijt M., de Bruijn I., Dekkers E., van der Voort M., Schneider J.H.M., Piceno Y.M., De Santis T.Z., Andersen G.L., Bakker P.A.H.M. and Raaijmakers J.M. Deciphering the rhizosphere microbiome for disease-suppressive bacteria. Science. 2011;**332**:1097–1100.

[94] Pal K.K. and McSpadden Gardener B. Biological control of plant pathogens. Plant Health Instr. 2006;**2006**:1–25.

[95] Uppal A.K., El Hadrami A., Adam L.R., Tenuta M. and Daayf F. Biological control of potato Verticillium wilt under controlled and field conditions using selected bacterial antagonists and plant extracts. Biol. Control. 2008;**44**:90–100.

[96] Cretoiu M.S., Korthals G.W., Visser J.H.M. and van Elsas J.D. Chitin amendment increases soil suppressiveness toward plant pathogens and modulates the actinobacterial and oxalobacteraceal communities in an experimental agricultural field. Appl. Environ. Microbiol. 2013;**79**(17):5291–301.

[97] Angelim A.L., Costa S.P., Farias B.C., Aquino L.F. and Melo, V.M. An innovative bioremediation strategy using a bacterial consortium entrapped in chitosan beads. J. Environ. Manage. 2013;**127**:10–17.

[98] Vasconcelos M.W. Chitosan and chitooligosaccharide utilization in phytoremediation and biofortification programs: current knowledge and future perspectives. Front. Plant Sci. 2014;**5**:616.

[99] White P.J. and Brown P.H. Plant nutrition for sustainable development and global health. Ann. Bot. 2010;**105**:1073–1080.

[100] Utsunomiya N., Kinai H., Matsui Y. and Takebaysshi T. The effects of chitosan oligosaccharides soil conditioner and nitrogen fertilizer on the flowering and fruit growth of purple passion fruit (*Passiflora edulis* Sims var. edulis). J. Japanese Soc. Hortic. Sci. 1998;**64**(4):567–571.

[101] Kowalski B., Jimenez Terry F., Herrera L. and Agramonte Peñalver D. Application of soluble chitosan *in vitro* and in the greenhouse to increase yield and seed quality of potato minitubers. Potato Res. 2006;**49**:167–176.

[102] Kamari A., Pulford I.D. and Hargreaves J.S. Binding of heavy metal contaminants onto chitosans – an evaluation for remediation of metal contaminated soil and water. J. Environ. Manage. 2011;**92**:2675–2682.

[103] Kamari A., Pulford I.D. and Hargreaves J.S. Metal accumulation in *Lolium perenne* and *Brassica napus* as affected by application of chitosans. Int. J. Phytorem. 2012;**14**:894–907.

Section 2

Marine Polysaccharides and Biological Activities

Chapter 3

Chitosan from Marine Crustaceans: Production, Characterization and Applications

Jimena Bernadette Dima, Cynthia Sequeiros and Noemi Zaritzky

Additional information is available at the end of the chapter

http://dx.doi.org/10.5772/65258

Abstract

Chitosan is a very useful marine polysaccharide that forms structural components in the exoskeleton of crustaceans. In this chapter, the production of chitosan (CH) and chitosan reticulated micro/nanoparticles (CHM) is described. Three case studies corresponding to different effective applications of chitosan are discussed: (i) the performance of CH to destabilize oil/water emulsion waste for water clarification. It was observed that as long as colloidal charge was maintained around zero, turbidity also showed low values and water clarification was achieved. However, when the applied doses were higher than the optimum, colloidal charge and turbidity both increased, showing emulsion restabilization. Emulsions treated with the optimum chitosan doses were clarified in very short periods; (ii) CH and CHM were used as effective antibacterial agents against three different pathogenic microorganisms that are problematic for aquaculture: *Vibrio alginolyticus* and *parahaemolyticus*, and *Lactococcus garvieae* and the minimum bactericidal concentrations were determined; and (iii) the removal of hexavalent chromium and the comparative performance of CH versus CHM. Results showed that at pH < 2, the adsorption capacity of CHM was higher because CH is unstable. Additionally, Cr(VI) was adsorbed on CH without further reduction; in contrast, Cr(VI) adsorbed on CHM was reduced to nontoxic Cr(III).

Keywords: chitosan, coagulation, flocculation, antimicrobial properties, aquaculture, hexavalent chromium removal, micro and nanoparticles

1. Introduction

The largest group of marine arthropods is the class crustacean, made up of approximately 30,000 species. Marine crustaceans include animals as shrimp, crabs, lobsters, etc. The crustaceans have particular biological characteristics; they have an exoskeleton made of the polysaccharide chitin and calcium. This external shell, in addition to being protective, gives rigid support for the attachment of the muscles. When crustaceans grow, their outer shell, the exoskeleton, does not grow with them, so they must regularly shed these shells in order to increase in size. This process is known as molting; it occurs periodically whenever the body is ready to increase in height and weight, and involves the detachment of the exoskeleton [1]. Chitin, poly β-(1-4)-N-acetyl-D-glucosamine, is a natural polysaccharide synthesized by an enormous number of living organisms. Considering the amount of chitin produced annually in the world, it is the most abundant polymer after cellulose. Chitin occurs in nature as ordered crystalline microfibrils forming structural components in the exoskeleton of arthropods, like crustaceans, or in the cell walls of fungi and yeasts and in the pens of squids [2]. Crustaceans are of great direct and indirect importance to humans. The larger crustaceans (shrimps, lobsters and crabs) are used as food throughout the world and are therefore important to human economies. However, seafood processing industry discards large amounts of crustacean shellfish wastes; exoskeletons are converted in a solid residue, which accumulate in landfills becoming an environmental pollutant. The crustacean processing industries over the world turn out more than 60,000 ton of waste every year [3]. The exoskeleton of the crustaceans represents approximately 50–60% of the total weight in crabs and between 35 and 50% in shrimps. These crustacean wastes contain about 10–25% of chitin on dry weight basis, depending on the species. The proper utilization of these shell wastes not only solves the problem of their disposal but also forms the basis for many potential products used in different fields such as textiles, photography, medicine, agriculture, food processing, etc. Despite the widespread occurrence of chitin, up to now the main commercial sources of chitin have been crab and shrimp shells [4]. Chitosan (poly β-(1-4)-D-glucosamine) is a cationic linear polysaccharide obtained by partial deacetylation of chitin (**Figure 1**). It is composed of randomly distributed β-(1-4)-linked D-glucosamine (deacetylated unit) and N-acetyl-D-glucosamine (acetylated unit). The cationic nature of chitosan is owed to the free amino groups left by partial removal of acetyl groups of chitin. Chitosan is a nontoxic, biocompatible and biodegradable polysaccharide with many biomedical, chemical, agriculture and wastewater treatment applications [3], and represents an attractive alternative to other biomaterials because of its physico-chemical characteristics, chemical stability, high reactivity and excellent chelation behavior. Chemical and physical modifications of chitosan have been used to increase the stability of the polymer and to improve its functionality. In the last years, the production of chitosan nanoparticles (CHM) was investigated in different scientific areas as carriers of drugs, antifungal and antibacterial agents and metal bioadsorbents [5–7].

The major global producers of chitosan are Japan and the US as well as China, India, Australia, Poland and Norway [8].

In the present chapter, chitosan obtained from shrimp exoskeletons and chitosan micro/nanoparticles were tested in different applications. Three case studies are described: destabilization and clarification of emulsified wastewater, antimicrobial activity in aquaculture and removal of heavy metals from residual water.

http://dx.doi.org/10.5772/65258

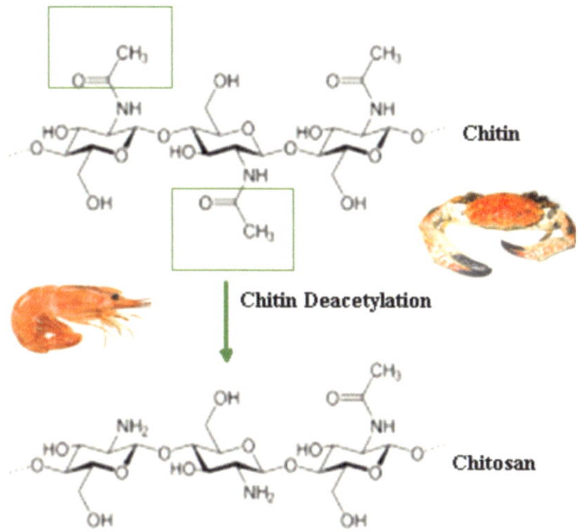

Figure 1. Chitin and chitosan.

2. Chitin and chitosan production and characterization

Shrimps shells (*Pleoticus muelleri*) were used for the extraction of chitin and chitosan. The shells were provided by the industry from Puerto Madryn, Patagonia, Argentina. Crude shrimp chitin was purified using acid and alkaline treatments, according to Dima et al. [9]. The exoskeleton powder was treated with HCl solutions to remove minerals and then treated with aqueous sodium hydroxide solution to remove proteins. The solid was washed with deionized water to reach neutral pH. Finally, the wet mass was dried at 65°C for 24 h; the product obtained was designated as chitin. Chitin was treated with concentrated sodium hydroxide solution at 120°C. After filtration, particles were washed with deionized water to neutral pH and dried at 65°C for 24 h, obtaining chitosan particles.

The yield of purified chitin was 24.8%, and this value was into the range reported by different authors [10, 11]. After N-deacetylation, the yield of shrimp chitosan represented the 76.9% of the initial crude chitin. The degree of N-deacetylation and the molecular weight were measured to characterize the obtained chitosan.

The N-deacetylation degree (DD%) of shrimp chitosan samples was determined using the potentiometric technique and Fourier transform infrared spectra (FTIR). The DD% by the potentiometric technique was determined according to Broussignac [12]. A sample of chitosan was mixed with 0.3 mol/L HCl. Potentiometric evaluation was carried out with a 0.1 mol/L NaOH solution, using a pH meter. The potentiometric curves were obtained by measuring the change in pH after each 2 mL addition of base solution. The titration curve shows two inflection points (**Figure 2a**); the difference between these points corresponds to the amount of acid required to protonate the amino groups of chitosan. The percentage of N-deacetylation was determined according to the following expression:

$$DD\ (\%) = \left(\frac{203M_{eq}}{1 + 42M_{eq}} \right) \tag{1}$$

where $M_{eq} = (N\Delta V)/w$; ΔV is the volume difference between both inflection points, N is the normality of NaOH solution, w is the chitosan mass used; 203 is the molar mass of glucosamine and 42 is the molar mass of the acetyl group.

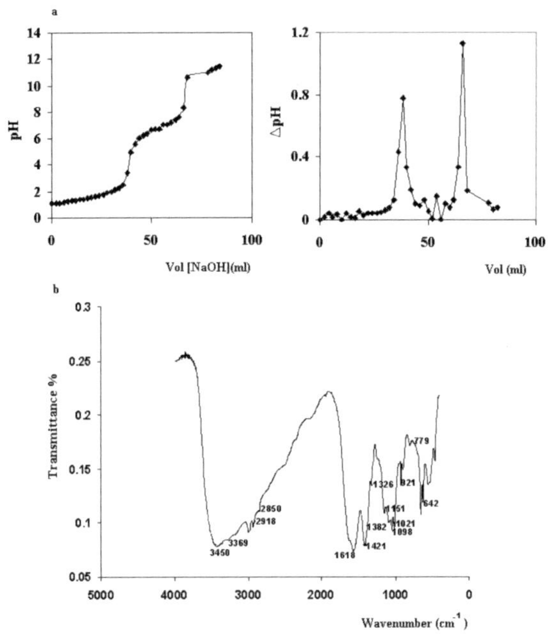

Figure 2. (a) Potentiometric titration curve for shrimp chitosan. Peaks correspond to the inflection points of the titration curve. (b) FTIR spectrum of CH obtained from shrimp.

For the determination of the degree of N-deacetylation by FTIR, chitosan was mixed with potassium bromide and compressed into pellets. The absorbances at 1320 and 1420 cm^{-1} were used to calculate the DD% according to the following equation proposed by Brugnerotto et al. [13]:

$$DD\ (\%) = 100 - \left(31.92 \times \frac{A_{1320\ cm^{-1}}}{A_{1420\ cm^{-1}}} \right) - 12.20 \tag{2}$$

Figure 2b shows the FTIR spectrum of shrimp chitosan; different characteristic bands can be observed: at 3450 cm^{-1} (N-H and O-H stretching vibrations), 2918 and 2852 cm^{-1} (tension group C-H), 1721 cm^{-1} (C=O carbonyl group), 1618 cm^{-1} (amine group), 1326 cm^{-1} (amide III), 1382 cm^{-1} ($-CH_3$ symmetrical deformation mode (scissoring) in amide group). The N-deacetylation degree calculated by the potentiometric technique was 90.2% and by FTIR was 86.3%; these values fall into the range of commercial chitosan. The degree of deacetylation necessary to obtain a chitosan product must be 65% or higher [14, 15].

The molecular weight of chitosan was determined by the viscometric method. The intrinsic viscosity [η] of shrimp chitosan was measured with an Ostwald capillary viscometer. The chitosan sample was dissolved in 0.3 M acetic acid, 0.2 M sodium acetate buffer. The viscosity average molecular weight of chitosan (Mv) was calculated by measuring the intrinsic viscosity [η] according to Mark-Houwink-Kuhn-Sakurada (MHKS) equation:

$$[\eta] = \mathrm{k}\ \mathrm{Mv}^{\mathrm{a}} \tag{3}$$

where k and a are viscometric coefficients, which depend on the polymer, the solvent used and the temperature; the parameters proposed by Rinaudo et al. [4] were adopted. In the case of chitosan with 12% degree of acetylation (deacetylation degree DD = 88%) in a solution 0.3 M acetic acid, 0.2 M sodium acetate buffer (at 25°C), the recommended values are k = 0.074 ml/g and a = 0.80 obtaining an average molecular weight of 1.46×10^5–1.52×10^5 Da; this value falls into the range reported by other authors, for different chitosan sources (1×10^5–5×10^5 Da) [4, 14, 15].

3. Chitosan as destabilizing agent of oil/water emulsion wastes for water clarification

Food processing plants generally discharge large volumes of wastewater. Emulsified oil in wastewater constitutes a severe problem in the different treatment stages. Oil in wastewaters has to be removed in order to: (1) prevent interference in water treatment units; (2) reduce fouling in process equipment; (3) avoid problems in the biological treatment stages; and (4) comply with water discharge requirements. Cationic polyelectrolytes such as chitosan can be used to coagulate and flocculate colloidal systems [16].

Chitosan is a natural linear polyelectrolyte at acidic pH; it has a high charge density, one charge per each glucosamine unit. The chain structure has positively charged amine functional groups which are responsible for the polyelectrolyte behavior. Chitosan can coagulate negatively charged material with its positively charged functional groups to give electric neutrality [17, 18].

The performance of chitosan as a destabilizing agent for emulsion wastes in order to clarify residual water was tested in the laboratory. Experiments were carried out on a model waste system consisted of sunflower oil/water emulsions with variable ionic strength (NaCl concentrations ranging between 10^{-3} and 10^{-1} mol/L). A biodegradable ionic surfactant (sodium dodecyl sulfate, SDS, molecular weight = 288.36) was added to each sample to stabilize the emulsions inhibiting the coalescence of the oil droplets. The presence of SDS produced a negatively charged emulsion. The emulsions were prepared in a colloidal mill at maximum speed with a stirring time of 15 min. Stability of the emulsions was verified by micrographs obtained by light microscopy over three days storage time. Sizes of the emulsion drops were measured by image analysis software. For comparison purposes, simultaneous experiments using another polyelectrolyte, a cationic polyacrylamide of high molecular weight (MW = 4.106), were carried out. Chitosan solutions (5 g/L) were prepared by dissolving chitosan in 1% (v/v) acetic acid solution during continuous agitation for several hours [16]. Cationic polyacrylamide (1 g/L) was prepared by dissolving the polyacrylamide in distilled water. The performance of both destabilizers was tested in terms of the doses and the time necessary to reach minimum turbidity in the system. The pH of the system changed from 4 to 8 by adding either NaOH or HCl. Flocculation experiments were carried out by adding the desired amount of the tested destabilizers to the emulsion with continuous agitation by a magnetic stirrer. To analyze the flocculation process, different techniques were used: electrophoretic mobility, colloid titration, jar test, turbidimetric method and light microscopy observation. Colloid titration was used to determine the colloidal charge and the isoelectric point of the system. A known excess amount of methyl glycol chitosan (MGC) was added in each test. MGC is a cationic polysaccharide which acts as a positively charged titrant over the entire pH range. The oppositely charged colloids react almost stoichiometrically, neutralizing the charge of the system. The remaining excess of MGC was back-titrated by potassium polyvinyl alcohol sulfate (PVSK), a negative colloid using toluidine blue (TB) as an indicator [16]. Electrophoretic mobility and colloidal titration methods gave equivalent information about the dose of polyelectrolyte to reach zero colloidal charge [18]. Microscopy observations were done on the emulsions and on the flocs. Flocculation assays were performed using the jar test with six stirrers having a maximum speed of 250 rpm. Different amounts of polyelectrolyte were added to 500 ml aliquots. Samples were stirred at high speed for 3 min and then at 50 rpm for 10 min. In the polyelectrolyte treatments, as long as colloidal charge was maintained around zero, turbidity showed the lowest values. However, when the applied doses were higher than the optimum, colloidal charge and turbidity both increased showing emulsion restabilization (**Figure 3**).

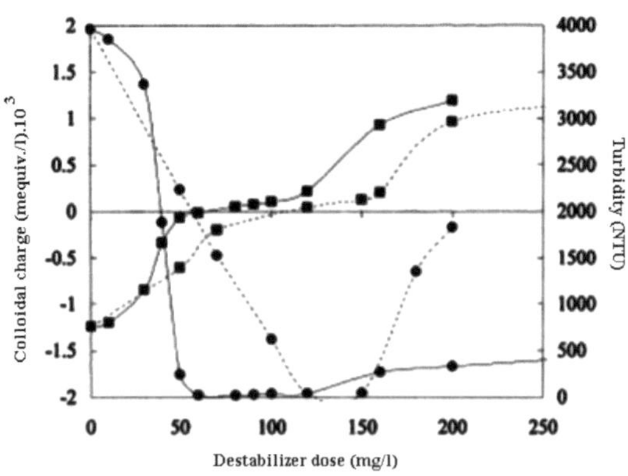

Figure 3. Colloidal charge (■) and turbidity (•) as a function of destabilizer dose: chitosan (—) and cationic polyacrylamide (----) for emulsions formulated with 5 g oil/L, 0.2 g SDS/L and 10^{-3} mol NaCl/L.

Nonsignificant differences between the doses to reach zero colloidal charge and those to reach minimum turbidity (optimum dose) were observed for both polyelectrolytes. Doses to reverse colloidal charge were lower for chitosan than for polyacrylamide; these results can be explained based on the higher charge density of chitosan, requiring lower doses to destabilize the emulsion. Differences in charge density between chitosan and polyacrylamide were detected by the titration of their electrical charge with potassium polyvinyl sulfate; results showed that charge density ratio, between chitosan and polyacrylamide, was 1.64 [16].

When NaCl concentration increased (higher ionic strength), the dose of destabilizer necessary to reach zero colloidal charge decreased. An increase in neutral salt concentration decreases the thickness of the double layer, and the electric potential falls markedly with distance. The repulsive forces diminish and the distance between colloidal particles is reduced, producing coagulation.

In the case of chitosan treatment, flocculation occurred immediately after its addition, requiring a shorter agitation time compared to that of polyacrylamide; pH values ranging between 4 and 8 had no significant effect ($p < 0.05$) on the optimum chitosan dose to destabilize the emulsion.

A linear relationship between initial colloidal charge and the chitosan dose necessary to reach zero colloidal charge was found and this result allows to determine the optimum dose of chitosan, knowing the initial physico-chemical variables of the colloidal system.

4. Synthesis and characterization of chitosan reticulated micro/nanoparticles

Chemical and physical cross-linking techniques are usually employed to modify chitosan. Chitosan chains can be chemically cross-linked with glutaraldehyde leading to quite stable matrixes; however, this reagent is toxic and a strong irritant. On the other hand, chitosan hydrogels can be obtained by ionic gelation, where micro- or nanoparticles are formed by electrostatic interactions between the positively charged chitosan chains and polyanions employed as physical cross-linkers. Chitosan reticulated microparticles (CHM) were prepared by ionic gelation of chitosan with a nontoxic reagent (tripolyphosphate), according to Calvo et al. [19] and modified by Dima et al. [9]. To obtain CHM, different concentrations and relative proportions of chitosan-TPP solutions were mixed under magnetic stirring; in each case, particle size was evaluated. When an opalescent suspension was detected, the presence of microparticles was observed (**Figure 4**). Microparticles were collected by centrifugation and were observed by SEM. Size distribution and zeta potential of CHM were determined by dynamic light scattering. For a concentration range of 1.00–1.50 g/L of TPP and 1.00–1.25 g/L chitosan, the obtained CHM size ranged between 88 and 140 nm (polydispersity <1), coinciding with that observed by SEM (**Figure 5**); pH has a marked effect on the electrokinetic behavior (Z-potential) and on particle size distribution. A high absolute value of Z-potential denotes stability of particles in suspension; Z-potential decreased with the amount of TPP added. The effect of pH and the amount of TPP added on particle size distribution and Z-potential are shown in **Figure 6**.

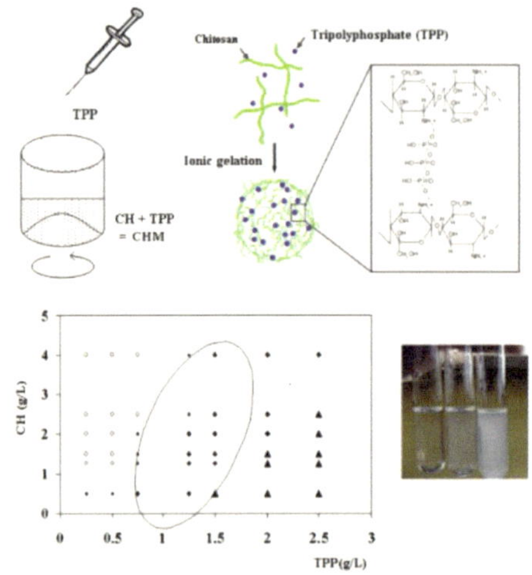

Figure 4. (a) Schematic procedure for the synthesis of micro/nanoparticles (CHM). (b) Effect of the relative concentrations of chitosan and tripolyphosphate on the formation of CHM. The photograph shows the aspect of the different zones in the graph: (♦) aggregates, (▲) clear solution, (○) opalescent suspension.

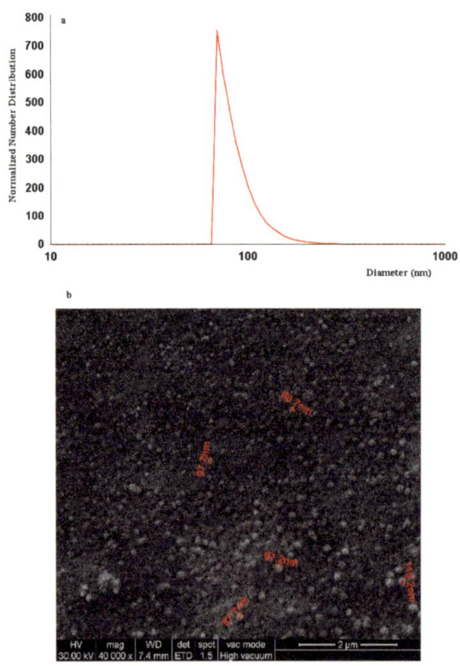

Figure 5. Particle size distribution determined by the Beckman Coulter equipment and SEM micrograph of chitosan micro- and nanoparticles for a ratio of CH:TPP of 1.25:1.50 (g/L).

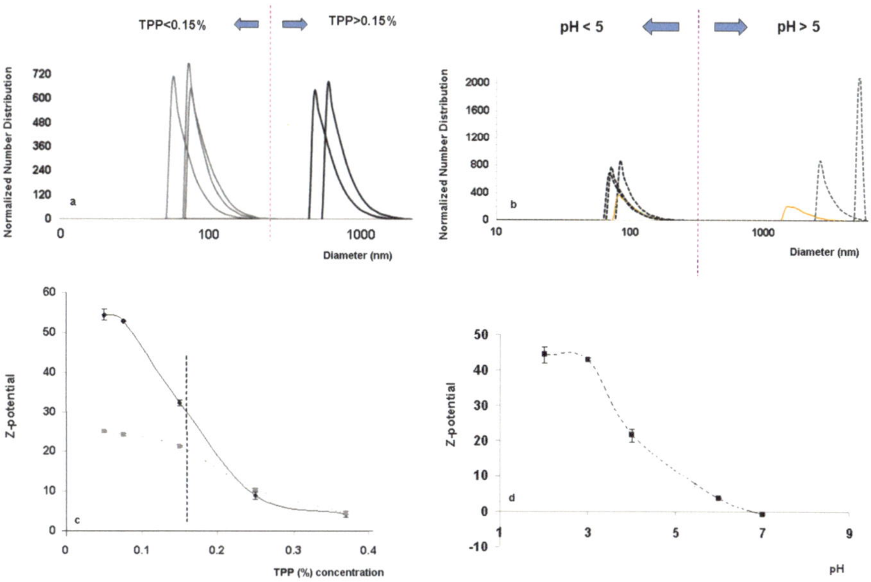

Figure 6. Characterization of chitosan-TPP micro/nanoparticles. Effect of TPP concentration on: (a) particle size distribution; (c) Z-potential (pH = 4): CHM in the original suspension (full line), CHM resuspended in water (dotted line). Effect of pH on: (b) particle size distribution; (d) Z-potential [Ratio of CH:TPP of 1.25:1.50 (g/L)].

5. Applications of chitosan and chitosan micro- and nanoparticles

5.1. Chitosan as antimicrobial agent in aquaculture

Aquaculture is the farming of aquatic organisms such as fish, crustaceans, mollusks and aquatic plants. It is one of the fastest growing food-producing sectors. This accelerated growth of finfish aquaculture has resulted in a series of developments detrimental to the environment and human health [20]. The use of a wide variety of antibiotics in large amounts, including nonbiodegradable antibiotics useful in human medicine, ensures that they remain in the aquatic environment, for long periods of time. This process has resulted in the emergence of antibiotic-resistant bacteria in aquaculture environments, in the increase of antibiotic resistance in fish pathogens, in the transfer of this resistance to bacteria of land animals and to human pathogens, and in alterations of the bacterial flora both in sediments and in water [20]. Chitosan as an antimicrobial agent can be considered an alternative to the use of these antibiotics. Chitosan and its derivatives have attracted considerable interest due to their antimicrobial and antifungal activity. It has been documented that chitosan itself has antimicrobial activity due to its cationic properties that cause a membrane-disrupting effect [20–23]. The antibacterial activity of chitosan is influenced by a number of factors that include the degree of chitosan polymerization and some of its other physicochemical properties. Chitosan exhibits higher antibacterial activity against Gram-positive than Gram-negative bacteria [23]. A number of chitosan derivatives with different modifications have been prepared to improve its antibacterial activity; in this way, chitosan micro/nanoparticles display unique physical and chemical features [6]. The nanoparticles may also penetrate inside the cell causing damage by interacting with phosphorus- and sulfur-containing compounds such as DNA and protein [21, 22].

Experiments were performed using chitosan and reticulated chitosan microparticles as antibacterial agents against different microorganisms that are problematic for aquaculture: *Lactococcus garvieae* (Gram +), *Vibrio parahaemolyticus* and *Vibrio alginolyticus*, both Gram −. These microorganisms have been implicated as the main bacterial pathogens of mariculture industry and are responsible of important economic losses in cultured fish and seafood worldwide [24].

The minimal bactericidal concentration (MBC) values were determined by a standard broth dilution method. For the analysis, a number of sterile test tubes containing 5 mL of TS broth (Tryptic Soy) with NaCl for vibrios and MRS broth for *L. garvieae* were used. The initial concentration of stock solution was 1% (m/v) for chitosan in acetic acid solution (pH 4.8) and 0.8% (m/v) for chitosan reticulated microparticles in distilled water. Serial twofold dilutions of stock solutions were performed for each culture broth. The tubes were inoculated under aseptic conditions with 50 μL of the freshly prepared bacteria suspension (*Vibrio alginolyticus*, 9.6×10^7 UFC/mL; *V. parahaemolyticus*, 1×10^8 UFC/mL and *L. garvieae*, 5×10^7 UFC/mL). MBC was determined transferring an aliquot of contents of each tube after 24 h to TS or MRS agar plates and incubated at 30°C for 24 h. The lowest concentration showing no revival strain on agar plates was considered as the MBC. For the selected pathogens, the MBC for reticulated chitosan was 0.20%, while for chitosan was 0.25% for *V. alginolyticus* and 0.125% for *V. parahaemolyticu* and *L. garvieae*. The antibacterial activity of chitosan particles and microparticles can be attributed to the disruption of cell membranes and the leakage of cytoplasm [21, 22, 25].

5.2. Removal of heavy metals from water using chitosan and chitosan reticulated micro/nanoparticles

Chitosan has been broadly used for the sorption of heavy metals; it combines with metal ions by three forms: ion exchange, adsorption and chelation [7]. This biopolymer has been shown to effectively remove metals such as chromium, copper, mercury and lead [7, 9, 25, 26] from aqueous solutions. In recent years, research has been performed on novel adsorbents to maximize their adsorptive capacity [27, 28]; chemical and physical modifications of chitosan have been used to increase the stability of the polymer and to improve its functionality. According to Schmuhl et al. [26], chitosan forms chelates with metal ions by releasing hydrogen ions, and hence, the adsorption of a metal ion on chitosan depends strongly on the pH of the solution. Chitosan is soluble in most dilute mineral acids (except in sulfuric acid solutions) and in dilute organic acids, such as acetic, propionic, formic and lactic acids [28]. Consequently, its chemical stability needs to be reinforced through treatments using cross-linking agents for its application in acidic media. These treatments induce new linkages between the chitosan chains allowing the polymer to be highly resistant to dissolution, even in solutions, such as hydrochloric acid [28, 29].

Cr(VI) is a toxic metal and must be removed from wastewater before it can be discharged. Cr(III) and Cr(VI) are the stable oxidation states for chromium in nature. Cr(III) is stable and less toxic or nontoxic and is considered an essential element for the good health and nutrition of many organisms. Cr(VI) is 500 times more toxic, mutagenic and carcinogenic than Cr(III). The United States Environmental Protection Agency has laid down the maximum contaminant level for Cr(VI) into inland surface waters as 0.1 mg/L and in domestic water supplies to be 0.05 mg/L [9, 26, 30].

The performance of chitosan and reticulated chitosan micro/nanoparticles in the adsorption process of Cr(VI) was analyzed in the laboratory [9]. Adsorption experiments were performed using different initial concentrations of Cr(VI) (50–400 ppm), contact times (30 min–2 h) and pH values (2–6). All the adsorption experiments of Cr(VI) ions onto CH or CHM were carried out in batch at 25°C, under constant stirring. Final concentrations of Cr(VI) were determined by flame atomic absorption spectrometry. The equilibrium adsorption capacity of Cr(VI) onto chitosan (Q_e) was calculated according to the following equation:

$$Q_e = \frac{(C_i - C_{eq})}{w} V \tag{4}$$

where Q_e (mg/g) is the amount of metal ions adsorbed by the CH or CHM, C_i and C_{eq} are the metal concentrations (mg/L) in the solution initially (time zero) and after equilibrium, respectively, V(L) is the volume of the solution and w is the mass (g) of adsorbent used.

The percentage of Cr(VI) removal was calculated according to:

$$(\%)\ \text{Removal} = \left(\frac{C_i - C_{eq}}{C_i} \right) \times 100 \tag{5}$$

The pH and the initial chromium concentration have a marked effect on the adsorption process. The optimum pH value for the adsorption of Cr(VI) was 4 for CH and 2 for CHM. The highest value of Q_e (equilibrium adsorption capacity) was 127.1 mg/g for CH (pH = 4), and in the case of CHM, a higher value of Q_e = 135.2 mg/g was observed at pH = 2. At very low pH, the adsorption capacities were higher for the CHM because chitosan is unstable at pH < 2.5, and thus cross-linking with TPP improved the adsorption performance of Cr(VI). The adsorption capacity increased from 31.4 to 230.2 mg/g for CH and from 35.5 to 71.4 mg/g for CHM, when the initial Cr(VI) concentration varied from 50 to 400 mg/L, at pH = 4. At pH = 2, the adsorption capacity of CHM increased and adsorption capacity of CH decreased. **Figure 7a**, **b** shows the simultaneous effect of initial Cr(VI) concentration and pH on the percentage of Cr(VI) removal [Eq. (5)] for CH and CHM, respectively. Contact time is an important parameter because this factor determines the adsorption kinetics of an adsorbent at a given initial concentration of the adsorbate. The adsorption kinetic curves of Cr(VI) (initial concentration of 100 mg/L, pH = 4) onto CH and CHM are shown in **Figure 7**. The curves show that equilibrium was reached after approximately 1 h for CH (**Figure 7c**) and after 2 h for CHM (**Figure 7d**). The maximum amounts of adsorbed chromium were produced after 3 h contact time obtaining equilibrium values of 66.9 mg/g for CH and 38.8 mg/g for CHM. After these periods, both systems remained almost unchanged until the end of the experiment.

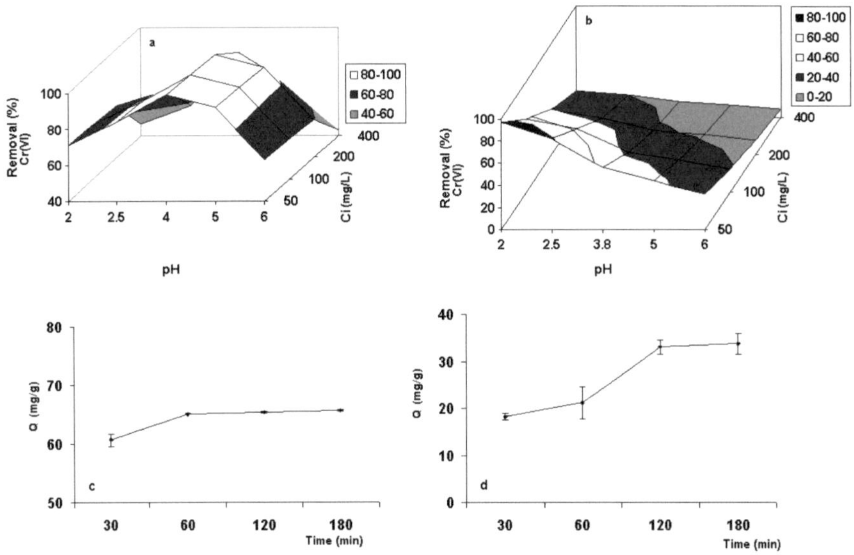

Figure 7. Effect of contact time on the adsorption capacity (Q) of chromium (VI) on: (a) CH; (b) CHM. Simultaneous effect of chromium (VI) initial concentration and pH on the percentage of Cr(VI) removal using; (c) CH; (d) CHM.

Other important observation was that the chemical analysis to determine the oxidation state of the adsorbed Cr showed that Cr(VI) was adsorbed on CH particles without further reduction; in contrast, Cr(VI) removed from the solution was reduced and bound to the CHM as Cr(III). The reduction in toxic Cr(VI) to the less or nontoxic Cr(III) by the reticulated chitosan micro/nanoparticles can be considered a very efficient detoxification technique for the treatment of Cr(VI) contaminated water [9].

Langmuir, Freundlich and Temkin equilibrium sorption isotherms [30] were used for the mathematical description of the adsorption equilibrium of Cr(VI) ions on CH or CHM adsorbents. The operating parameters were: T = 25°C, pH = 4, time = 3 h. Adsorption was evaluated by the determination of Cr(VI) concentration in the solution under equilibrium conditions reached at the end of the experiments (3 h, pH = 4). The Langmuir isotherm is given by the equation:

$$\frac{C_{eq}}{Q_e} = \frac{C_{eq}}{Q_m} + \frac{1}{K_L Q_m} \tag{6}$$

where Q_e is the amount adsorbed per unit weight of adsorbent at equilibrium (mg/g); C_{eq} is the equilibrium concentration of adsorbate in solution after adsorption (mg/L). K_L is the

Langmuir constant (g/L) related to the affinity of the binding sites; Q_m is the maximum monolayer adsorption capacity (mg/g).

The Freundlich isotherm is an empirical equation that assumes that the adsorption process takes place on heterogeneous surfaces and adsorption capacity is related to the concentration of Cr(VI) at equilibrium. It is defined as follows:

$$\ln Q_e = \ln K_f + \left(\frac{1}{n}\right)\ln C_{eq} \tag{7}$$

where K_f is the Freundlich constant or capacity factor (mg/g) and 1/n is the Freundlich exponent; n is the heterogeneity factor related to adsorption intensity.

The Temkin isotherm in its linear form is given by the equation:

$$Q_e = B_t \ln(K_t) + B_t \ln(C_{eq}) \tag{8}$$

The obtained parameters from Freundlich, Langmuir and Temkin adsorption equations and the corresponding R^2 values (coefficients of determination) are shown in **Table 1**. The regression values indicate that the adsorption data for Cr(VI) removal fitted well to Langmuir isotherm. This equation is representative of monolayer adsorption occurring on an energetically uniform surface on which the adsorbed molecules are not interactive. For CHM, adsorption at pH = 2 was more effective than at pH = 4.

	Langmuir			**Freundlich**			**Temkin**		
	Qm (mg/g)	**K_L (l/mg)**	**R^2**	**1/n**	**K_f (mg/g)**	**R^2**	**B_t**	**K_t (l/mg)**	**R^2**
Chitosan (CH) pH = 4	250	0.018	0.999	0.43	44.70	0.941	45.01	2.76	0.997
Reticulated microparticles (CHM) pH = 4	68.9	0.014	0.990	0.36	7.02	0.983	13.95	1.76	0.989
Reticulated micro/nanoparticles (CHM) pH = 2	124	0.086	0.990	0.12	60.42	0.977	22.55	1.46	0.939

Table 1. Parameters of the equilibrium isotherms for Cr(VI) adsorption upon chitosan and chitosan microparticles.

The kinetic data were analyzed using the pseudo-first-order, pseudo-second-order kinetic models and Elovich equation [31]. Kinetic analysis is required to get an insight into the rate of adsorption and the limiting step of the transport mechanism, which are primarily used in the modeling and design of the process [32]. The pseudo-first-order kinetic model of Lagergren has been widely used to predict the metal adsorption kinetics and is given by:

$$\frac{dQ}{dt} = k_1(Q_e - Q) \tag{9}$$

where Q is the amount of metal adsorbed at any time (mg/g), Q_e is the amount of metal adsorbed at equilibrium time (mg/g) and k_1 is the pseudo-first-order rate constant (min^{-1}). Integrating Eq. (9) becomes:

$$\ln\left(\frac{Q_e}{Q_e - Q}\right) = k_1 t \tag{10}$$

The adsorption kinetic data can be further analyzed using the pseudo-second-order kinetics, which is represented by:

$$\frac{dQ}{dt} = k_2 (Q_e - Q)^2 \tag{11}$$

where k_2 is the pseudo-second-order rate constant. Integrating Eq. (11) gives:

$$\frac{t}{Q} = \frac{1}{k_2 Q_e^2} + \frac{1}{Q_e} t \tag{12}$$

The Elovich or Roginsky-Zeldovich equation is satisfied in chemical adsorption processes and is suitable for systems with heterogeneous adsorbing surfaces. This equation has been widely used in adsorption kinetics, which describes chemical adsorption mechanism. It is generally expressed as follows:

$$\frac{dQ}{dt} = \alpha \exp(-\beta Q) \tag{13}$$

where Q is the amount of metal adsorbed by chitosan at a time t, α is the initial adsorption rate (mg g^{-1} h^{-1}), and β is the Elovich constant. To simplify the Elovich equation, Chien and Clayton [33] assumed $\alpha\beta t \gg 1$, and on applying the initial condition $Q = 0$ at $t = 0$, the equation becomes:

$$Q = \beta \ln(\alpha\beta) + \beta \ln t \tag{14}$$

	Pseudo-first order		Pseudo-second order		Elovich		
	$k_1(h^{-1})$	R^2	k_2 (g/mgh)	R^2	β (g/mg)	α (mg/gh)	R^2
Chitosan (CH)	2.71	0.871	0.76	0.999	0.37	6.0×10^{10}	0.750
Chitosan micro/nanoparticles (CHM)	1.94	0.932	0.078	0.946	0.10	116.8	0.877

Table 2. Kinetic parameters for Cr(VI) adsorption on chitosan flakes (CH) and chitosan reticulated microparticles (CHM).

The parameters obtained by the different tested kinetic equations with the corresponding correlation coefficients are given in **Table 2**.

For CH, the kinetic studies indicated a rapid removal of chromium from aqueous solutions. The large value of α in Elovich equation for chitosan particles indicates a very high initial adsorption rate in comparison with the microparticles. The kinetic analysis of chromium adsorption showed that pseudo-second-order kinetic model fitted successfully the experimental data.

6. Conclusions

Chitosan is a marine polysaccharide with multiple applications in different fields. Chitin and chitosan were obtained from shrimps shells (*Pleoticus muelleri*) from Patagonia, Argentina. The degree of N-deacetylation and the molecular weight were measured to characterize the obtained chitosan. The N-deacetylation degree (DD%) of shrimp CH samples was determined using the potentiometric technique and Fourier transform infrared spectra. The molecular weight of CH was determined by the viscometric method measuring the intrinsic viscosity and using the Mark-Houwink-Kuhn-Sakurada equation.

Chitosan is a cationic polyelectrolyte that was applied to coagulate and flocculate colloidal systems; oil/water emulsion wastes were destabilized for water clarification. Results showed that as long as colloidal charge was maintained around zero, turbidity showed the lowest values and water clarification was achieved. However, when the applied doses were higher than the optimum, colloidal charge and turbidity both increased showing emulsion restabilization. Emulsions treated with the optimum CH doses were clarified in very short periods.

Chitosan reticulated micro/nanoparticles were prepared by ionic gelation of chitosan with a nontoxic reagent (tripolyphosphate, TPP). Microparticles were observed by SEM; size distribution and zeta potential were determined by dynamic light scattering. Chitosan and reticulated microparticles were used as effective antibacterial agents against different pathogenic microorganisms that are problematic for aquaculture: *Vibrio alginolyticus* and *parahaemolyticus*, and *L. garvieae* and the minimum bactericidal concentrations were determined.

Finally, the performance of CH and reticulated chitosan micro/nanoparticles in the adsorption process of Cr(VI), a very toxic metal, was analyzed. At very low pH, the adsorption capacity was higher for the microparticles because CH is unstable at $pH < 2.5$. Chitosan cross linking with TPP improved the adsorption performance of hexavalent chromium at low pH. In addition, Cr(VI) was reduced and bound to the microparticles as Cr(III). The reduction of toxic and carcinogenic Cr(VI) to the less soluble and less toxic Cr(III) by the reticulated chitosan microparticles can be considered a very efficient detoxification technique. Therefore, the use of marine polysaccharides such as chitosan is a potential tool for innovative technologies to improve environmental quality and sustainability.

Author details

Jimena Bernadette Dima[1,2], Cynthia Sequeiros[2] and Noemi Zaritzky[1,3*]

*Address all correspondence to: zaritkynoemi@.gmail.com

1 CIDCA – Center of Research and Development in Food Cryothechnology (CONICET-UNLP-CIC), La Plata, Argentina

2 Center for the Study of Marine Systems-Patagonian National Center (CECIMAR-CENPAT-CONICET), Puerto Madryn, Argentina

3 Faculty of Engineering, National University of La Plata (UNLP), La Plata, Argentina

References

[1] Petriella AM, Boschi EE. Growth of decapod crustaceans: results of research made on Argentine species. Investigaciones marinas. 1997; 25: 135–157.

[2] Shepherd R, Reader S, Falshaw A. Chitosan functional properties. Glycoconjugate Journal. 1997; 14(4): 535–542.

[3] Thirunavukkarasu N, Shanmugam A. Extraction of chitin and chitosan from mud crab Scylla tranquebarica (Fabricius, 1798). International Journal on Applied Bioengineering. 2009; 4(2): 31–33.

[4] Rinaudo M. Chitin and chitosan: properties and applications. Progress in Polymer Science. 2006: 31(7): 603–632.

[5] Rodríguez-Hamamura N, Valderrama-Negrón A, Alarcón-Cavero H, López-Milla A. Preparation of chitosan particles cross-linked with tripolyphosphate and modified with polyethylene glycol. Revista de la Sociedad Química del Perú. 2010; 76(4): 336–354.

[6] Du WL, Niu SS, Xu YL, Xu ZR, Fan CL. Antibacterial activity of chitosan tripolyphosphate nanoparticles loaded with various metal ions. Carbohydrate Polymers. 2009; 75(3): 385–389.

[7] Wu FC, Tseng RL, Juang RS. A review and experimental verification of using chitosan and its derivatives as adsorbents for selected heavy metals. Journal of Environmental Management. 2010; 91(4): 798–806.

[8] Hoell, I A., Vaaje-Kolstad G, Eijsink VGH. Structure and function of enzymes acting on chitin and chitosan. *Biotechnology and Genetic Engineering Reviews* 2010; 27:331–366.

[9] Dima JB, Sequeiros C, Zaritzky NE. Hexavalent chromium removal in contaminated water using reticulated chitosan micro/nanoparticles from seafood processing wastes. Chemosphere. 2015; 141: 100–111.

[10] Abdou ES, Nagy KSA, Elsabee MZ. Extraction and characterization of chitin and chitosan from local sources. Bioresources Technology. 2008; 99: 1359–1367.

[11] Cocoletzi HH, Almanza EA, Agustin OF, Nava EV, Cassellis ER. Obtención y caracterización de quitosano a partir de exoesqueletos de camarón. Surfaces and Materials. 2009; 22(3): 57–60.

[12] Broussignac P. Chitosan: a natural polymer not well known by the industry. Chimie et Industrie, Genie Chimique. 1968; 99: 1241–1247.

[13] Brugnerotto J, Lizardi J, Goycoolea FM, Argüelles-Monal W, Desbrieres J, Rinaudo M. An infrared investigation in relation with chitin and chitosan characterization. Polymer. 2001; 42(8): 3569–3580.

[14] Ravi Kumar MN. A review of chitin and chitosan applications. Reactive and Functional Polymers. 2000; 46(1): 1–27.

[15] Yen MT, Yang JH, Mau JL. Antioxidant properties of chitosan from crab shells. Carbohydrate Polymers. 2008; 74(4): 840–844.

[16] Pinotti A, Zaritzky N. Effect of aluminum sulfate and cationic polyelectrolytes on the destabilization of emulsified wastes. Waste Management. 2001; 21(6): 535–542.

[17] Ahmad AL, Sumathi S, Hameed BH. Coagulation of residue oil and suspended solid in palm oil mill effluent by chitosan, alum and PAC. Chemical Engineering Journal. 2006; 118(1): 99–105.

[18] Pinotti A, Bevilacqua A, Zaritzky N. Optimization of the flocculation stage in a model system of a food emulsion waste using chitosan as polyelectrolyte. Journal of Food Engineering. 1997; 32: 69–81.

[19] Calvo P, Remuñan-López C, Vila-Jato JL, Alonso MJ. Chitosan and chitosan/ethylene oxide-propylene oxide block copolymer microparticles as novel carriers for proteins and vaccines. Pharmaceutical Research. 1997; 14(10): 1431–1436.

[20] Cabello FC. Aquaculture and public health: the emergence of diphyllobothriasis in Chile and the world. Revista médica de Chile. 2007; 135(8): 1064–1071.

[21] Rabea EI, Badawy MET, Stevens CV, Smagghe G, Steurbaut W. Chitosan as antimicrobial agent: applications and mode of action. Biomacromolecules. 2003; 4(6): 1457–1465.

[22] Wei D, Sun W, Qian W, Ye Y, Ma X. The synthesis of chitosan-based silver nanoparticles and their antibacterial activity. Carbohydrate Research. 2009; 344(17): 2375–2382.

[23] Chaiyakosa S, Charernjiratragul W, Umsakul K, Vuddhakul V. Comparing the efficiency of chitosan with chlorine for reducing *Vibrio parahaemolyticus* in shrimp. Food Control. 2007; 18(9): 1031–1035.

[24] Toranzo AE, Magariños B, Romalde JL. A review of the main bacterial fish diseases in mariculture systems. Aquaculture. 2005; 246(1): 37–61.

[25] Qi L, Xu Z, Jiang X, Hu C, Zou X. Preparation and antibacterial activity of chitosan nanoparticles. Carbohydrate Research. 2004; 339(16): 2693–2700.

[26] Schmuhl R, Krieg HM, Keizer K. Adsorption of Cu(II) and Cr(VI) ions by chitosan: kinetics and equilibrium studies. Water SA. 2001; 27(1): 1–8.

[27] Yu K, Ho J, McCandlish E, Buckley B, Patel R, Li Z, Shapley NC. Copper ion adsorption by chitosan nanoparticles and alginate microparticles for water purification applications. Colloids Surface A. 2013; 425: 31–41.

[28] Laus R, CostaTG, Szpoganicz B, Fávere VT. Adsorption and desorption of Cu(II), Cd(II) and Pb(II) ions using chitosan crosslinked with epichlorohydrin-triphosphate as the adsorbent. Journal of Hazardous Materials. 2010; 183(1): 233–241.

[29] Beppu MM, Vieira RS, Aimoli CG, Santana CC. Crosslinking of chitosan membranes using glutaraldehyde: effect on ion permeability and water absorption. Journal of Membrane Science. 2007; 301(1): 126–130.

[30] Hena S. Removal of chromium hexavalent ion from aqueous solutions using biopolymer chitosan coated with poly 3-methyl thiophene polymer. Journal of Hazardous Materials. 2010; 181(1): 474–479.

[31] Sağ Y, Aktay Y. Kinetic studies on sorption of Cr(VI) and Cu(II) ions by chitin, chitosan and Rhizopus arrhizus. Biochemical Engineering Journal. 2002; 12(2): 143–153.

[32] Ho YS. Removal of copper ions from aqueous solution by tree fern. Water Research. 2003; 37: 2323–2330.

[33] Chien SH, Clayton WR. Application of Elovich equation to the kinetics of phosphate release and sorption in soils. Soil Science Society of America Journal. 1980; 44(2): 265–268.

Chapter 4

Alginate and Sericin: Environmental and Pharmaceutical Applications

Thiago Lopes da Silva,
Jacyara Moreira Martins Vidart,
Meuris Gurgel Carlos da Silva,
Marcelino Luiz Gimenes and
Melissa Gurgel Adeodato Vieira

Additional information is available at the end of the chapter

http://dx.doi.org/10.5772/65257

Abstract

Alginate is a polysaccharide that, for commercial purposes, is extracted exclusively from marine brown algae. In this chapter, we discuss the main sources of alginate and sodium alginate manufacturing, its chemical structure and physicochemical properties, the alginate modifications, and blend formation. We also present applications of alginate and sericin blend in the pharmaceutical and environmental fields as well as case studies.

Keywords: alginate, alginate production, alginate properties, alginate modification, alginate pharmaceutical applications, alginate environmental applications

1. Introduction

"Alginate" is the term usually used for the salts of alginic acid (carboxylic salts), but it can also refer to all the derivatives of alginic acid and alginic acid itself [1]. Alginates are biopolymers that have two main sources: bacteria and seaweed (brown algae). This biomaterial is a natural polysaccharide that occurs as structural components in the cell wall of marine brown algae (*Phaeophyceae*) as well as capsular polysaccharides in some bacteria (*Azotobacter* and *Pseudomonas*). Commercial alginates are extracted exclusively from marine algae sources although the microbial fermentation is technically feasible for this biopolymer production [2].

Alginate is composed of blocks of mannuronic acid residues (M-blocks), blocks of guluronic acid residues (G-blocks), and blocks with alternating M and G residues (MG-blocks). The source of brown seaweed, growth, location, tissue used in the alginate extraction, age of the tissue used for alginate preparation, and season of the year, among other conditions, are important factors that contribute to vary the chemical composition and sequence of M and G units what implies in different alginate properties, like viscosity, gelation, solubility, among others [3, 4].

Alginate, polymer with polyelectrolyte nature, is considered low or nontoxic, nonimmunogenic, biocompatible, and biodegradable. The industrial applications of alginate are linked to its ability to retain water, and its gelling, viscosifying, and stabilizing properties [5, 6]. As biopolymer with important properties, it has large uses in several industrial fields, including textiles, food industry, agri-foods, pharmaceuticals, cosmetics, paper [6], and medical supplies, among others. The physical properties significantly control the stability of the gels, the rate of drug release from gels, and the phenotype and function of cells encapsulated in alginate gels [7].

The gelling property, which is the ability of alginate to form gels in the presence of multivalent cations (e.g., Ca^{2+}), is one of its main biofunctional properties [5, 6, 8, 9]. Emerging biotechnological applications are based on this unique property, which allow specific biological effects of the alginate molecule. The gel formation and the almost temperature-independent sol/gel transition in the presence of multivalent cations make alginate suitable for the development of biomaterial that can be used in cell immobilization, tissue engineering, drug delivery, controlled release, immobilization of microorganism, and matrix for living cell, among other applications [10].

2. Alginate sources

At present, commercial alginates are exclusively extracted from marine brown algae found in coastal waters around the globe [7].

The brown algae are an important assemblage of plants that are classified in about 265 genera with more than 1500 species. They derive their characteristic color from the large amounts of the carotenoid fucoxanthin (which yields a brown color) contained in their chloroplasts and the presence of various pheophycean tannins. Brown algae flourish in temperate to subpolar regions where they exhibit the greatest diversity in species and morphological expression [11].

Brown algae belong to Phaeophyta class. Typical algal cell walls of Phaeophyta are composed of a fibrillar skeleton (made from cellulose material) and as amorphous embedding matrix. The Phaeophyta algal-embedding matrix is predominately alginic acid or alginate (the salt of alginic acid) with a smaller amount of sulfated polysaccharide (fucoidan) [11, 12].

The main commercial sources are species of *Ascophyllum, Durvillaea, Ecklonia, Laminaria, Lessonia, Macrocystis, Sargassum,* and *Turbinaria* [13]. Of these, the most important species are *Laminaria, Macrocystis,* and *Ascophyllum*. The alginate content in algae can reach values of 40% of its dry

http://dx.doi.org/10.5772/65257

weight. On a dry weight basis, the alginate contents are 22–30% for *A. nodosum*, 25–44% for *L. digitata*, and 17–33 and 25–30%, respectively, for the fronds (leaves) and stems of *L. hyperborea* [14].

Algae species	Monomer M (mannuronic acid) (%)	Monomer G (guluronic acid) (%)	M/G ratio
Ascophyllum nodosum	60.0	40.0	1.50
Laminaria digitate	59.0	41.0	1.43
Laminaria japonica	69.3	30.7	2.26
Macrocystis pyrifera	61.0	39.0	1.56
Laminaria hyperborean, fronds	56.0	44.0	1.28
Laminaria hyperborean, stems	30.0	70.0	0.43
Sargassum filipendula	16.0	84.0	0.19
S. polycystum	17.4	82.6	0.21
S. muticum	23.7	76.3	0.31
S. oligocystum	38.3	61.7	0.62
S. horneri	39.0	61.0	0.64
S. miyabei	43.2	56.8	0.76
S. thunbergii	43.8	56.2	0.78
S. henslowianum	45.1	54.9	0.82
S. hemiphyllum	51.5	48.5	1.06
S. siliquastrum	53.1	46.9	1.13
S. vulgare (high viscosity)	53.9	46.1	1.17
S. fluitans (Cuba)	34.2	65.8	0.52
S. fluitans (Florida)	54.1	45.9	1.18
S. pallidum	55.8	44.2	1.26
S. maclurei	59.5	40.5	1.47
S. tenerrimum	60.5	39.5	1.53
S. vulgare (low viscosity)	60.9	39.1	1.56
S. patens	61.4	38.6	1.59
Turbinaria ornata	47.1	52.9	0.89

Table 1. Percentages of monomers M and G (mannuronic acid and guluronic acid) and M/G ratio of some alginate extracted from brown seaweeds [14, 18, 19].

Chemical composition and sequence of M and G units in alginate may vary widely among species and even in different parts of the algae [2]. The time of the year (season) when the algae is harvested, the location of growth, and the age of the tissue used for alginate preparation also influence the composition and sequence of the unit monomer of the polysaccharide [6, 15–17].

The ratio M/G is an important factor because the properties of alginate solution, gels, and its produced biomaterials depend on the G and M contents. **Table 1** shows the M and G contents of alginate extracted from common species of brown seaweed.

From **Table 1**, it can be seen that same species can present different composition of M and G monomers depending on the local harvest (*S. fluitans* – Cuba and Florida) and the tissue utilized in alginate extraction (*L. hyperborean* – fronds and stems). The ratio M/G varies largely ranging from 0.19 to 2.26 indicating wide differences in species composition. Depending on the alginate material that must be developed, this is a critical feature in choosing the raw alginate source.

Alginates with more extreme compositions containing up to 100% mannuronate can be isolated from bacteria. Alginates with a very high content of guluronic acid can be prepared from special algal tissues such as the outer cortex of old stipes of *L. hyperborea*, by chemical fractionation, or by enzymatic modification in vitro using mannuronan C-5 epimerases from bacteria [5].

3. Sodium alginate manufacturing

Alginates occur in brown algae in the intracellular matrix as gels containing sodium, calcium, magnesium, strontium, and barium ions, such that the counterion composition is determined by the ion-exchange equilibrium with seawater [20]. The extraction process of sodium alginate is relatively simple and can be divided into two categories: calcium alginate process and alginic acid process. At first, the key intermediate products formed are calcium alginate and alginic acid, while the second only alginic acid is formed. The calcium alginate process has the advantage of easy separation of both calcium alginate and the alginic acid that are precipitates in fibrous form. Furthermore, although the process of alginic acid has one step less as compared to the calcium alginate process, it should be noted that the overall losses of alginic acid in this process are greater than in the calcium alginate process, due to the fact that the alginic acid precipitated forms a gelatinous precipitate which is very difficult to separate [13].

Figure 1 shows the steps involved in the manufacture of sodium alginate. Initially, the ions (Na^{2+}, Ca^{2+}, Mg^{2+}, Sr^{2+}, and Ba^{2+}) are removed by protons exchange by adding a dilute mineral acid, such as HCl, which will result in the formation of insoluble salts of alginic acid. Treatment with formaldehyde is carried out for the removal of phenolic compounds and also to bleach the material. Then, an alkaline extraction is performed by adding Na_2CO_3 or NaOH, yielding soluble sodium alginate and insoluble seaweed residue. Alkaline extraction is the main step as it corresponds to the extraction phase itself [21]. A separation process is employed to separate the sodium alginate solution of the extraction residue. From the sodium alginate solution, it is possible to employ the precipitation method which generates both calcium alginate and alginic acid as intermediates (calcium alginate process) or generate only alginic acid as an intermediate (alginic acid process). In calcium alginate process, $CaCl_2$ is added to the sodium alginate solution to form insoluble calcium alginate, which is converted into insoluble alginic acid by the addition of dilute mineral acid (e.g., HCl). Alginic acid is finally converted to sodium alginate by the addition of Na_2CO_3 or NaOH. In the case of alginic acid process, sodium alginate solution is treated with dilute mineral acid, giving rise to the insoluble alginic acid. Alginic acid is suspended in alcohol (ethanol or methanol), and NaOH or Ca_2CO_3 solution is added for obtaining sodium alginate [13, 20–23].

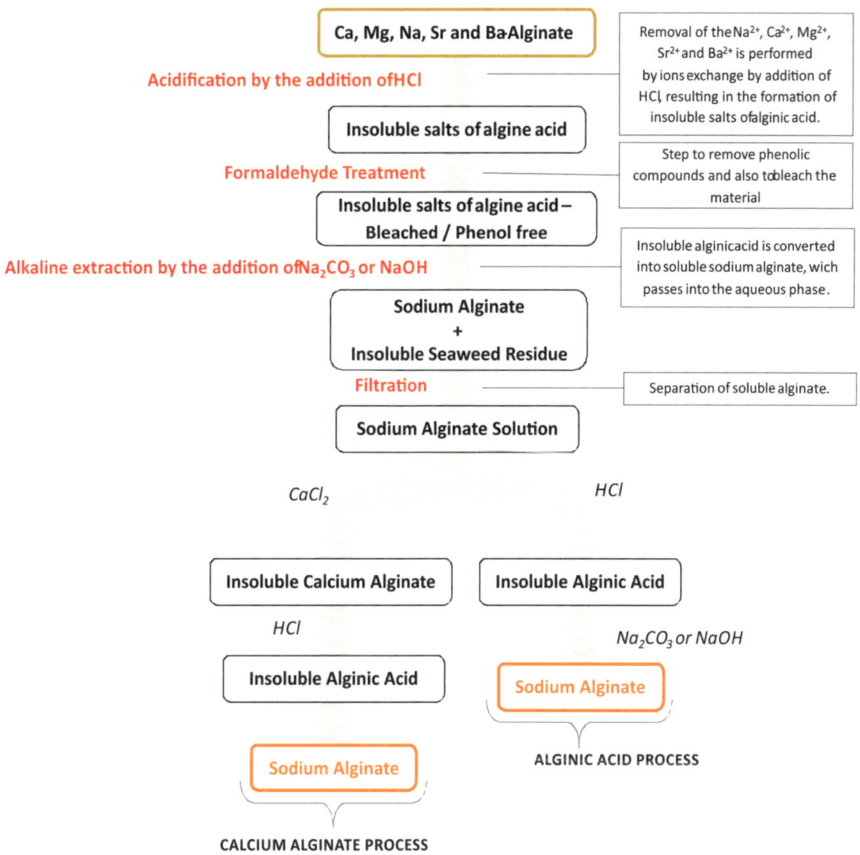

Figure 1. Scheme of sodium alginate production from brown algae.

4. Chemical structure

Alginates constitute a family of linear binary unbranched copolymers composed of 1,4-linked β-D-mannuronic acid (monomer M) and α-L-guluronic acid (monomer G) residues [3, 6]. These two acid residues (saccharide unit) present stereochemically differences at C-5 [5, 14]. Alginate presents a number of free hydroxyl and carboxyl groups distributed along the backbone which allow reactions and chemical functionalization [10].

The alginate polymer accepts different conformation of M and G saccharides in its chain. The chain can be composed of homopolymeric regions of β-ᴅ-mannuronic acid residues (M-blocks: MMMMMM), homopolymeric region of β-ᴅ-mannuronic acid residues (G-blocks: GGGGGG), and heteropolymeric regions where G and M exist in alternating sequence (MG-block: MGMGMG) [24, 25]. **Figure 2** presents both M and G monomers and the chain conformation of alginate.

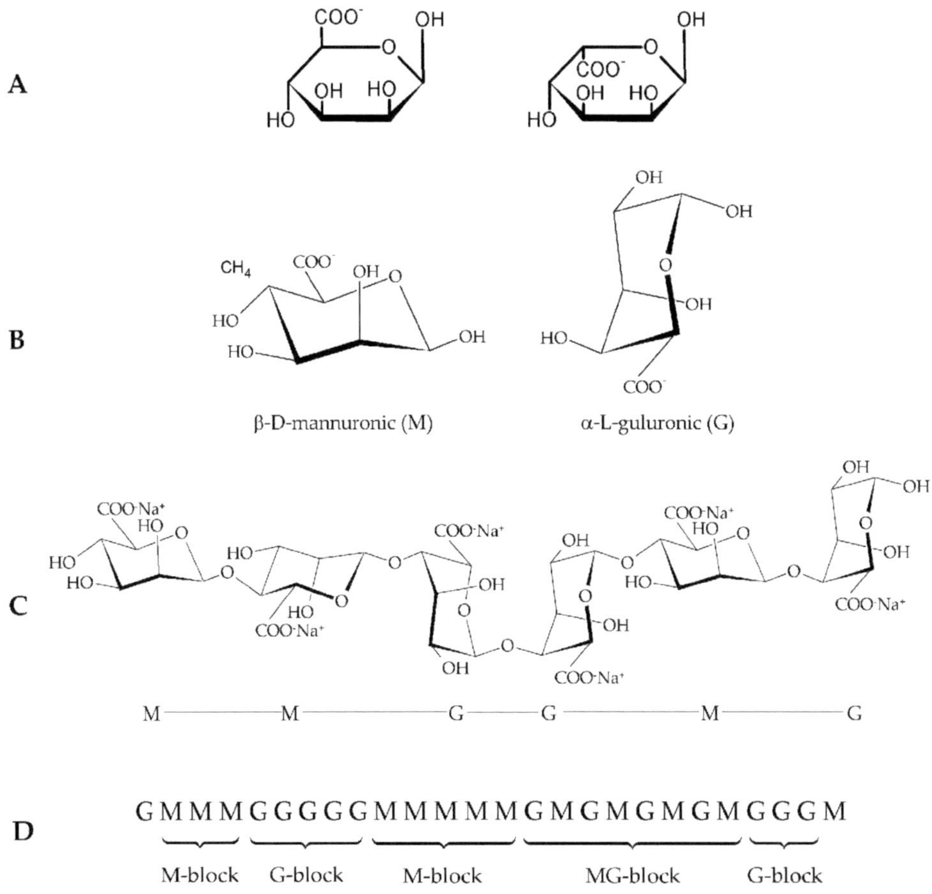

Figure 2. Structural characteristics of alginates: (A) alginate monomers, (B) chemical structure of monomers, (C) chain conformation, and (D) block distribution.

Because of differences in chemical structure of alginates, the properties of alginate can vary depending on the source of brown seaweed. The proportion and sequential arrangement of M and G (uronic acid residues) in the alginate chain, that is, the proportion of the three types of blocks, leads to differences in the physical properties of the respective alginate products [6].

http://dx.doi.org/10.5772/65257

Properties of alginates depend on the relative proportion of three types of uronic blocks; for industrial utilization of any particular alginate, it is quite important to quantify the relative proportions of the uronic acids. Methods such as H NMR and C NMR (proton nuclear magnetic resonance spectroscopy) have been developed to measure the ratio M/G, as well the MM, GG, and MG/GM contents [3, 6, 26]. The ratio of mannuronic acid to guluronic acid, although the number and size of blocks are not provided, provides a practical estimate to evaluate the composition and quality of the alginate to a particular use.

Alginates extracted from different sources differ in M and G contents as well as the size of each block, and nowadays more than 200 different alginates are currently being manufactured [7].

Alginates have no regular repeating distribution of the monomers along the polymer chain. That is why this distribution cannot be described by Bernoullian statistics, which implies that the knowledge of the monomeric composition is not sufficient to determine the sequential structure of alginates [2, 5]. It was suggested that a second-order Markov model would be required for a general approximate description of the monomer sequence in alginates [5].

Alginate contains all four possible glycosidic linkages within the alginate molecule: diequatorial linkages connect mannuronic acid residues in M-blocks, diaxial linkages connect guluronic acid residues in G-blocks, and equatorial-axial (MG) and axial-equatorial (GM) glycosidic bonds connect both uronic residues in MG blocks [5, 24]. Due to this kind of linkages, the M-bock is a relatively straight polymer, such as a flat ribbon, while G-block presents a buckled arrangement. The conformation of linkages and chain alginate is reported in **Figure 2**.

The diaxial linkage in G-blocks results in a large hindered rotation around the glycosidic linkage, which combined with the polyelectrolyte nature of the alginate molecule may account for its stiff and extended nature [2]. G-blocks are stiffer than alternating blocks, which in turn are more soluble at low pH [27]. In the uronic blocks, the rigidity decreases along the series GG > MM > MG [28]. The electrostatic repulsion between the charged groups on the polymer chain also will increase the chain extension and hence the intrinsic viscosity [5].

5. Physicochemical properties

5.1. Ionic cross-linking

The most important feature of alginate properties is its ability to form hydrogels with divalent cations. The alginate chelation with multivalent cations is the basis for gel formation. Selective binding of earth metal ions increases significantly with the increase of G content in the alginate backbone chain.

Gel formation is driven by the interactions between G-blocks, which associate to form tightly held junctions in the presence of divalent cations [20]. The divalent cations, such as Ca^{2+}, act as cross-links between the functional groups of alginate chain [10], "zipping" the G-blocks in alginate chain, that is, the G-block of one polymer forms junctions with the G-block of adjacent polymer chain through interactions with the carboxylic groups in the sugars, which leads to the formation of a gel network. Because of the structural form of the G-block, the metal chelation-binding chain is called the egg-box model of cross-linking. **Figure 3** shows the egg-box model for alginate gel formation.

It was believed that only G-blocks of alginate participate in intermolecular cross-linking with divalent cations to form hydrogel, but some researches indicate that MG-blocks also participate in this process [7, 29]. The participation of MG-block is less important to hydrogel formation because these blocks form weak junctions [20]. The linkage of long alternating sequences in secondary MG/GM junctions is suggested to account for the shrinking of alginate gels in view of its dependence on the length of the MG-blocks [29]. Gels prepared from alginate with a high content of G residues (high M/G ratio) exhibit higher stiffness than those with a low amount of G residues [7].

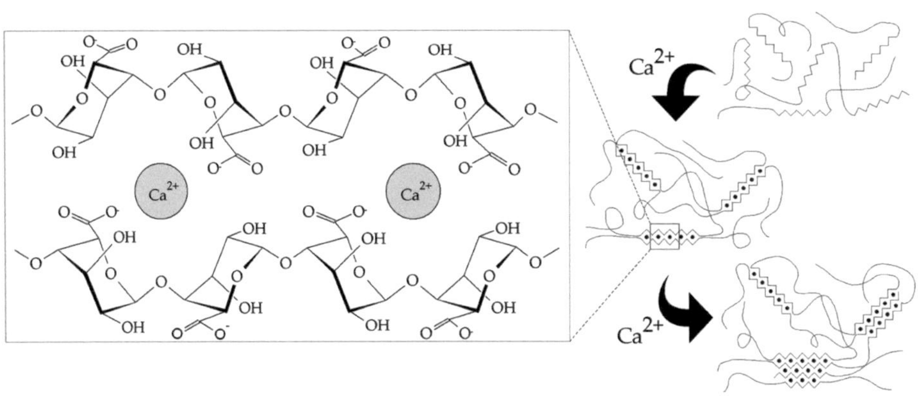

Figure 3. The "egg-box" model of gelation of alginate by calcium.

Alginate's affinity toward the different divalent ions has been shown to decrease in the following order: Pb > Cu > Cd > Ba > Sr > Ca > Co, Ni, Zn > Mn. Since the composition and block structure varies greatly in different types of alginates, it follows that both the gel and ion-binding properties of alginate are influenced by the choice of alginate material and cross-linking ion [30]. Despite the variety of cations, because of the cost and no toxicity, Ca^{2+} is the most used ion to produce alginate gel.

Concerning alginate particles, the preparation method of calcium alginate particles also interferes with their physical properties, such as the porosity, volume of water, sphericity, and elasticity [31]. Methods such as atomization [32], emulsification [31], and dripping [33] are also employed. Calcium cross-linking of alginates can be performed by mainly two methods: diffusion method and internal setting method. In the "diffusion" method, the ions diffuse into the alginate solution from an outside reservoir. In the "internal setting" method, the ion source is located within the alginate solution and a controlled trigger (typically pH or solubility of the ion source) sets off the release of cross-linking ions into the solution. The diffusion method yields gels having a Ca^{2+} ion concentration gradient across the thickness, while internal setting gives gels with uniform ion concentrations throughout [2, 20].

The gelation rate is an important factor that affects the uniformity and strength of hydrogels. Lower rates can be achieved by temperature control (lower temperature implies slower ionic cross-linking), alginate composition (high G content implies higher stiffness), pH control, and Ca^{2+} concentration of calcium solution source.

Calcium chloride ($CaCl_2$) is one of the most frequently used agents to ionically cross-linking alginate. However, it typically leads to rapid and poorly controlled gelation due to its high solubility in aqueous solutions [7]. The fast gelation rate with $CaCl_2$ results in varying cross-linking densities and a polymer concentration gradient within the gel bead. By contrast, the use of $CaCO_3$ and $CaSO_4$, at internal setting method, which has very low solubility in pure water, allows its uniform distribution in alginate solution before gelation occurs [34].

The ionic gelation process (by diffusion method) to produce alginate beads usually is performed by dripping a sodium alginate solution into a $CaCl_2$ bath. This process has been used in both drug delivery and cell encapsulation [34], and to produce adsorbent beads from a blend of sericin-protein/alginate [33, 35]. **Figure 4** shows the schematic process for producing alginate beads.

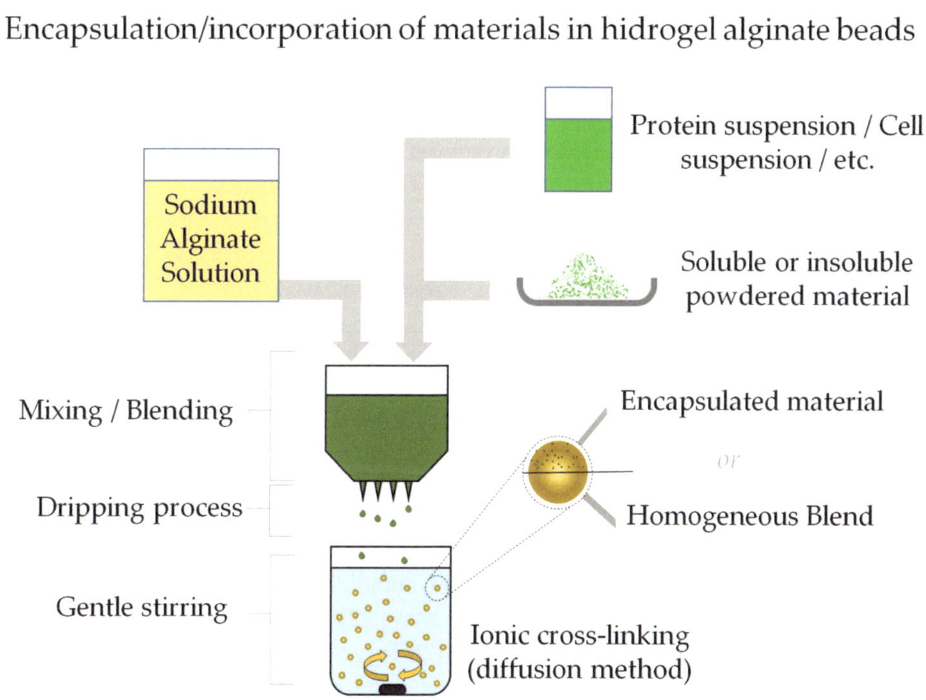

Figure 4. Alginate beads preparation by diffusion setting method.

5.2. Covalent cross-linking

Alginate hydrogels have been attractive for a variety of biomedical applications, but they possess limited mechanical properties when ionically cross-linked with divalent cations [36]. Covalent cross-linking is applied in order to improve physical properties of alginate gels compared with those ionically cross-linked ones [37]. This kind of cross-linking is typically formed by the reaction between carboxylic groups in alginate chains and a cross-linking molecule possessing primary diamines [38]. The use of cross-linking reagents must be carefully investigated because many of them can be toxic and unreacted cross-linkers need to be thoroughly removed from gels [37].

The stress-relaxation behavior of hydrogels is strongly affected by how the polymers are cross-linked. In gels with ionic cross-links, stress relaxes mainly through breaking and subsequent reforming of the ionic cross-links, while in gels with covalent cross-links, stress relaxes mainly through migration of water [38]. The ionic cross-linking stresses lead to plastic deformation of the alginate gel, while the covalent cross-linking leads to a stress relaxation allowing a significant elastic deformation [7].

Covalent cross-linking of alginate with poly(ethylene glycol) (PEG)-diamines of various molecular weights (MWs) generates hydrogels with a range of mechanical properties. Hydrogels with a range of elastic moduli could be generated by controlling either the chain length of the cross-linking molecule or the cross-linking density. The elastic modulus increased gradually with an increase in cross-linking density or weight fraction of PEG in the hydrogel [36].

The introduction of hydrophilic cross-linking molecules as second macromolecules (such as PEG) compensates for the loss of hydrophilic groups in the alginate backbone [37].

Maiti and Sa [39] used ionotropic gelation method for the preparation of ibuprofen-loaded calcium alginate (CALG) and ethylenediamine (EDA)-treated calcium alginate (EDA-CALG) microspheres to investigate drug delivery and the retard of the drug release to some extent. The reduction in drug entrapment efficiency by a maximum of 44.60% for EDA-CALG microspheres compared to untreated CALG microspheres was observed. EDA-CALG microspheres released almost all of its contents within 7 h in pH 6.8 phosphate buffer; however, CALG microspheres were found to release the same within 3 h.

5.3. Alginic acid gels

Alginates may also gel following a third and ion-independent way in that they form acid gels at pH values below the pKa values of the uronic residues [2, 40]. When the pH of alginate solutions is lowered below the pKa of the uronic acids in a highly controlled fashion, acid gels are formed. Such gels, often called "acid gels," are stabilized by an intermolecular hydrogen-bonding network [20]. With the exception of some pharmaceutical uses, the number of applications of acid gels is rather limited to date [2]. Two methods are generally used to make acid gels: In the first method, a slowly hydrolyzing lactone such as glucono delta-lactone (GDL) is added to a solution of Na-alginate, and in the second method, preformed Ca-alginate gels are converted to acid gels by proton exchange [40].

http://dx.doi.org/10.5772/65257

6. Physical properties

6.1. Molecular weight (MW)

Alginates, like polysaccharides in general, are polydisperse with respect to MW. In this aspect, they resemble synthetic polymers rather than other biopolymers such as proteins and nucleic acids. Because of this polydispersity, the MW of an alginate is an average over the whole distribution of MW [5].

The MW distribution can have implications for the uses of alginates, as low-molecular-weight fragments containing only short G-blocks may not take part in gel-network formation and consequently do not contribute to the gel strength. Furthermore, in some high-tech applications, the leakage of mannuronate-rich fragments from alginate gels may cause problems, and a narrow molecular-weight distribution therefore is recommended [5].

The MW of commercially available sodium alginates ranges between 32,000 and 400,000 g/mol [7]. Usually, a higher MW would result in a higher level of interchain bonding and a greater mechanical gel strength and viscosity [14].

The use of alginate with high MW can improve the physical properties of its gels. However, an alginate solution formed from high-MW polymer becomes greatly viscous, which is often undesirable in processing [7, 41].

For the successful use of hydrogels as cell immobilization/delivery vehicles, a key property that must be satisfied is the maintenance of the viability of the cells through the gel-preparation process. The high viscosity may not be desirable in terms of maintaining cell viability during the pre-gel/cell-mixing process, as a high-solution viscosity would lead to cells being exposed to high shear forces during the mixing. Cell membranes are highly labile to shear forces, and the mixing can lead to damage or cell death [42].

Manipulation of the MW and its distribution can independently control the pre-gel solution viscosity and postgelling stiffness. The elastic modulus of gels can be increased significantly, while the viscosity of the solution minimally raises, by using a combination of high- and low-MW alginate polymers [7, 8].

6.2. Viscosity

Alginates can be prepared with a wide range of molecular weights (50–100,000 kDa), and aqueous solutions of alginates have non-Newtonian characteristics, that is, the viscosity decreases with increasing shear rate (shear thinning). The viscosity of an alginate solution depends on the concentration of the polymer, the MW distribution [27], pH, and G- and M-residues content of alginate.

The viscosity of the alginate solution increases as the molecular weight increases making it difficult to dissolve a high concentration of alginate in a given amount of water. Alginate manufacturers can control the MW (or the degree of polymerization, DP) by varying the severity of the extraction conditions to produce products with viscosities in a 1% solution ranging from 10 to 1000 mPa, with a DP range of 100–1000 units [14].

The viscosity of the alginate solution increases sharply as the concentration increases. However, high-solution viscosities make it difficult to remove bubbles in the solution brought in during the mixing process. For practical purposes, aqueous solutions of alginate have a maximum content of 5–6% of sodium alginate. The temperature of solution also interferes in the viscosity solution. The viscosity decreases as temperature increases (at a rate of 2.5% per degree Celsius). Since viscosity drops sharply on heating, it is useful to heat a solution during the dissolution process. Also, the heating is beneficial since the reduced viscosity helps the bubbles to rise from the solution. However, if alginate solutions are maintained above 50°C for several hours, depolymerization may occur, giving a permanent loss of viscosity and MW [14].

The viscosity of alginate solutions is unaffected over the range pH = 5–11. Below pH = 5, the free $-COO^-$ ions in the chain start to become protonated, to –COOH, and as the electrostatic repulsion between chains is reduced, they are able to come closer and form hydrogen bonds, producing higher viscosities. When the pH is further reduced, a gel will form, usually between pH = 3 and 4. Above pH = 11, slow depolymerization occurs on the storage of alginate solutions, giving a fall in the viscosity [14].

7. Alginate modification

One of the most effective ways to design high-performance biomaterials is by chemically reacting the functional groups available on the alginate backbone [43]. Alginate has the ease of chemical functionalization due to the presence of free hydroxyl and carboxyl groups distributed along the backbone. By forming alginate derivatives through functionalizing available hydroxyl and carboxyl groups, the properties such as solubility, hydrophobicity, and physicochemical and biological characteristics may be modified. Techniques such as oxidation, sulfation, esterification, and amidation can be employed to perform chemical modification of alginate [10]. These chemical modifications allow tailored physical and chemical properties in the modified alginate.

7.1. Oxidation

Alginates may form physical gels under specific conditions, and their functional groups (–OH and –COOH) also allow different chemical and physical modifications. The oxidation on its groups can be performed and the features of new material present important responses. The oxidized alginate presents more reactive groups implying on the modification on its properties, such as a faster degradation, which is important when these ones are used in support for drug-controlled delivery, for instance [44].

Although alginate is an attractive material due to its biocompatibility and ability to form hydrogels, its slow and uncontrollable degradation can be an undesirable feature [45, 46]. Ionically cross-linked alginate hydrogels exhibit a remarkably slow degradation rate, which is typically months to years for their complete removal from injection sites. The alginate oxidation can accelerate the degradation rate improving its property for medical purposes [47].

Oxidized alginates present more reactive groups and a faster degradation when these ones are used in supports for drug-controlled delivery, for example [44].

The periodate oxidation has been used for alginate oxidation and extensively reviewed in literature. Periodate-oxidized alginates are highly susceptible to biodegradation, and therefore oxidized alginates have the potential to be used in a number of biomedical applications wherein biocompatibility and biodegradability are important criteria. Oxidized alginates could also function as potential nontoxic and biodegradable cross-linking agents for proteins in the preparation of hydrogels [48].

When alginate is oxidized by reacting with sodium periodate, the carbon-carbon bonds of the cis-diol groups in the uronate residues are cleaved and changed to dialdehyde groups. Varying the oxidation degree, the degradation rate can be controlled increasing the vulnerability of alginate hydrogels to hydrolysis [47]. The oxidation reaction with sodium periodate on –OH groups at C-2 and C-3 positions of the uronic units of sodium alginate is presented in **Figure 5**.

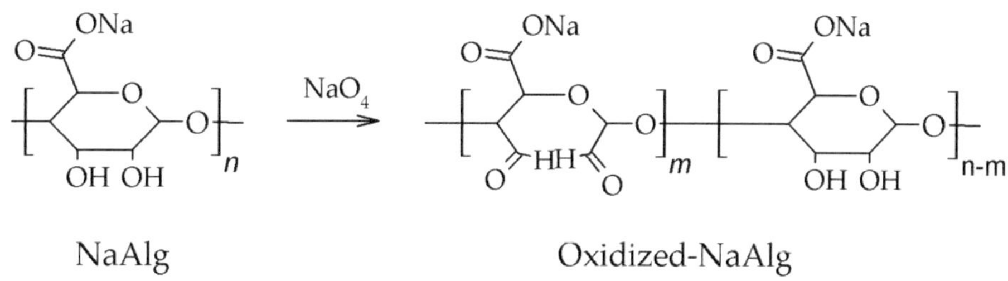

Figure 5. Oxidation of sodium alginate.

The periodate oxidation reactions on –OH groups of the uronic units of alginate leads in a rupture of carbon-carbon bond, to the formation of two aldehyde groups in each oxidized monomeric unit. Therefore, larger rotational freedom and new reactive groups along the backbone are obtained [10].

Gomez et al. [44] conducting a study about the characterization of the oxidized derivatives of sodium alginate (alginate oxidation by sodium periodate) found that the molar mass decreases rapidly until an oxidation of 10 mol% and then remains nearly constant. In addition, the polymers with a degree of oxidation higher than 10 mol% were no more able to form gels with calcium ions. A decrease in the cooperative interactions between calcium ions and carboxylate groups was observed due to the decrease in molar mass and the number of unreacted G units.

7.2. Sulfation

Sulfation of polysaccharides, both enzymatically in nature and by chemical methods, is known to provide blood compatibility and anticoagulant activity [20]. When alginate is sulfated, it will show high blood compatibility because of the structural similarity to that of heparin, which has been widely used for anticoagulant therapy [10]. **Figure 6** presents the sulfation of sodium alginate using chlorosulfonic acid in formamide.

After sulfated modification, the sodium alginate would contain sulfate and carboxyl groups, as the nearest structural analogs of the natural blood anticoagulant heparin [49]. Heparin from animal sources had the potential to induce disease-affecting mammals, such as the avian influenza virus and bovine spongiform encephalopathy [50]. These reasons strongly motivated the necessity to find new anticoagulants and antithrombotics to replace heparin [49].

Ronghua et al. [51] reported the sulfation of sodium alginate using chlorosulfonic acid in formamide. The in vitro coagulation assay of human plasma containing the sulfates indicated that alginate sulfates had considerably high anticoagulant activity especially to the intrinsic coagulation pathway.

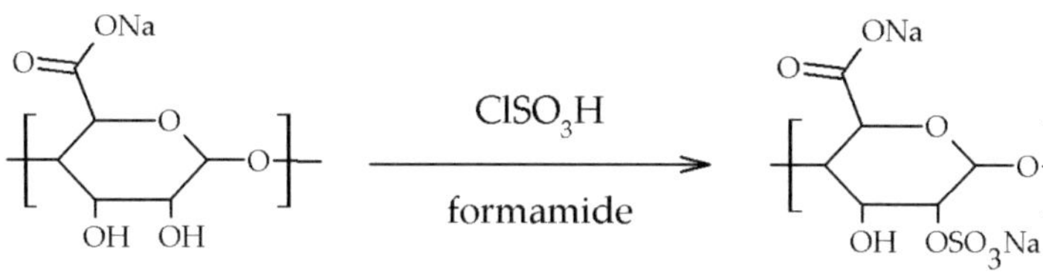

Figure 6. Sulfation of sodium alginate.

7.3. Esterification

Esterification involves the reaction of –COOH groups with –OH groups producing an ester as final product. As shown in **Figure 7**, alginate can be modified by direct esterification with several alcohols in the presence of catalyst and the alcohol is present in excess to ensure that the equilibrium is in favor of product formation. This method was successfully used by researchers to modify native alginate, increasing its hydrophobic nature by the addition of alkyl groups to the backbone of the native alginate [10].

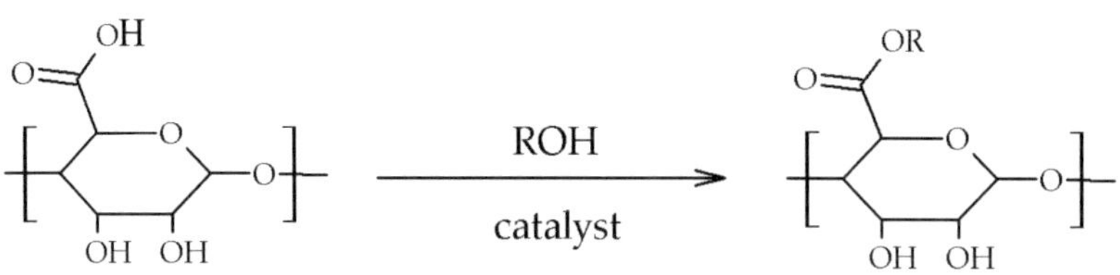

Figure 7. Esterification of alginate.

The only synthetic derivative of alginic acid to find wide use, and acceptance as a food additive, is propylene glycol alginate. This is formed by reacting propylene oxide with moist alginic acid. Esterification occurs at the carboxylic acid groups on the alginate chain, mainly with the primary hydroxyl group of propylene glycol. Depending on the reaction conditions, such as reaction temperature and ratios of propylene oxide to alginic acid, varying degrees of esterification can be achieved. A product with about 60–70% esterification is satisfactory for most purposes but up to about 90% esterification can be achieved and this type of product (80–90%) is useful in very acidic, short-term applications [13].

7.4. Ugi reaction

Pure alginate has its inherent drawbacks, such as rigid backbone, poor mechanical strength, uncontrolled degradation, and extensive water uptake properties, which restricts its practical applications. As a result of available hydroxyl and carboxyl groups, chemical modification of alginate could be achieved mostly at the two secondary hydroxyl positions (C-2 and C-3) or the one carboxyl (C-6) position via the acetylation, phosphorylation, sulfation oxidation, esterification, amidation, and Ugi reaction [52].

The Ugi reaction is an important reaction used in combinatorial chemistry, and it is a multi-component reaction in organic chemistry involving a ketone or an aldehyde, an amine, an isocyanide, and a carboxylic acid to form a bisamide [10, 53]. Among the chemical modification methods, the Ugi four-component condensation reaction is the most effective and unique that could endow alginate with specific property without the aid of the catalyst [52]. The Ugi multicomponent reactions lead a hydrophobic behavior to the modified alginate [10].

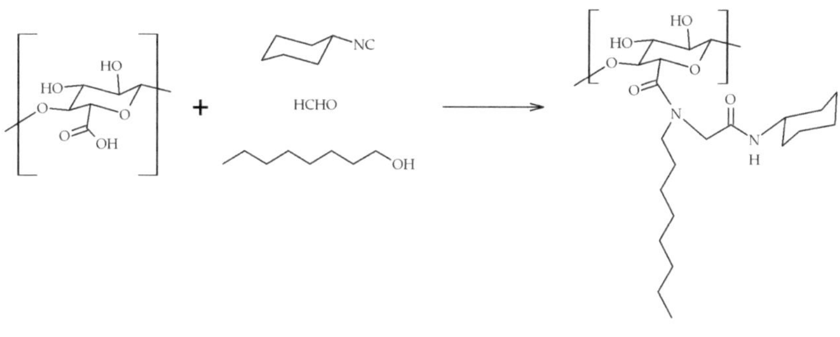

Figure 8. Ugi four-component reaction.

Amphiphilic alginate derivative modified by introducing the hydrophobic groups onto its hydrophilic backbone could enhance its affinity for the hydrophobic drug, thus making it an attractive candidate for drug delivery [52].

The Ugi reaction procedure involves the preparation of aqueous alginate solution followed by acidification (HCl, pH 3.6) to allow the Ugi reaction. The n-octylamine groups are added with respect of molar amount of carbohydrate monomers. Formaldehyde, n-octylamine, and cyclohexyl isocyanide are added to the solution successively. After the addition, the solution is stirred for 24 h. Unreacted monomers and other low-molecular weight impurities are removed from solution by dialysis, and thereafter the solution is freeze-dried. **Figure 8** presents the Ugi four-component reaction.

8. Blend formation

Alginates are established among the most versatile biopolymers, used in a wide range of applications [54]. It can be easily modified in any form, such as hydrogels, microspheres, microcapsules, sponges, foams, and fiber [55], and can cross-link, copolymerize, and blend with other polymers due its polar side chain made of hydroxyl and carboxyl groups [33, 56].

The possibility of alginate modifications (including blending) can increase the applications of alginate in various fields such as tissue engineering, drug delivery [57], environmental [33], and others. The combination of favorable properties of each constituent polymer results in a new hybrid system with the properties that are often significantly improved or substantially different from those of the individual polymers [58]. Alginate beads can be prepared easily through simple and economic procedures; these beads suffer from low drug encapsulation and poor mechanical property in the intestinal pH, which lead to rapid drug release [59]. Therefore, modified alginate beads using sodium alginate and natural polysaccharide blends have been investigated to improve drug encapsulation, swelling, mucoadhesion, drug release, and others. Okra (*Hibiscus esculentus*) is an annual plant cultivated throughout the tropical and subtropical areas of the world, and it is chemically inert, nonirritant, biodegradable, and biocompatible. Okra gum is already investigated as useful excipients in the development of pharmaceutical formulations, and its highly viscous property leads to the usefulness of it as a drug-release-retarding polymer [60]. Sinha et al. [60] developed controlled drug release Zn^{2+}-ion-induced alginate-okra gum blend beads by ionic-gelation-cross-linking method using zinc sulfate ($ZnSO_4$) as a cross-linker in an aqueous environment. These beads exhibited sustained in vitro drug release over a prolonged period of 8 h and followed controlled-release pattern. The swelling and degradation of the optimized beads was influenced by the pH of test media, which might be suitable for intestinal drug delivery.

Alginate application in wastewater treatment is limited due its tendency to swell in water and other mechanical weakness. In this way, alginate can be blended with other polymers to overcome the drawbacks of alginate and to combine the good characteristics of both polymers [61]. Chitosan is a polysaccharide biopolymer derived from chitin [62]. It is well established as an excellent natural adsorbent due to the presence of the amino ($-NH_2$) and hydroxyl ($-OH$) groups, and has also other useful features such as being polycationic, nontoxic, biodegradable, and antibacterial properties. However, due to its weak mechanical property, chemical and physical modifications are carried out on chitosan [61]. Alginate can be easily blended with chitosan by the strong electrostatic interaction between the amino groups of chitosan and the carboxyl groups of alginate [63]. Dubey et al. [64] developed chitosan-alginate nanoparticles using microemulsion method for the removal of Hg (II) ions from aqueous solution. The results obtained in this study proved that the prepared biopolymer nanomaterial could be an effective and economically viable adsorbent for the removal of Hg (II) ions. Moreover, the nanoparticles can be regenerated and reused subsequently for the metal removal.

Sericin is a water-soluble globular protein that is easily soluble in hot or boiling water and is extracted from the silkworm *Bombyx mori* cocoons [65]. Most of the sericin is removed during the silk processing and it is usually discarded in the wastewater [66]. This fact leads to environmental contamination due to the high oxygen demand for its degradation by microbes. Therefore, finding viable means to recover sericin would bring environmental and economic benefits [67]. Sericin isolated presents weak structural properties, but the presence of polar side chain made of hydroxyl, carboxyl, and amino groups in this protein enables easy cross-linking, copolymerization, and blending with other polymers to obtain biodegradable products with better properties [33, 56]. The use of blends provides an improvement in the physical characteristics of the materials produced with the protein, and the use of alginate has an advantage because it has the affinity for a variety of cations [33]. Polymers such as silk sericin and alginate can undergo a sol-gel transition by thermal cross-linking (by heating) [68], photo-cross-linking, and with pH variation, depending on β-sheet content in case of sericin [69].

Other applications of the alginate and sericin blend will be discussed in the following topics.

9. Applications of alginate and sericin blend in the pharmaceutical and environmental fields

In the work of Khandai et al. [70], sericin was evaluated as release retardant along with sodium alginate in aceclofenac-sustained release mucoadhesive microsphere formulation. Microparticulate drug delivery of aceclofenac was prepared by gelation technique using a blend of sodium alginate and sericin as release retardant. All the formulations developed showed diffusion type of release mechanism in a sustained manner. Thus, the aceclofenac microsphere helped to increase the patient compliance, decrease the dosing frequency, and also prevent gastric hemorrhage which is commonly found to be associated with conventional dosage form.

Khampieng et al. [71] assessed the anti-inflammatory efficacy and preparation of silk sericin-loaded alginate nanoparticles that were prepared by the emulsification method followed by internal cross-linking. This study confirms the hypothesis that the topical application of silk sericin-loaded alginate nanoparticle gel can inhibit inflammation induced by carrageenan.

Sericin-alginate microbeads were developed by Nayak et al. [72] via ionotropic gelation under high voltage, and the beads were coated with chitosan and cross-linked with genipin, for the purpose of encapsulating hepatocytes for advanced cellular functions. This study suggests that the developed sericin-alginate-chitosan microcapsule contributes toward the development of cell encapsulation model. It also offers to generate enriched population of metabolically and functionally active cells for the future therapeutics especially for hepatocytes transplantation in acute liver failure.

Silva et al. [33] studied the adsorption of copper and zinc by particles produced from silk sericin and alginate blend. The results obtained suggest the potential use of sericin-alginate particles, cross-linked by ionic gelation and by heat, to adsorption processes of toxic metals, zinc and copper.

10. Case study

10.1. Case study 1: incorporation and release of diclofenac sodium in sericin and alginate blend

10.1.1. Drug incorporation and preparation of particles

In this study, sericin was extracted from silkworm cocoons (*B. mori*), and the concentration of the sericin solution (SS) extracted was adjusted to 2.5% (w/v). In order to prepare the sericin/alginate blend, the sodium alginate was added to the sericin solution and the mixture was stirred at 4000 rpm. The drug incorporation was performed by adding diclofenac sodium (DS) to the sericin/alginate blend and dispersed with an Ultraturrax® (T18, IKA, USA) at 8000 rpm until homogeneity was obtained. The ionic gelation method was used to prepared particles with different compositions [73]. For this, a mixture of sericin, alginate, and DS was added dropwise to a calcium chloride solution (3% w/v) and stirred continuously. After dripping, the particles were stirred at 100 rpm for 30 min, and then washed with deionized water and dried at room temperature.

10.1.2. Determination of incorporation efficiency

In order to determine the incorporation efficiency, accurately weighed 0.1 g of dried particles was added to 500 mL of phosphate buffer (pH 6.8), and kept overnight. Hence, the suspension was subjected to agitation for 15 min in a sonicator (1510RMTH, Branson, USA) and filtered through a 0.45-μm filter. The DS content in the filtrate was determined by spectrophotometer (UVmini1240, Shimadzu, Japan) at 276 nm. All determinations were carried out in triplicate. The incorporation efficiency was calculated by

$$\text{Incorporation} = \frac{\text{practical DS content}}{\text{theoretical DS content}} \times 100 \tag{1}$$

Table 2 shows the effect of sericin and alginate concentration in the DS incorporation efficiency of the formulations developed. It was noticed that the incorporation efficiency increased by increasing sericin proportion in the blend; therefore, sericin significantly contributes to the incorporations of the DS.

Formulation	Sericin (% w/v)	Alginate (% w/v)	DS (% w/v)	Incorporation efficiency (%)
F1	2.5	1.25	2.0	91.1 ± 2.4
F2	2.5	2.60	2.0	82.5 ± 3.6
F3	2.5	3.30	2.0	77.9 ± 2.1
F4	–	4.00	2.0	75.5 ± 2.1

Table 2. Effect of blend composition in the DS incorporation efficiency [73].

The surface morphology of the sericin/alginate/DS particles (F1, F2, and F3) and alginate/DS particles (F4) was visualized by scanning electron microscopy (SEM) with the magnification of 150× and is presented in **Figure 9**. By analyzing the F1 micrograph, it was found that it has no clear-cut boundaries and has poorly defined shape, which does not favor the reproducibility of these particles. Spherical particles with a rough and rugged surface were observed in F2 and F3 micrographs. F4 micrographs indicated oval particles and a very rough and rugged surface too. Thus, it was evident from the SEM micrographs that balanced concentrations of sericin and alginate, as in F2 and F3, favor the sphericity of the particles. Besides, it is verified that roughness increased by increasing alginate proportion in the blend. It can be inferred that the F4 particles (alginate/DS) possibly release the drug in its dissolution medium (gastric or enteric) more rapidly when compared to particles containing sericin in its composition, since the surface has greater roughness and therefore its contact surface is higher. Consequently, particles containing sericin can contribute to the sustained release of the DS present in the matrix and maybe reduce the drug side effects.

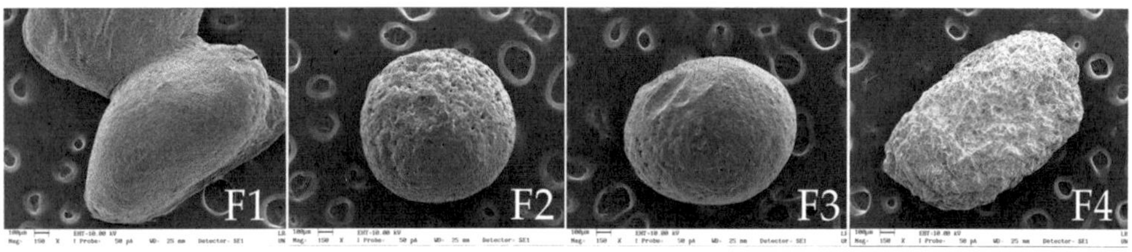

Figure 9. Micrographs of formulations evaluated (as shown in **Table 2**).

10.1.3. In vitro drug release study

According to the methodology recommended by the US Pharmacopeia, the *in vitro* release of DS was tested using a dissolution apparatus. The dissolution was measured at 37 ± 1°C under 50 rpm speed. Accurately weighed quantities of sericin/alginate/DS particles equivalent to 50 mg DS were added to 900 mL of dissolution medium. The test was carried out in 0.1 M HCL (pH 1.2) for 2 h, and then continued in phosphate buffer (pH 6.8) for the next 10 h. Aliquots of 5 mL were collected at regular time intervals, and the same amounts of fresh dissolution medium were replaced into a dissolution vessel. The collected aliquots were filtered, and suitably diluted to determine the absorbance using an ultraviolet-visible infrared spectroscopy (UV-VIS) spectrophotometer at 276 nm.

Table 3 shows the percentage of DS released after 2 h of the dissolution test in acid medium, for F2, F3, and F4. Whereas F1 is nonreproductive, dissolution tests were not conducted for this formulation.

It was found that all formulations were resistant to the gastric medium. Among all studied formulations, F4 presented the lowest drug release in gastric medium. Because F4 has only alginate and DS in its formulation, the low drug release observed in other formulations can be attributed to the presence of this polysaccharide in all composition.

Formulation	DS release (%)
F2	2.63 ± 0.66
F3	1.92 ± 0.36
F4	1.49 ± 0.26

Table 3. Diclofenac sodium release in acid medium, pH 1.2.

Figure 10 shows the diclofenac sodium dissolution profiles for F2, F3, and F4. It can be seen that the presence of sericin in the formulations causes a prolonged drug release compared to the formulation without sericin (F4), which releases all incorporated drug at 45 min. Formulations containing sericin in the composition (F2 and F3) showed the total drug release between 240 and 300 min of dissolution in simulating enteric medium (phosphate buffer, pH 6.8).

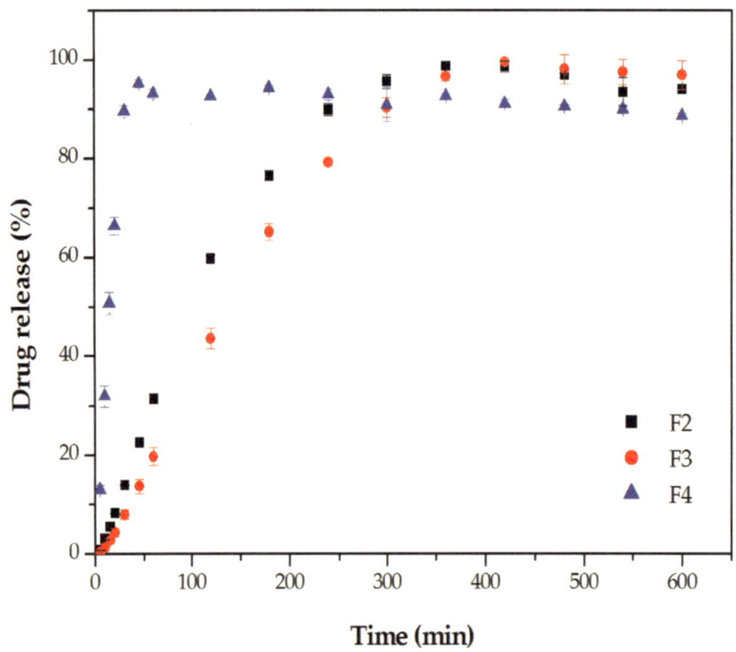

Figure 10. Diclofenac sodium dissolution profile in pH 6.8.

10.2. Case study 2: metallic affinity of toxic and noble metals by particles produced from blends of sericin, alginate and poly(ethylene glycol) diglycidyl

In this study, particles were produced from sericin-alginate blend, by diffusion method, and the evaluation of metallic affinity toward toxic and noble metals by particles was performed. Also, the influence of the presence of the poly(ethylene glycol) diglycidyl ether, as cross-linking agent, in the blend was investigated. The particles were produced dripping the blends in aqueous and alcoholic (ethanol) solutions of $CaCl_2$ and $Ca(NO_3)_2$, providing different sources of divalent ion, and then the particles were dried first at 40°C (24 h) and then at 100°C (24 h). The metallic affinity for toxic metals: copper (Cu^{2+}), nickel (Ni^{2+}), cadmium (Cd^{2+}), zinc (Zn^{2+}), lead (Pb^{2+}), and chromium (Cr^{3+}), and noble metals: palladium (Pd^{2+}), platinum (Pt^{4+}), gold (Au^{3+}), and silver (Ag^{+}) were investigated in this work [74].

10.2.1. Particles preparation

The sericin was obtained by autoclave extraction (40 min, 1 kgf/cm²) from silkworm cocoons. All cocoons were cleaned, washed (in deionized water), and cut into pieces about 1 cm² before the extraction procedure (the ratio of cocoons and ultrapure water was 40 g of cocoons to 1000 mL of water). The extracted sericin solution, still hot, was filtered to remove the fibers of fibroin and stored in a sealed bottle at room temperature for 12 h [35, 74]. After this period, the SS solution was frozen for, at least, 24 h in a conventional refrigerator (-4°C) and then it was thawed at room temperature. This procedure was performed in order to separate the sericin from diluted solution. The freezing procedure promotes the precipitation of sericin, which can be recovered by filtration. The sericin was heated in autoclave (120°C for 10 min) to solubilize again the protein and then the concentration was adjusted by dilution to 25 g/L.

Particle	Blend formulation (g/L) [sericin/alginate/PEG]	Calcium solution
1	25/20/0	Aqueous solution of $CaCl_2$
2	25/20/0	Alcoholic solution of $CaCl_2$
3	25/20/0	Aqueous solution of $Ca(NO_3)_2$
4	25/20/0	Alcoholic solution of $Ca(NO_3)_2$
5	25/20/5	Aqueous solution of $CaCl_2$
6	25/20/5	Alcoholic solution of $CaCl_2$
7	25/20/5	Aqueous solution of $Ca(NO_3)_2$
8	25/20/5	Alcoholic solution of $Ca(NO_3)_2$

Table 4. Formulations and methods of preparations of the produced particles.

The sodium alginate (Sigma-Aldrich brand) was added in adjusted SS in a concentration of 20 g/L. The blend was mixed at 5000 rpm (Ultraturrax® T18—USA) until it was homogeneous. In the particle formulations containing PEG, the cross-linking agent was added in the blend in a concentration of 5 g/L, being mechanically agitated for 30 min. The particles were prepared by ionic gelation technique (diffusion method) where the blend was dripped with a peristaltic pump in alcoholic (ethanol) and aqueous solutions of $CaCl_2$ (3% m/V) and $Ca(NO_3)_2.4H_2O$ (6.4% m/V) (same concentration of Ca^{2+} in both solutions). The particles were agitated in Jar test for 12 h in Ca^{2+} solution. The produced particles were rinsed in deionized water and dried at a continuous flow oven at 40°C, and then submitted to thermal cross-linked at 100°C for 24 h [33]. **Table 4** presents the blend formulation and the respective calcium solution used to produce particles.

10.2.2. Metal affinity tests of single components

Metal solutions of 1 mmol/L of toxic metals: copper (Cu^{2+}), nickel (Ni^{2+}), cadmium (Cd^{2+}), zinc (Zn^{2+}), lead (Pb^{2+}), and chromium (Cr^{3+}), and noble metals: palladium (Pd^{2+}), platinum, gold, and silver (Ag^{+}) were prepared and for the respective affinity tests 0.5 g of each particle formulated was immersed in 50 mL of each metal solution. The particles were maintained in contact with metal solutions under agitation (200 rpm) for 24 h at 25°C. The metal concentrations, before and after adsorption process, were measured at atomic absorption spectroscopy (AAS – 7000A – Shimadzu) according to equipment instructions.

The percentage of metal removal (%*R*) of each was determined by Eq. (2)

$$\%R = \frac{C_0 - C_e}{C_0} \times 100 \tag{2}$$

10.2.3. Adsorption results

From **Table 5**, it can be seen that the adsorption process to toxic metal, Cu, Cd, Pb, and Cr, presents high percentage reduction. Obtained results of low-removal values indicate that the adsorption of nickel and zinc by sericin/alginate and sericin/alginate/PEG particles is not very effective. With exception to silver, which results were slightly lower than the general results observed for Cu, Cd, Pb, and Cr metal ions, the results observed to noble metals (Pd, Pt, and Au) showed greater values than the ones observed to toxic metals.

Particle	Pd	Pt	Au	Ag	Cu	Cd	Ni	Pb	Zn	Cr
1	87.1	73.1	98.9	–	65.0	62.2	14.8	65.9	15.6	73.2
2	88.7	71.8	99.2	–	73.9	70.7	15.5	82.0	21.9	71.0
3	88.6	66.4	99.4	61.3	74.4	72.1	23.5	80.0	24.1	74.4
4	88.9	70.3	99.2	60.4	71.4	73.6	14.0	83.2	28.5	72.0
5	86.8	74.4	99.3	–	74.9	70.7	19.7	82.3	17.7	73.3
6	86.9	73.7	99.7	–	72.8	70.9	12.0	81.8	22.3	71.8
7	88.9	71.6	99.7	63.4	73.1	79.6	23.7	82.8	27.8	73.7
8	89.0	73.2	99.7	61.0	74.3	79.6	13.3	83.3	30.2	69.6

Table 5. Percentage removal (%*R*) of the metal concentration in solution by sericin/alginate/PEG particles.

In general, the particles containing PEG in its formulation (i.e., particles 5–8), exhibit a slightly higher range of %*R*. For example, copper is removed within the observed range (65.0–73.9%) when particles without PEG were used and comparatively within the range (72.9–74.9%) for the particles with PEG. PEG possesses as amphipathic behavior, and the introduction of this cross-linking agent develop modifications in particle structure. Thus, the modifications promoted by PEG in the structure of the particles seem to improve the capacity of the adsorbent.

10.2.4. Considerations

Sericin consists of 17–18 kinds of amino acids with large amount of polar side chains made of hydroxyl, carboxyl, and amino groups. Alginate presents the polar groups hydroxyl and carboxyl along the chain backbone. The presence of these kinds of groups enables this polymer to ionic and covalent cross-linking, allowing interactions and linkages between the proteins and polysaccharide chains, and also allows interactions with pollutants (such as toxic metal) and pharmaceutical.

Despite the interesting properties that sericin and alginate present, there are few works dedicated to study biomaterials that use the blend of both polymers as raw material.

Concerning pharmaceutical field, the development of a new pharmaceutical form of diclofenac sodium-modified release based on alginate-sericin blends is intended to solve current problems related to its therapeutic administration. These problems are related to the development of gastric irritations and decrease of therapeutic effects by partial degradation in acid medium that occurs in the usual forms.

As a bioadsorbent material, or a matrix to drug incorporation or to improve the therapeutic effects of some pharmaceuticals, there are many potential uses for the sericin-alginate blend.

Acknowledgements

The authors thank the BRATAC Company for providing the silkworm cocoons, Geolab® Indústria Farmacêutica S/A for providing the drugs, CNPq (Proc. 470615/2013-3, 473808/2012-9 and 300986/2013-0), CAPES, FAPESP (Proc. 2015/13505-9, 2014/26355-2 and 2011/51824-8), and Fundação Araucária (Proc. 22597-257/12) for financial support.

Author details

Thiago Lopes da Silva[1], Jacyara Moreira Martins Vidart[1], Meuris Gurgel Carlos da Silva[1], Marcelino Luiz Gimenes[2] and Melissa Gurgel Adeodato Vieira[1*]

*Address all correspondence to: melissagav@feq.unicamp.br

1 Department of Process and Products Design, Chemical Engineering School, University of Campinas, Campinas, SP, Brazil

2 Department of Chemical Engineering, State University of Maringá, Maringá, PR, Brazil

References

[1] McHugh DJ. A guide to the seaweed industry. FAO Fisheries Technical Paper. 2003. 105 p.

[2] Draget KI, Taylor C. Chemical, physical and biological properties of alginates and their biomedical implications. Food Hydrocoll. 2011;25(2):251–6. http://dx.doi.org/10.1016/j.foodhyd.2009.10.007.

[3] Grasdalen H, Larsen B, Smidsrød O. 13C NMR studies of alginate. Carbohydr Res. 1977;56:C11–5. DOI: 10.1016/S0008-6215(00)83369-8.

[4] Haug A, Larsen B, Smidsrød O. A study of the constitution of alginic acid by partial acid hydrolysis. Acta Chem Scand. 1966;20:183–90. DOI: 10.3891/acta.chem.scand.20-0183.

[5] Draget K, Smidsrød O, Skjåk-Bræk G. Alginates from algae. Biopolym Online. 2005;6:1–30. http://onlinelibrary.wiley.com/doi/10.1002/3527600035.bpol6008/full.

[6] Grasdalen H, Larsen B, Smidsrød O. A p.m.r. study of the composition and sequence of uronate residues in alginates. Carbohydr Res. 1979;68(1):23–31. DOI: 10.1016/S0008-6215(00)84051-3.

[7] Lee KY, Mooney DJ. Alginate: properties and biomedical applications. Prog Polym Sci. 2012;37(1):106–26. http://dx.doi.org/10.1016/j.progpolymsci.2011.06.003.

[8] Kong H-J, Lee KY, Mooney DJ. Decoupling the dependence of rheological/mechanical properties of hydrogels from solids concentration. Polymer. 2002;43(23):6239–46. http://dx.doi.org/10.1016/S0032-3861(02)00559-1.

[9] Grant GT, Morris ER, Rees DA, Smith PJC, Thom D. Biological interactions between polysaccharides and divalent cations: The egg-box model. FEBS Lett. 1973;32(1):195–8. http://dx.doi.org/10.1016/0014-5793(73)80770-7.

[10] Yang JS, Xie YJ, He W. Research progress on chemical modification of alginate: A review. Carbohydr Polym. 2011;84(1):33–9. http://dx.doi.org/10.1016/j.carbpol.2010.11.048.

[11] Davis TA, Volesky B, Mucci A. A review of the biochemistry of heavy metal biosorption by brown algae. Water Res. 2003;37(18):4311–30. http://dx.doi.org/10.1016/S0043-1354(03)00293-8.

[12] Kim SK. Handbook of marine macroalgae: Biotechnology of applied phycology. 2nd ed. John Wiley & Sons; 2011. 736 p.

[13] McHugh DJ. Production and utilization of products from commercial seaweeds. (288) ed. FAO Fisheries Technical Paper, editor. 1987. 189 p.

[14] Qin Y. Alginate fibers: An overview of the production processes and applications in wound management. Polym Int. 2008;57:171–80. DOI: 10.1002/pi.2296.

[15] Latifi AM, Nejad ES, Babavalian H. Comparison of extraction different methods of sodium alginate from brown alga *Sargassum* sp. Localized in the southern of Iran. J Appl Biotechnol Rep. 2015;2(2):251–5.

[16] Sanchez-Machado DI, Lopez-Cervantes J, Lopez-Hernandez J, Paseiro-Losada P, Simal-Lozano J. Determination of the uronic acid composition of seaweed dietary fibre by HPLC. Biomed Chromatogr. 2004;18(2):90–7. http://dx.doi.org/10.1002/bmc.297.

[17] Haug A, Larsen B, Fykse O, Block-Bolten A, Toguri JM, Flood H. Quantitative determination of the uronic acid composition of alginates. Acta Chem Scand. 1962;16:1908–18. DOI: 10.3891/acta.chem.scand.16-1908.

[18] Torres MR, Sousa APA, Silva Filho EAT, Melo DF, Feitosa JPA, de Paula RCM, et al. Extraction and physicochemical characterization of Sargassum vulgare alginate from Brazil. Carbohydr Res. 2007;342(14):2067–74. http://dx.doi.org/10.1016/j.carres.2007.05.022.

[19] Minghou J., Yujun W., Zuhong X, It I. Studies on the M:G ratios in alginate. Hydrobiologia. 1984;116(1): 554–6. DOI:10.1007/BF00027745.

[20] Pawar SN, Edgar KJ. Alginate derivatization: A review of chemistry, properties and applications. Biomaterials. 2012;33(11):3279–305. http://dx.doi.org/10.1016/j.biomaterials.2012.01.007.

[21] Vauchel P, Kaas R, Arhaliass A, Baron R, Legrand J. A new process for the extraction of alginates from *Laminaria digitata*: Reactive extrusion. Food Bioprocess Technol. 2008;1(3):297–300. DOI: 10.1007/s11947-008-0082-x.

[22] Ruvinov E, Cohen S. Alginate biomaterial for the treatment of myocardial infarction: Progress, translational strategies, and clinical outlook. From ocean algae to patient bedside. Adv Drug Deliv Rev. 2016;96:54–76. http://dx.doi.org/10.1016/j.addr.2015.04.021

[23] Bertagnolli C, Uhart A, Dupin JC, da Silva MGC, Guibal E, Desbrieres J. Biosorption of chromium by alginate extraction products from *Sargassum filipendula*: Investigation of adsorption mechanisms using X-ray photoelectron spectroscopy analysis. Bioresour Technol. 2014;164:264–9. http://dx.doi.org/10.1016/j.biortech.2014.04.103.

[24] Daemi H, Barikani M. Synthesis and characterization of calcium alginate nanoparticles, sodium homopolymannuronate salt and its calcium nanoparticles. Sci Iran. 2012;19(6): 2023–8. http://dx.doi.org/10.1016/j.scient.2012.10.005.

[25] Morris ER, Rees AD, Thom D. Characterisation of alginate composition and block structure by circular dichroism. Carbohydr Res. 1980;81:305–14. DOI:10.1016/S0008-6215(00)85661-X.

[26] Haug A, Larsen B, Smidsrød O. Uronic acid sequence in alginate from different sources. Carbohydr Res. 1974;32(2):217–25. DOI:10.1016/S0008-6215(00)82100-X.

[27] Augst AD, Kong 1 HJ, Mooney DJ. Alginate hydrogels as biomaterials. Macromol Biosci.2006;6(8):623–33. DOI: 10.1002/mabi.200600069.

[28] Jang LK, Harpt N, Grasmick D, Vuong LN, Geesey G. A two-phase model for determining the stability constants for interactions between copper and alginic acid. J Phys Chem. 1990;94(1):482–8. DOI: 10.1021/j100364a083.

[29] Donati I, Holtan S, Mørch YA, Borgogna M, Dentini M, Skjåk-Bræk G. New hypothesis on the role of alternating sequences in calcium-alginate gels. Biomacromolecules. 2005;6(2):1031–40. DOI: 10.1021/bm049306e.

[30] Mørch ÝA, Donati I, Strand BL, Skjåk-Bræk G. Effect of Ca^{2+}, Ba^{2+}, and Sr^{2+} on alginate microbeads. Biomacromolecules. 2006;7(5):1471–80. DOI: 10.1021/bm060010d.

[31] Silva MGC, Canevesi RLS, Welter RA, Vieira MGA, da Silva EA. Chemical equilibrium of ion exchange in the binary mixture Cu^{2+} and Ca^{2+} in calcium alginate. Adsorption. 2015;21(6):445–58. DOI: 10.1007/s10450-015-9682-8.

[32] Tu J, Bolla S, Barr J, Miedema J, Li X, Jasti B. Alginate microparticles prepared by spray coagulation method: Preparation, drug loading and release characterization. Int J Pharm. 2005;303(1):171–81. http://dx.doi.org/10.1016/j.ijpharm.2005.07.008.

[33] Silva TL, da Silva AC, Vieira MGA, Gimenes ML, da Silva MGC. Biosorption study of copper and zinc by particles produced from silk sericin–alginate blend: Evaluation of blend proportion and thermal cross-linking process in particles production. J Clean Prod. 2015;1–9. http://dx.doi.org/10.1016/j.jclepro.2015.05.067.

[34] Kuo CK, Ma PX. Ionically crosslinked alginate hydrogels as scaffolds for tissue engineering: Part 1. Structure, gelation rate and mechanical properties. Biomaterials. 2001;22(6):511–21. http://dx.doi.org/10.1016/S0142-9612(00)00201-5.

[35] Silva TL, da Silva AC, Vieira MGA, Gimenes ML, da Silva MGC. Production and physicochemical characterization of microspheres made from sericin and alginate blend. Chem Eng Trans. 2014;39:643–8. DOI: 10.3303/CET1439108.

[36] Eiselt P, Lee KY, Mooney DJ. Rigidity of two-component hydrogels prepared from alginate and poly(ethylene glycol)-diamines. Macromolecules. 1999;32(17):5561–6. DOI: 10.1021/ma990514m.

[37] Reis RL, Neves NM, Mano JF, Gomes ME, Marques AP, Helena S. Azevedo. Natural-based polymers for biomedical applications. Woodhead Publishing Series in Biomaterials; 2008. 832 p.

[38] Zhao X, Huebsch N, Mooney DJ, Suo Z. Stress-relaxation behavior in gels with ionic and covalent crosslinks. J Appl Phys. 2010;107(6):1–5. DOI: 10.1063/1.3343265.

[39] Maiti S, Sa B. Preparation and characterization of ibuprofen-loaded alginate microspheres using ethylenediamine as a crosslinker. Orient Pharm Exp Med. 2008;8(2):178–86. DOI 10.3742/OPEM.2008.8.2.178.

[40] Draget KI, Bræk GS, Smidsrød O. Alginic acid gels: The effect of alginate chemical composition and molecular weight. Carbohydr Polym. 1994;25(1):31–8. DOI: 10.1016/0144-8617(94)90159-7.

[41] LeRoux MA, Guilak F, Setton LA. Compressive and shear properties of alginate gel: Effects of sodium ions and alginate concentration. J Biomed Mater Res. 1999;47(1):46–53.

[42] Kong HJ, Smith MK, Mooney DJ. Designing alginate hydrogels to maintain viability of immobilized cells. Biomaterials. 2003;24(22):4023–9. http://dx.doi.org/10.1016/S0142-9612(03)00295-3.

[43] Pawar SN, Edgar KJ. Alginate esters via chemoselective carboxyl group modification. Carbohydr Polym. 2013;98(2):1288–96. http://dx.doi.org/10.1016/j.carbpol.2013.08.014.

[44] Gomez CG, Rinaudo M, Villar MA. Oxidation of sodium alginate and characterization of the oxidized derivatives. Carbohydr Polym. 2007;67(3):296–304. http://dx.doi.org/10.1016/j.carbpol.2006.05.025.

[45] Gao C, Liu M, Chen J, Zhang X. Preparation and controlled degradation of oxidized sodium alginate hydrogel. Polym Degrad Stab. 2009;94(9):1405–10. http://dx.doi.org/10.1016/j.polymdegradstab.2009.05.011.

[46] Boontheekul T, Kong HJ, Mooney DJ. Controlling alginate gel degradation utilizing partial oxidation and bimodal molecular weight distribution. Biomaterials. 2005;26(15):2455–65. DOI: 10.1016/j.biomaterials.2004.06.044.

[47] Jeon O, Alt DS, Ahmed SM, Alsberg E. The effect of oxidation on the degradation of photocrosslinkable alginate hydrogels. Biomaterials. 2012;33(13):3503–14. DOI: 10.1016/j.biomaterials.2012.01.041.

[48] Balakrishnan B, Lesieur S, Labarre D, Jayakrishnan A. Periodate oxidation of sodium alginate in water and in ethanol-water mixture: A comparative study. Carbohydr Res. 2005;340(7):1425–9. http://dx.doi.org/10.1016/j.carres.2005.02.028.

[49] Fan L, Jiang L, Xu Y, Zhou Y, Shen Y, Xie W, et al. Synthesis and anticoagulant activity of sodium alginate sulfates. Carbohydr Polym. 2011;83(4):1797–803. http://dx.doi.org/10.1016/j.carbpol.2010.10.038.

[50] Mendes SF, Santos O dos, Barbosa AM, Vasconcelos AFD, Aranda-Selverio G, Monteiro NK, et al. Sulfonation and anticoagulant activity of botryosphaeran from *Botryosphaeria rhodina* MAMB-05 grown on fructose. Int J Biol Macromol. 2009;45(3):305–9. http://dx.doi.org/10.1016/j.ijbiomac.2009.06.004.

[51] Ronghua H, Yumin D, Jianhong Y. Preparation and in vitro anticoagulant activities of alginate sulfate and its quaterized derivatives. Carbohydr Polym. 2003;52(1):19–24. http://dx.doi.org/10.1016/S0144-8617(02)00258-8.

[52] Yan H, Chen X, Li J, Feng Y, Shi Z, Wang X, et al. Synthesis of alginate derivative via the Ugi reaction and its characterization. Carbohydr Polym. 2016;136:757–63. http://dx.doi.org/10.1016/j.carbpol.2015.09.104.

[53] Ugi I. The α-addition of immonium ions and anions to isonitriles accompanied by secondary reactions. Angew Chem. 1962;I(1):8–21. DOI: 10.1002/anie.196200081.

[54] Tønnesen HH, Karlsen J. Alginate in drug delivery systems. Drug Dev Ind Pharm. 2002;28(6):621–30. http://dx.doi.org/10.1081/DDC-120003853.

[55] Sun J, Tan H. Alginate-based biomaterials for regenerative medicine applications. Materials. 2013;6(4):1285–309. DOI:10.3390/ma6041285.

[56] Dash BC, Mandal BB, Kundu SC. Silk gland sericin protein membranes: Fabrication and characterization for potential biotechnological applications. J Biotechnol. 2009;144(4):321–9. http://dx.doi.org/10.1016/j.jbiotec.2009.09.019.

[57] Venkatesan J, Bhatnagar I, Manivasagan P, Kang K, Kim S. Alginate composites for bone tissue engineering: A review. Int J Biol Macromol. 2014;72C:269–81. http://dx.doi.org/10.1016/j.ijbiomac.2014.07.008.

[58] Zhang Y, Liu J, Huang L, Wang Z, Wang L. Design and performance of a sericin-alginate interpenetrating network hydrogel for cell and drug delivery. Sci Rep. 2015;5(4):12374. DOI:10.1038/srep12374

[59] Singh B, Sharma V, Chauhan D. Gastroretentive floating sterculia-alginate beads for use in antiulcer drug delivery. Chem Eng Res Des. 2010;88(8):997–1012. http://dx.doi.org/10.1016/j.cherd.2010.01.017.

[60] Sinha P, Ubaidulla U, Hasnain MS, Nayak AK, Rama B. Alginate-okra gum blend beads of diclofenac sodium from aqueous template using $ZnSO_4$ as a cross-linker. Int J Biol Macromol. 2015;79:555–63. http://dx.doi.org/10.1016/j.ijbiomac.2015.04.067.

[61] Ngah WSW, Fatinathan S. Adsorption of Cu(II) ions in aqueous solution using chitosan beads, chitosan-GLA beads and chitosan-alginate beads. Chem Eng J. 2008;143(1):62–72. http://dx.doi.org/10.1016/j.cej.2007.12.006.

[62] Gotoh T, Matsushima K, Kikuchi KI. Preparation of alginate-chitosan hybrid gel beads and adsorption of divalent metal ions. Chemosphere. 2004;55(1):135–40. http://dx.doi.org/10.1016/j.chemosphere.2003.11.016.

[63] Vijaya Y, Popuri SR, Boddu VM, Krishnaiah A. Modified chitosan and calcium alginate biopolymer sorbents for removal of nickel (II) through adsorption. Carbohydr Polym. 2008;72(2):261–71. http://dx.doi.org/10.1016/j.carbpol.2007.08.010.

[64] Dubey R, Bajpai J, Bajpai AK. Chitosan-alginate nanoparticles (CANPs) as potential nanosorbent for removal of Hg (II) ions. Environ Nanotechnol Monit Manag. 2016;6:32–44. http://dx.doi.org/10.1016/j.enmm.2016.06.008

[65] Cao T-T, Zhang Y-Q. Processing and characterization of silk sericin from Bombyx mori and its application in biomaterials and biomedicines. Mater Sci Eng C. 2016;61:940–52. http://dx.doi.org/10.1016/j.msec.2015.12.082.

[66] Zhang YQ. Applications of natural silk protein sericin in biomaterials. Biotechnol Adv. 2002;20(2):91–100. http://dx.doi.org/10.1016/S0734-9750(02)00003-4.

[67] Aramwit P, Siritientong T, Srichana T. Potential applications of silk sericin, a natural protein from textile industry by-products. Waste Manag Res. 2012;30(3):217–24. DOI: 10.1177/0734242X11404733.

[68] Mohammed ZH, Hill SE, Mitchell JR. Covalent crosslinking in heated protein systems. J Food Sci. 2000;65(2):221–6. http://dx.doi.org/10.1111/j.1365-2621.2000.tb15983.x.

[69] Gasperini L, Mano JF, Reis RL. Natural polymers for the microencapsulation of cells. J R Soc Interface. 2014;11(100):20140817. http://dx.doi.org/10.1098/rsif.2014.0817.

[70] Khandai M, Chakraborty S, Sharma A, Pattnaik S, Dinda SC, Sen KK, et al. Preparation and evaluation of algino-sericin mucoadhesive: An approach for sustained drug delivery. J Adv Pharm Res. 2010;1:48–60.

[71] Khampieng T, Aramwit P, Supaphol P. Silk sericin loaded alginate nanoparticles: Preparation and anti-inflammatory efficacy. Int J Biol Macromol. 2015;80:636–43. http://dx.doi.org/10.1016/j.ijbiomac.2015.07.018.

[72] Nayak S, Dey S, Kundu SC. Silk sericin-alginate-chitosan microcapsules: Hepatocytes encapsulation for enhanced cellular functions. Int J Biol Macromol 2014;65:258–66. http://dx.doi.org/10.1016/j.ijbiomac.2014.01.042.

[73] Vidart JMM, Nakashima M, da Silva TL, Rosa PCP, Gimenes ML, Vieira MGA, et al. Sericin and alginate blend as matrix for incorporation of diclofenac sodium. Chem Eng Trans. 2016;52 (*In Press*).

[74] Silva TL, Meinerz VH, Vidart JM, Gimenes ML, Vieira MGA, Silva MGC. Metallic affinity of toxic and noble metals by particles produced from sericin, alginate and poly (ethylene glycol). Chem Eng Trans. 2016;52 (*In press*).

Chapter 5

Chitosan, Chitosan Derivatives and their Biomedical Applications

Gustavo Adolfo Muñoz Ruiz and
Hector Fabio Zuluaga Corrales

Additional information is available at the end of the chapter

http://dx.doi.org/10.5772/66527

Abstract

Chitosan is one of the most studied polysaccharides nowadays. Because of its biocompatibility, biodegradability and abundance in nature, it has had a wide number of applications. In this chapter, an overview of chitosan including its physicochemical properties and characterization methods is presented. Subsequently, the main chitosan chemical modifications via the hydroxyl and amino groups are discussed. These chemical modifications improve chitosan physical properties and expand its range of applications especially in the biomedical field which will also be studied.

Keywords: chitin, chitosan, hydrogels, chitosan derivatives, biomedical applications

1. Introduction

Chitin is the second most abundant polysaccharide in nature after cellulose and can be found in the exoskeletons of crustaceans and mollusks, insect cuticles and fungi [1]. Due to its low solubility in water as well as most organic solvents, chitin is usually converted to chitosan by deacetylation process, obtaining a soluble material in aqueous acid medium. Considering its nontoxicity, biocompatibility, biodegradability and indirect abundance in nature, chitosan has attracted much research interest and has found potential applications in pharmaceutical, textile, paper and food industries, as well as in agriculture and medicine [2, 3]. On the other hand, chitosan structure can be modified through its amino group and the hydroxyl groups. These chemical modifications improve chitosan mechanical properties and its solubility or bring new functional properties and promising applications. In this chapter, the generalities about chitosan will be discuss, including both chemical and enzymatic isolation processes, the characterization techniques and its physicochemical properties. Subsequently, we will focus particularly on the current state of knowledge of chitosan functionalization methods, including carboxymethylation, cross-linking, copolymer grafting, among others. Finally, the most recent advances in the biomedical applications of chitosan and its derivatives, especially tissue engineering, drug delivery systems, gene therapy systems and wound healing will be discussed.

2. Chitin

Chitin is a linear polysaccharide composed by units of 2-acetamide-2-deoxy-β-D-glucopyranose linked through β(1, 4) bonds. This polysaccharide has a structure similar to cellulose but instead of having a hydroxyl group at carbon number 2, chitin has the N-acetyl group as can be seen in **Figure 1**.

Chitin was isolated in 1811 by French chemist and botanist Henri Braconnot, who found in some fungi an insoluble material in alkaline conditions [4, 5]; chitin was the first polysaccharide identified by man, even before cellulose which was discovered in 1838 by French chemist Anselme Payen [6]. In 1823, Antoine Odier found a similar material described by Braconnot while studying the cuticle of some insects and called it chitin, from the Greek "chiton" which refers to a robe or wrap. In 1824, Children extracted chitin from May bug elytra and found nitrogen by elemental analysis. Payen, Fischer and Leuchs also noted the presence of nitrogen in chitin in 1843. Later, Karrer and Zechmeister carried out experiments with chitin and found the N-acetylglucosamine as its main component. Finally, Meyer and Pankow confirmed the structure of chitin through X-ray diffraction studies in the first half of the twentieth century [7].

Chitin has average molecular weight (MW) between 1.03×10^6 and 2.5×10^6 Da and is a polysaccharide insoluble in water as well as in most organic solvents. Chitin can be dissolved in solutions with high ionic strength as hexafluoroacetone or dichloroethane with mineral

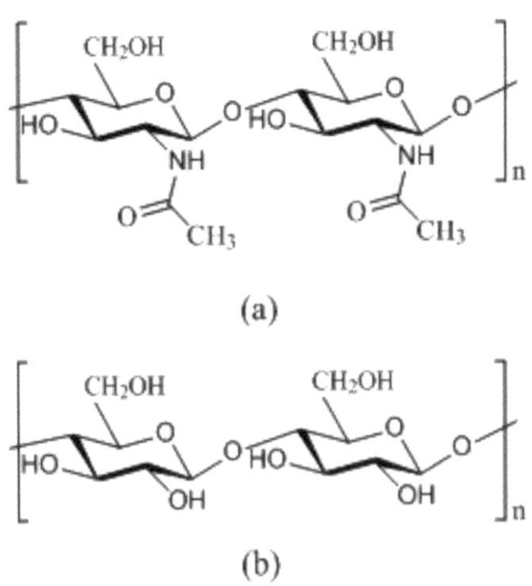

Figure 1. Chemical structure of: (a) chitin and (b) cellulose.

acids, and 5% lithium chloride in dimethylacetamide [2]. In 2006, Tamura and coworkers solubilized chitin in saturated methanol with dihydrate calcium chloride [8]. The insolubility of chitin is associated with a high degree of crystallinity due to the formation of different hydrogen bonds between the N-acetyl groups (N-H...O=C) at C-2 and hydroxyl group at C-3 and C-6 with N-acetyl group (O-H...O=C), keeping chitin chains tightly bound. There are three crystalline forms for this polysaccharide, depending on the source of chitin. α-Chitin is the most abundant form and is present in the shell of shellfish and fungi cell walls. In this crystalline form, chitin chains are organized in an antiparallel configuration, allowing an orthorhombic crystal to form and giving rigidity to the polymer. In β-chitin form, the chains are aligned in parallel, as do cellulose chains, forming monoclinic crystals. In this case, there are more intramolecular hydrogen bonds rather than intermolecular interactions. β-Chitin is commonly associated with proteins in squid and diatomaceae and has weak packing. Finally, there is a γ-chitin form, which is a mixture of the α and β forms. γ-Chitin combines the properties of both polymorphisms and swells when in contact with water [9].

3. Chitosan

Chitosan is obtained via partial or total chitin deacetylation (**Figure 2**) and can be classified as a copolymer of 2-amino-2-deoxy-β-D-glucopyranose (glucosamine) and 2-acetamide-2-deoxy-β-D-glucopyranose (N-acetylglucosamine). In general, when the content of N-acetyl groups is >50% is considered chitin, while for lower values is considered chitosan. It has nitrogen content of 6.80% or higher and is characterized by molecular weights between 1×10^5 and 5×10^5 Da. The discovery of chitosan is attributed to Rouget, who in 1859 found that when chitin was heated in alkaline medium, a material that was soluble in organic acids was obtained [4]. In 1894, Hoppe-Seyler called this material chitosan; however, only until 1950, its chemical structure was elucidated [5].

As a biopolymer, chitosan has potential biomedical applications, since it is biocompatible and biodegradable. Due to its solubility in acidic aqueous medium, many applications at industrial level can be found for chitosan; its solubility is related to the degree of acetylation, molecular weight, and distribution of the acetyl and amino groups along the chain. Additionally, antimicrobial activity is attributed to chitosan when the amino groups are in cationic form, which means that antimicrobial activity of chitosan is higher at low pH [10]. Chitosan has a

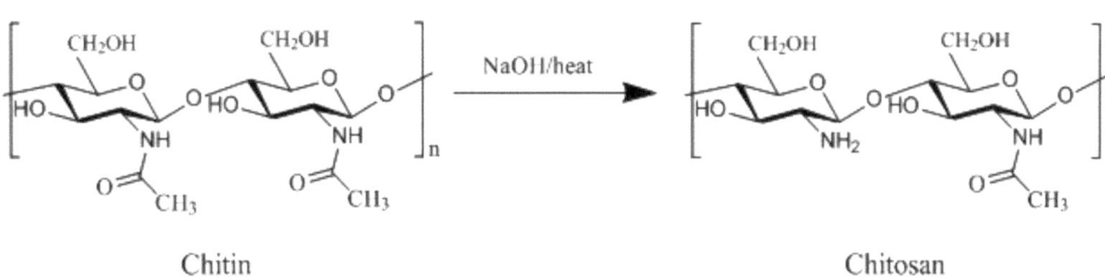

Figure 2. Chitin deacetylation process to produce chitosan.

broad-spectrum antimicrobial activity against both Gram-positive and Gram-negative bacteria [11]. Due to this property, chitosan is a natural antimicrobial agent with potential application in agriculture, food, biomedical and biotechnology fields [12].

4. Global distribution of chitin and chitosan in nature

Chitin is the second most abundant polysaccharide in nature after cellulose [13], it is widely distributed in fungi, diatomaceae, arthropods and other large number of animals and plants; today, a worldwide production of 10^{11} tons of chitin per year is estimated [14, 15]. The commercial exploitation of chitin has been mainly focused on products derived from the marine food industry, such as shrimp, crabs, lobsters and squid [16]. Each year 6–8 million tons of waste crab, shrimp and lobster shells are produced globally [17]. The major useful components and its percentage on a dry weight basis in commercial crustacean wastes are as follows: chitin 15–40%, protein 20–40% and minerals like calcium carbonate 20–50%. The production of chitin from seafood is reported in countries such as India, Poland, Japan, China, United States, Norway and Australia [3]. On the other hand, the production of chitosan from fungal mycelia has increased in recent decades; the study of different types of fungi as a source of chitosan have been reported, with yields ranging from 1 to 8%, average molecular weights between 5.6×10^4 and 1.6×10^5 Da and degree of deacetylation (DD) between 41 and 70% [18–20].

5. Isolation of chitosan

The production of chitosan is based on a hydrolysis of the acetamide group as shown in **Figure 2**. When fungi are used to produce chitosan, the alkaline treatment removes the protein and deacetylates chitin simultaneously. When the shells of crustaceans are used as source of chitosan, two pretreatment are required, one to remove traces of organic material and another to remove calcium carbonate. Nowadays, there are chemicals and enzymatic methods to produce chitosan. In this section, both methods will be discussed showing the advantage and disadvantage of each of them.

5.1. Chemical method

During chemical isolation of chitosan, after obtaining the shells from different sources they are washed, dried and size reduced by grounding into a powder [21]. Then demineralization is carried out using acidic treatment with HCl, H_2SO_4, HNO_3, among others. Subsequently, the deproteination of shells requests an alkaline treatment, which usually is carried out with NaOH. Finally, a deacetylation process of chitin is performed using NaOH again but in a higher concentration, with heating, to produce chitosan. This product is dissolved in acidic medium such as diluted acetic acid, formic acid or lactic acid and finally precipitated with dilute sodium hydroxide [2, 22].

The advantages of chemical production of chitosan include the short processing time and the applicability of this methodology at the industrial scale. However, some disadvantages are found:

1. This method is environmentally unfriendly, due to the large amount of alkaline waste and organic material.
2. There is a high cost because of the effluent treatment generated after acid and alkaline reagents.
3. The continuous hydrolysis of the polymer during the alkaline treatment causes a decrease in the molecular weight and therefore its mechanical properties.

5.2. Enzymatic method

Enzymatic treatments offer an alternative way to extract chitosan from crustacean shells. Lactic acid-producing bacteria have been used for demineralization of crustacean shells instead of acidic treatment. The obtained lactic acid reacts with calcium carbonate yielding calcium lactate, which can be precipitated and removed [23]. For the deproteination of crustacean shells, proteases from bacteria are used. The treatment consists of a fermentation of the crustacean by different species of bacteria such as *Pseudomonas aeruginosa* K-187, *Serratia marcescens* FS-3, and *Bacillus subtilis* [24]. Commercial proteases have previously been used to produce chitosan [25]. Chitin acetyl groups are removed by chitin deacetylase [25]. This enzyme was first found in Mucor rouxii. However, Serratia sp. and Bacillus sp. are bacteria that also produce chitin deacetylase and can be used to generate chitosan [23].

The main advantages of using enzymatic methods are the following:

1. Acidic and alkali treatments, which could be a source of environmental problems, are avoided.
2. Decreasing chitosan molecular weight and mechanical properties is avoided, because selective enzymes are used in each step.

The disadvantage of this biological method is its high cost, which limits its use only to laboratory scale.

6. Physicochemical properties of chitosan

Polysaccharides were considered a source of metabolic energy at first; however, throughout their existence, they have found many applications, as they are nontoxic, biodegradable and abundant in nature. Polysaccharides can be classified according to their acid-base character; there are neutral polysaccharides such as glycogen, cellulose and dextran. Pectin, alginic acid and agar are examples of acidic polysaccharides. Finally, there are some highly basic polysaccharides such as chitin and chitosan [2].

As mentioned previously, chitosan is a linear polysaccharide that contains copolymers of D-glucosamine and N-acetyl-D-glucosamine linked by β(1, 4) glycosidic bonds. The degree of deacetylation (DD) is defined as the glucosamine/N-acetylglucosamine ratio; in other words, DD is the percentage of glucosamine units present in the copolymer chain. Under acidic pH, the amino groups in the chitosan chain become protonated and the polymer dissolves in aqueous media. The protonation constants p*K*a of chitosan change depending on the molecular weight (MW) and DD; p*K*a of 6.51 and 6.39 were found when MW was 1370 and

60 kDa, respectively. On the other hand, p*K*a value was increased from 6.17 to 6.51 when DD decreased from 94.6 to 73.3% [26].

7. Characterization of chitosan

The molecular weight and deacetylation degree are the most important properties of chitosan and dictate the quality and applications of these biomaterials. **Table 1** summarizes the most common methods used to characterize chitosan.

Chitosan property	Characterization methods
Deacetylation degree (DD)	Potentiometric titration [27, 28]
	Elemental analysis [29]
	Fourier transform infrared (FTIR) [30–33]
	Nuclear magnetic resonance (NMR) [34–36]
Molecular weight (MW)	Viscometry [37, 38]
	Gel permeation chromatography (GPC) [37, 39, 40]

Table 1. Chitosan characterization methods.

8. Functionalization of chitosan

Chitosan is a biomaterial that after chemical modifications can find specific biomedical applications. The presence of amino and hydroxyl groups is considered an interesting functionality to modify or improve desired properties. The chemical reactions like cross-linking, carboxymethylation, etherification, graft copolymerization and esterification are the most common modification carried out with chitosan [41].

8.1. Cross-linking

A hydrogel is a natural or synthetic polymeric system, where the chains are cross-linked through covalent and noncovalent bonds, resulting in three-dimensional networks (**Figure 3**). These systems are able to swell rapidly, retain large volumes of water, while maintaining their original shape [42].

Physical or chemical chitosan hydrogels can be formed, sensitive to external stimuli such as pH or ionic strength. In an acidic medium, free chitosan amino groups are protonated generating electrostatic repulsion that promotes the swelling of the polymeric material [43]. To carry out the formation of chemical hydrogels, chitosan can use both hydroxyl groups at C-3 and C-6, as well as the amino group at C-2. Due to the nucleophilic character of amino groups, they are used widely to react with the cross-linking agent. Therefore, chitosan chains are typically cross-linked with dialdehydes: glyoxal and glutaraldehyde [29]. **Figure 4** shows the scanning electron microscopy (SEM) for a chitosan hydrogels cross-linked with glyoxal and

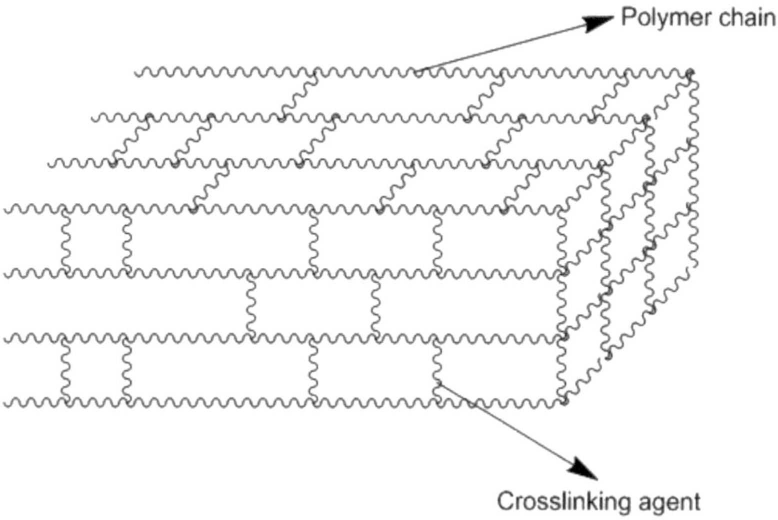

Figure 3. Three-dimensional representation of the polymeric network in a hydrogel.

glutaric acid. The cross-linking of chitosan with glutaraldehyde and sulfosuccinic acid has been used to generate membranes with a good proton conductivity, which makes them useful in fuel cells [44]. Genipin, a compound extracted from the fruit of Gardenia jasminoides and American Genipa has also been used as cross-linker agent [45]. Bodnar and coworkers reported the preparation of chitosan nano-hydrogels, using dicarboxylic and tricarboxylic acids as nontoxic cross-linking agents, which contribute to the biocompatibility of the material [46]. Succinic, glutaric and adipic acid were used recently to prepare chitosan hydrogels [47]. The degree of cross-linking affects hydrogels features such as pore size, mechanical strength and percent swelling. It is clear that small amounts of cross-linking agent increase the absorption capacity. Increasing the cross-link density reduces the swelling due to decreased mobility

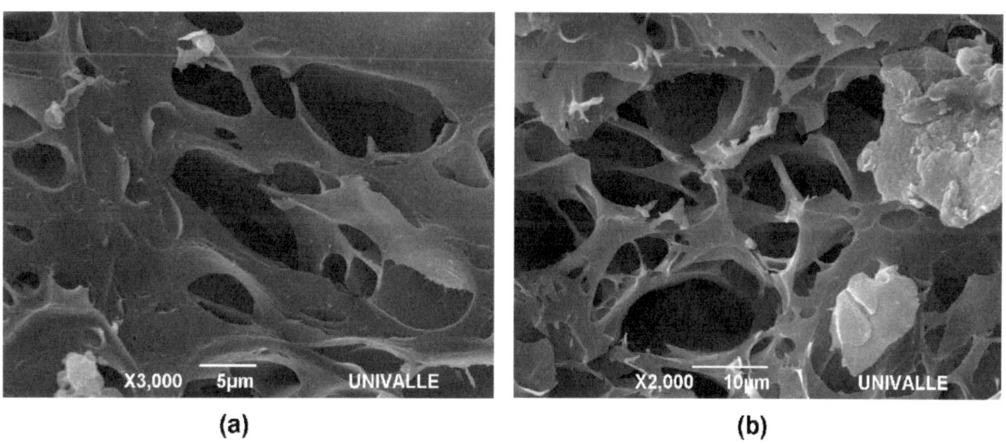

Figure 4. SEM of chitosan hydrogels cross-linked with: (a) glyoxal and (b) glutaric acid. From Laboratorio de Polímeros, Universidad del Valle.

of the polymer chains. Chitosan hydrogels are also used as controlled release systems, both as epidermal and intracorporeal implants, because they are able to maintain a constant drug concentration in a particular environment for a prolonged time [48, 49]. Due to the biocompatibility of chitosan, hydrogels are not only used for controlled release of drugs such as insulin [50], but also used to produce hemodialysis membranes, biodegradable sutures, artificial substituents skin burns healing agents, immobilizing enzymes and cells, among others.

8.2. Graft copolymerization

There are various techniques being used to tailor the surface characteristics of chitosan to introduce or improve desired properties. However, graft copolymerization of synthetic polymers with chitosan is of the most importance. Chitosan and carboxymethyl chitosan have been grafted with metacrylic acid by using ammonium persulfate (APS) as initiator in aqueous media. After grafting, the chitosan derivatives had much improved water solubility (**Figure 5a**) [51, 52]. The copolymerization of aniline with chitosan in the presence of APS has been carried out, producing films and fibers, which have shown protonic conductivity (**Figure 5b**) [53]. Different initiators such as, potassium persulfate (PPS), ceric ammonium nitrate (CAN), thiocarbonationpotassium bromate (TCPB), potassium diperiodatocuprate (III) (PDC), 2,2-azobisisobutyronitrile (AIBN) and ferrous ammonium sulfate (FAS) have been used to initiate grafting copolymerization [54].

Due to its solubility in both water and organic solvents, low toxicity, good biocompatibility and biodegradability, poly(ethylene glycol) (PEG) is a suitable graft-forming polymers. PEG grafted with chitosan have been synthesized using the hydroxyl groups of chitosan (**Figure 5c**). The new material was soluble in water and aqueous solutions of wide pH range [55].

8.3. Carboxymethylation

The low solubility of chitosan is a disadvantage in many of its potential applications. Thus, the preparation of chitosan derivatives has been carried out to improve chitosan solubility in aqueous media. One modification widely used is the carboxymethylation. The procedure consists in dispersing chitosan in isopropanol and NaOH aqueous solution under magnetic stirring and room temperature. Then, a monochloroacetic acid/isopropanol mixture is added to the suspension [56]. Carboxymethyl-chitosan is an amphoteric polymer, and the solubility depends on pH. Although during the reaction O- and N-carboxymethylation may occur simultaneously (**Figure 6**), under controlled reaction conditions (like temperature and stoichiometry) is possible to yield only one of them [37].

8.4. Etherification

Hydroxypropyl chitosan was prepared from chitosan and propylene epoxide under alkaline conditions. This functionalization allows to carry out graft copolymerization, for example, maleic acid was grafted onto hydroxypropyl chitosan to improve the antimicrobial applications, showing good inhibition effect against *S. aureus* and *E. coli* [51].

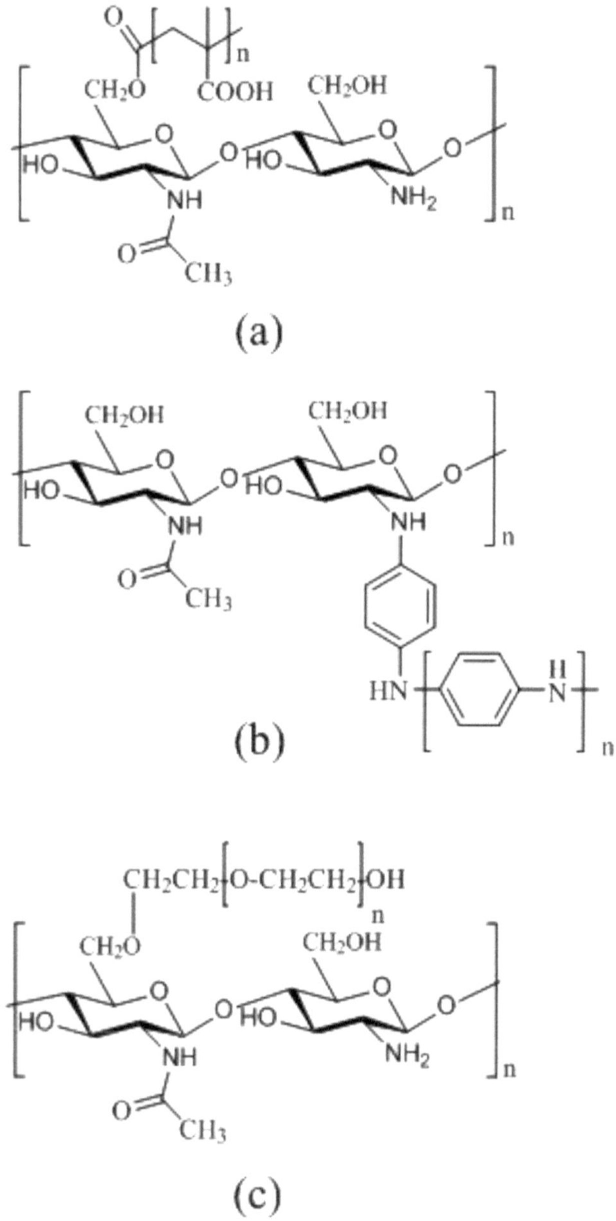

Figure 5. Chemical structure of graft copolymers: (a) chitosan-PMMA, (b) chitosan-PANI and (c) chitosan-PEG.

8.5. Esterification

Other chemical modification that is widely documented in the literature is the esterification of chitosan. Special attention has been paid to the preparation of N,O-acyl chitosan using acetyl chloride in $MeSO_3H$ as solvent (**Figure 7**). Under these conditions, O-acetylated chitosan is the

Figure 6. N- and O-carboxymethylation of chitosan.

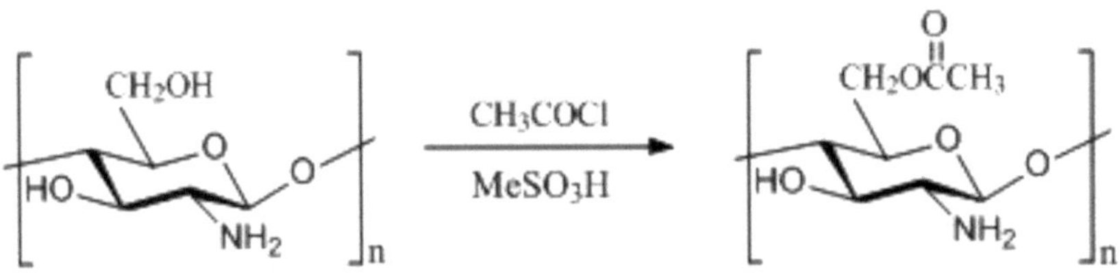

Figure 7. O-acylation of chitosan.

major product compared to the N-acetylated chitosan [57]. It has been shown that acetylation of chitosan substantially improve its antifungal activity [58].

8.6. Phosphorylation

There are some methods to obtain phosphorylated derivatives of chitosan. These derivatives are important due to their interesting biological and chemical properties. They could exhibit bactericidal and osteoinductive properties. Phosphorylated chitosan can be prepared by heating chitosan with orthophosphoric acid in N,N-dimethylformamide (DMF). Another way to prepare phosphorylated chitosan is by the reaction of chitosan with phosphorous pentoxide in methanesulphonic acid, this method is considered very efficient (**Figure 8**) [59]. The methods described here promote phosphorylation of the hydroxyl groups in carbons 3 and 6 of chitosan.

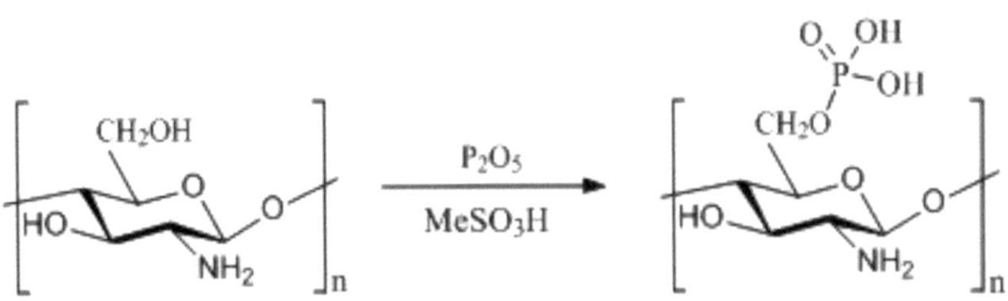

Figure 8. Phosphotylation of chitosan using phosphorous pentoxide.

9. Biomedical applications of chitosan

Biomedical applications of chitosan are related with its most important properties, biocompatibility and biodegradability. Biocompatibility is defined as the ability of a material to be biologically inert in the human body or any other guest while performing certain function (ASTM F2475-11). In other words, biocompatibility is directly related to cytotoxicity that may present the material. The cell culture method is the most simple and widely used way to study both the toxicity and the chitosan-cell interaction. L929 fibroblasts studies supported on chitosan have shown biocompatibility of this biopolymer trough no interference with cell growth, on the contrary, favoring its proliferation [60, 61].

On the other hand, biodegradability is defined as the ability of a material to decompose into low molecular weight molecules like carbon dioxide, methane and water by an enzymatic process (ASTM D-5488-944). Biodegradation of chitosan is related to its depolymerization, oligomers produced are subsequently decomposed to the corresponding monomers glucosamine and N-acetylglucosamine. To carry out this process of depolymerization, nature uses the bacterial enzyme known as chitinase, which is present in the digestive tract of vertebrates and many invertebrates such as fish, lizards, birds and mammals such as pigs. For humans, biodegradation of chitosan is mainly associated with the enzyme lysozyme, widely available in macrophages [62].

9.1. Tissue engineering

Tissue engineering involves the use of scaffolds for the formation of new viable tissue for a medical purpose. To generate tissue, it is necessary to use a supporting material including natural and synthetic polymers. Polylactic acid (PLA) and polyglycolic acid (PGA) are commonly used for fibroblasts growth at laboratory scale [63, 64]. However, chitosan is now one of the leading cell culture media due to its biocompatibility and because it accelerates the growth process. In many of these studies, chitosan, PLA and PGA are mixed to form films, porous structures or beads [65].

Electrospinning is the production of fiber using electric field to draw charged threads of polymer solutions. This method allows to produce scaffolds with tissue engineering applications from different polymeric materials. Due to the repulsive forces between ionic groups within polymer backbone in chitosan, it is restricted the formation of continuous fibers during the electrospinning process. However, some fiber of chitosan have been prepared successfully using trifluoroacetic acid, which can destroy the rigid interactions between the chitosan molecules, improving the electrospinning process and the diameter of the chitosan fibers [66, 67]. **Figure 9** shows PLA nanofibers with a chitosan coating produced by coaxial electrospinning. These nanofibers nets are used to grow osteoblasts and promote the regeneration of bone tissue.

9.2. Drug delivery systems

Like agar, glucomannan and pectin, chitosan with low molecular weight can be used as carrier for solid drug formulations because of their easy bioabsorption, with the advantage that it acts as an antacid, preventing stomach irritation [2]. Chitosan has been studied for colon-targeted delivery of a drug, due to its pH sensitivity and its complete digestion by the colonic bacteria [68].

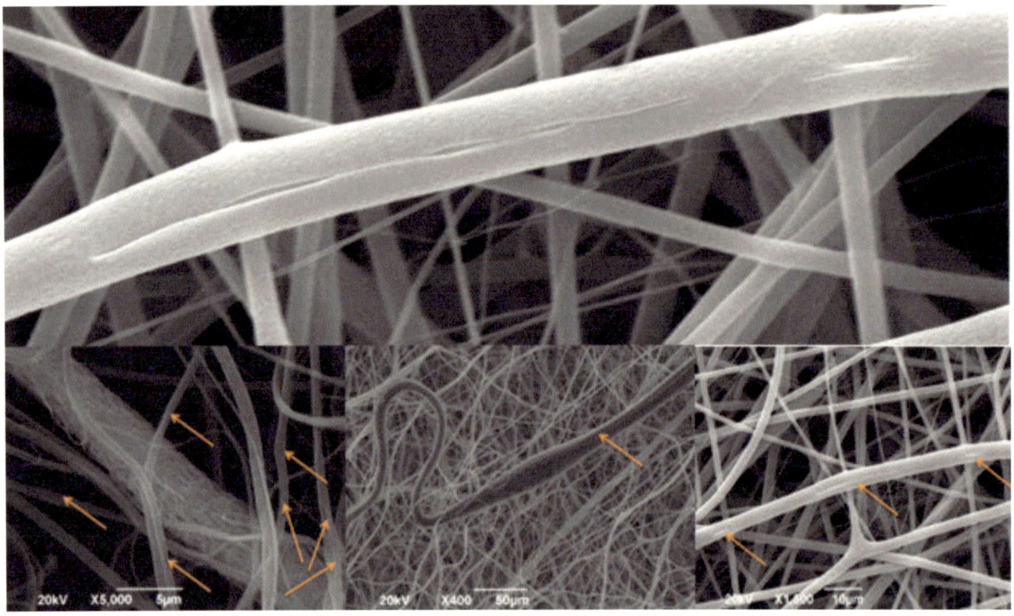

Figure 9. Nanofibers of PLA and chitosan produced by coaxial electrospinning. From Escuela de Ingeniería de Materiales, Universidad del Valle.

9.3. Gene therapy systems

Delivery systems using DNA as a kind of therapeutic agent are considered gene therapy systems, which are a powerful tool for curing many hereditary diseases and treating acquired diseases such as multigenetic disorders [69]. Chitosan is a cationic polyelectrolyte whose positive charge provides a strong electrostatic interaction with negatively charged DNA and protects it from nuclease degradation [70, 71]. For that reason, chitosan is used for nonviral gene delivery. Nanoparticles of folic acid-chitosan-DNA have been used to improve the targeting and transfection rates in gene therapy. These nanoparticles have lower cytoxicity and good DNA condensation. In this case, folic acid is attached to chitosan for targeting cell membrane with folate receptor overexpressed on human cancer cells [72]. Deoxycholic acid-modified chitosan were prepared and characterized to yield self-aggregates/DNA complexes in aqueous media. This chitosan self-aggregates were used as a delivery carrier for the transfection of genetic material in mammalian cells [69]. Reverse microemulsion was used as a template to fabricate chitosan-alginate coreshell nanoparticles encapsulated with enhanced green fluorescent protein (EGFP)-encoded plasmids. These chitosan-alginate coreshell nanoparticles endocytosed by NIH 3T3 cells causes swelling of transport vesicles which renders gene escaping before entering digestive endolysosomal compartment and promotes gene transfection rate [73, 74].

9.4. Wound healing

Wound healing is a biological process related with growth and tissue regeneration. The wound healing process has five important stages: homeostasis, inflammation, migration, proliferation and maturation [75]. A number of studies have reported the use of chitosan membranes to produce wound healing with potential application in patients with deep burns, wounds, etc. PVA have been mixed with chitosan for the preparation of composites and blends films. PVA-chitosan membranes have shown an increasing of the mechanical properties [41]. Chitosan films with oleic acid and glycerol at 1% were prepared previously. These chitosan films exhibited suitable morphology, and in vivo studies with Wistar® rats showing that implanted chitosan films were biocompatible and bioabsorbable leaving the tissue healthy. For all this, chitosan films may allow their use for wound healing [76]. Due to the importance of antibacterial resistance in the field of wound care to avoid complications such as infection and delayed wound healing, nanoparticles of silver combine with chitosan has been used as an antibacterial agent [75, 77].

9.5. Antioxidant and antimicrobial properties

Chitosan is a polysaccharide with antimicrobial properties [78]. The antimicrobial mechanism of chitosan is currently unknown. However, it has been suggested that the polycationic nature below pH 6.5 is a decisive factor. The positive charge in the backbone chain interacts with negatively charged components in microbial cell membranes, altering their barrier properties, and thereby preventing the entry of nutrients or causing the leakage of intracellular contents [79]. Due to its ability to form films, chitosan is employed to produce food packaging as a potential food preservative. However, the limited solubility of chitosan restricts its applicability. At the same time, chitosan have been considered a natural antioxidant [80–82]. Thus, different functionalizations have been carried out to improve both the antimicrobial and antioxidant activities of chitosan.

Functionalization of chitosan with epigallocatechin gallate (EGCG) by a free radical-induced grafting procedure improves both its antimicrobial and its antioxidant activity [83]. Chitosan antioxidant activity can be enhanced by grafting on it protocatechuic acid which is a natural phenolic antioxidant [84]. Chitosan modified with monomethyl fumaric acid presented antibacterial activity against Gram-positive and Gram-negative bacteria [85]. Other chitosan derivatives with antimicrobial activity included the following: chitosan modified with thioglycolic acid [86] and chitosan with polyethylene glycol diacrylate (PEGDA) [87].

10. Summary

Chitin is the second most abundant polysaccharide in nature after cellulose. However, due to its solubility problems, scientific research has focused more on studying its main derivative, chitosan. Chitosan is a compound of great interest due to its biocompatibility and biodegradability. At the same time, antimicrobial and antioxidant properties have been found in chitosan. For all these properties, chitosan has potential application at industrial level as well as biomedical applications. The chemical modifications of chitosan through its functional groups have extended its range of applications, improving its mechanical properties, its solubility, among others. Chitosan is a promising material because in addition to various applications, it is a product that is produced from waste seafood, offsetting the negative impact they may have. For all this, we are about to find chitosan commercially available in many everyday products.

Author details

Gustavo Adolfo Muñoz Ruiz and Hector Fabio Zuluaga Corrales*

*Address all correspondence to: hector.zuluaga@correounivalle.edu.co

University of Valle, Cali, Colombia

References

[1] Kaur S., Dhillon G.S. The versatile biopolymer chitosan: potential sources, evaluation of extraction methods and applications. Critical Reviews in Microbiology. 2014;**40**(2):155–175. doi:10.3109/1040841X.2013.770385

[2] Ravi Kumar M.N.V. A review of chitin and chitosan applications. Reactive and Functional Polymers. 2000;46(1):1–27. DOI: 10.1016/S1381-5148(00)00038-9

[3] Dutta P.K., Dutta J., Tripathi V.S. Chitin and chitosan: chemistry, properties and applications. Journal of Scientific and Industrial Research. 2004;**63**(1):20–31.

[4] Roberts G.A.F. Chitin Chemistry. London: The Macmillan Press Ltd.; 1992. 339 p. doi:10.1007/978-1-349-11545-7

[5] Khor E. Chitin: Fulfilling a Biomaterials Promise. Elsevier Ltd.; 2001. 136 p. doi:10.1016/B978-008044018-7/50001-4

[6] Zugenmaier P. Crystalline Cellulose and Derivatives. Springer Berlin Heidelberg; 2008. 285 p. doi:10.1007/978-3-540-73934-0

[7] Meyer K.H., Pankow G.W. Sur la constitution et la structure de la chitine. Helvetica Chimica Acta. 1935;**18**(1):589–598. doi:10.1002/hlca.19350180177

[8] Tamura H., Nagahama H., Tokura S. Preparation of chitin hydrogel under mild conditions. Cellulose. 2006;**13**(4):357–364. doi:10.1007/s10570-006-9058-z

[9] Minke R., Blackwell J. The structure of α-chitin. Journal of Molecular Biology. 1978;**120**(2):167–181. doi:10.1016/0022-2836(78)90063-3

[10] Goy R.C., De Britto D., Assis O.B. A review of the antimicrobial activity of chitosan. Polímeros. 2009;**19**(3):241–247. doi:10.1590/S0104-14282009000300013

[11] Kanatt S.R., Rao M.S., Chawla S.P., Sharma A. Effects of chitosan coating on shelf-life of ready-to-cook meat products during chilled storage. LWT—Food Science and Technology. 2013;**53**(1):321–326. doi:10.1016/j.lwt.2013.01.019

[12] Hosseinnejad M., Jafari S.M. Evaluation of different factors affecting antimicrobial properties of chitosan. International Journal of Biological Macromolecules. 2016;**85**:467–475. doi:10.1016/j.ijbiomac.2016.01.022

[13] Synowiecki J. Al-Khateeb N.A. Production, Properties, and Some New Applications of chitin and its derivaties. Critical Reviews in Food Science and Nutrition. 2003;43(2):145–171. DOI: dx.doi.org/10.1080/10408690390826473

[14] Kurita K. Chitin and chitosan: functional biopolymers from marine crustaceans. Marine Biotechnology. 2006;**8**:203–226. doi:10.1007/s10126-005-0097-5

[15] Revathi M., Saravanan R., Shanmugam A. Production and characterization of chitinase from Vibrio species, a head waste of shrimp Metapenaeus dobsonii (Miers, 1878) and chitin of Sepiella inermis Orbigny, 1848. Advances in Bioscience and Biotechnology. 2012;**3**(4):392–397. doi:10.4236/abb.2012.34056

[16] Kim S. K., editor. Chitin, Chitosan, Oligosaccharides and Their Derivatives: Biological Activities and Applications. Boca Raton: CRC Press; 2011. 633 p.

[17] Food and Agriculture Organization of the United Nations. The State of World Fisheries and Aquaculture. Rome: FAO; 2014. 243 p.

[18] White S.A., Farina P.R., Fulton I. Production and isolation of chitosan from Mucor rouxii. Applied and Environmental Microbiology. 1979;**38**(2):323–328.

[19] Chatterjee S., Adhya M., Guha A.K., Chatterjee B.P. Chitosan from Mucor rouxii: production and physico-chemical characterization. Process Biochemistry. 2005;**40**(1):395–400. doi:10.1016/j.procbio.2004.01.025

[20] Nwe N., Furuike T., Osaka I., Fujimori H., Kawasaki H., Arakawa R., et al. Laboratory scale production of 13C labeled chitosan by fungi Absidia coerulea and Gongronella butleri grown in solid substrate and submerged fermentation. Carbohydrate Polymers. 2011;**84**(2):743–750. doi:10.1016/j.carbpol.2010.06.023

[21] Abdou E.S., Nagy K.S., Elsabee M.Z. Extraction and characterization of chitin and chitosan from local sources. Bioresource Technology. 2008;**99**(5):1359–1367. doi:10.1016/j.biortech.2007.01.051

[22] Bajaj M., Winter J., Gallert C. Effect of deproteination and deacetylation conditions on viscosity of chitin and chitosan extracted from Crangon crangon shrimp waste. Biochemical Engineering Journal. 2011;**56**(1–2):51–62. doi:10.1016/j.bej.2011.05.006

[23] Hamed I., Özogul F., Regenstein J.M. Industrial applications of crustacean by-products (chitin, chitosan, and chitooligosaccharides): a review. Trends in Food Science and Technology. 2016;**48**:40–50. doi:10.1016/j.tifs.2015.11.007

[24] Jo G.H., Jung W.J., Kuk J.H., Oh K.T., Kim Y.J., Park R.D. Screening of protease-producing *Serratia marcescens* FS-3 and its application to deproteinization of crab shell waste for chitin extraction. Carbohydrate Polymers. 2008;**74**(3):504–508. doi:10.1016/j.carbpol.2008.04.019

[25] Cai J., Yang J., Du Y., Fan L., Qiu Y., Li J., Kennedy J.F. Enzymatic preparation of chitosan from the waste Aspergillus niger mycelium of citric acid production plant. Carbohydrate Polymers. 2006;**64**(2):151–157. doi:10.1016/j.carbpol.2005.11.004

[26] Wang Q.Z., Chen X.G., Liu N., Wang S.X., Liu C.S., Meng X.H., Liu C.G. Protonation constants of chitosan with different molecular weight and degree of deacetylation. Carbohydrate Polymers. 2006;**65**(2):194–201. doi:10.1016/j.carbpol.2006.01.001

[27] Dos Santos Z.M., Caroni A.L., Pereira M.R., Da Silva D.R., Fonseca J.L. Determination of deacetylation degree of chitosan: a comparison between conductometric titration and CHN elemental analysis. Carbohydrate Research. 2009;**344**(18):2591–2595. doi:10.1016/j.carres.2009.08.030

[28] Jiang X., Chen L., Zhong W. A new linear potentiometric titration method for the determination of deacetylation degree of chitosan. Carbohydrate Polymers. 2003;**54**(4):457–463. doi:10.1016/j.carbpol.2003.05.004

[29] Muñoz G., Valencia C., Valderruten N., Ruiz-Durántez E., Zuluaga F. Extraction of chitosan from *Aspergillus niger* mycelium and synthesis of hydrogels for controlled release of betahistine. Reactive and Functional Polymers. 2015;**91–92**:1–10. doi:10.1016/j.reactfunctpolym.2015.03.008

[30] Sajomsang W., Ruktanonchai U.R., Gonil P., Nuchuchua O. Mucoadhesive property and biocompatibility of methylated N-aryl chitosan derivatives. Carbohydrate Polymers. 2009;**78**(4):945–952. doi:10.1016/j.carbpol.2009.07.020

[31] Urreaga J.M., De la Orden M.U. Chemical interactions and yellowing in chitosan-treated cellulose. European Polymer Journal. 2006;**42**(10):2606–2616. doi:10.1016/j.eurpolymj.2006.05.002

[32] Nwe N., Stevens W.F., Tokura S., Tamura H. Characterization of chitosan and chitosan–glucan complex extracted from the cell wall of fungus Gongronella butleri USDB 0201 by enzymatic method. Enzyme and Microbial Technology. 2008;**42**(3):242–251. doi:10.1016/j.enzmictec.2007.10.001

[33] Kasaai M.R. A review of several reported procedures to determine the degree of N-acetylation for chitin and chitosan using infrared spectroscopy. Carbohydrate Polymers. 2008;**71**(4):497–508. doi:10.1016/j.carbpol.2007.07.009

[34] Heux L., Brugnerotto J., Desbrières J., Versal M.F., Rinaudo M. Solid state NMR for determination of degree of acetylation of chitin and chitosan. Biomacromolecules. 2000;**1**(4):746–751. doi:10.1021/bm000070y

[35] Hirai A., Odani H., Nakajima A. Determination of degree of deacetylation of chitosan by 1H NMR spectroscopy. Polymer Bulletin. 1991;**26**(1):87–94. doi:10.1007/BF00299352

[36] Lavertu M., Xia Z., Serreqi A.N., Berrada M., Rodrigues A., Wang D., Buschmann M.D., et al. A validated 1H NMR method for the determination of the degree of deacetylation

of chitosan. Journal of Pharmaceutical and Biomedical Analysis. 2003;**32**(6):1149–1158. doi:10.1016/S0731-7085(03)00155-9

[37] Rinaudo M. Chitin and chitosan: properties and applications. Progress in Polymer Science. 2006;**31**(7):603–632. doi:10.1016/j.progpolymsci.2006.06.001

[38] Kasaai M.R. Calculation of mark–houwink–sakurada (MHS) equation viscometric constants for chitosan in any solvent–temperature system using experimental reported viscometric constants data. Carbohydrate Polymers. 2007;**68**(3):477–488. doi:10.1016/j.carbpol.2006.11.006

[39] Zielinska K., Shostenko A.G., Truszkowski S. Analysis of chitosan by gel permeation chromatography. High Energy Chemistry. 2014;**42**(2):72–75. doi:10.1134/S0018143914020143

[40] De Benedictis V.M., Soloperto G., Demitri C. Correction of MHS viscosimetric constants upon numerical simulation of temperature induced degradation kinetic of chitosan solutions. Polymers. 2016;**8**(6):1–17. doi:10.3390/polym8060210

[41] Rafique A., Zia K.M., Zuber M., Tabasum S., Rehman S. Chitosan functionalized poly(vinyl alcohol) for prospects biomedical and industrial applications: a review. International Journal of Biological Macromolecules. 2016;**87**:141–154. doi:10.1016/j.ijbiomac.2016.02.035

[42] Shibayama M., Tanaka T. Volume phase transition and related phenomena of polymer gels. In: Dusek K, editor. Responsive Gels: Volume Transitions I. 1st ed. Springer Berlin Heidelberg; 1993. pp. 1–62. doi:10.1007/3-540-56791-7_1

[43] Gil E.S., Hudson S.M. Stimuli-reponsive polymers and their bioconjugates. Progress in Polymer Science. 2004;**29**(12):1173–1222. doi:10.1016/j.progpolymsci.2004.08.003

[44] Dashtimoghadam E., Hasani-Sadrabadi M.M., Moaddel H. Structural modification of chitosan biopolymer as a novel polyelectrolyte membrane for green power generation. Polymers for Advanced Technologies. 2010;**21**(10):726–734. doi:10.1002/pat.1496

[45] Muzzarelli R.A. Genipin-crosslinked chitosan hydrogels as biomedical and pharmaceutical aids. Carbohydrate Polymers. 2009;**77**(1):1–9. doi:10.1016/j.carbpol.2009.01.016

[46] Bodnar M., Hartmann J.F., Borbely J. Preparation and characterization of chitosan-based nanoparticles. Biomacromolecules. 2005;**6**(5):2521–2527. doi:10.1021/bm0502258

[47] Valderruten N.E, Valverde J.D., Zuluaga F., Ruiz-Durántez E. Synthesis and characterization of chitosan hydrogels cross-linked with dicarboxylic acids. Reactive and Functional Polymers. 2014;**84**:21–28. doi:10.1016/j.reactfunctpolym.2014.08.006

[48] Hoarea T.R., Kohane D.S. Hydrogels in drug delivery: progress and challenges. Polymer. 2008;**49**(8):1993–2007. doi:10.1016/j.polymer.2008.01.027

[49] Peppas N.A, Bures P., Leobandung W., Ichikawa H. Hydrogels in pharmaceutical formulations. European Journal of Pharmaceutics and Biopharmaceutics. 2000;**50**(1):27–46. doi:10.1016/S0939-6411(00)00090-4

[50] Sarmento B., Ribeiro A., Veiga F., Sampaio P., Neufeld R., Ferreira D. Alginate/chitosan nanoparticles are effective for oral insulin delivery. Pharmaceutical Research. 2007;**24**(12):2198–2206. doi:10.1007/s11095-007-9367-4

[51] Xie W., Xu P., Wang W., Liu Q. Preparation and antibacterial activity of a water-soluble chitosan derivative. Carbohydrate Polymers. 2002;**50**(1):35–40. doi:10.1016/S0144-8617(01)00370-8

[52] Sun T., Xie W., Xu P. Superoxide anion scavenging activity of graft chitosan derivatives. Carbohydrate Polymers. 2004;**58**(4):374–382. doi:10.1016/j.carbpol.2004.06.042

[53] Yang S., Tirmizi S.A., Burns A., Barney A.A., Risen Jr W.M. Chitaline materials: soluble chitosan-polyaniline copolymers and their conductive doped forms. Synthetic Metals. 1989;**32**(2):191–200. doi:10.1016/0379-6779(89)90841-2

[54] Olteanu C.E. Applications of functionalized chitosan. Scientific Study and Research: Chemistry and Chemical Engineering, Biotechnology, Food Industry. 2007;**8**(3):227–256.

[55] Gorochovceva N., Makuška R. Synthesis and study of water-soluble chitosan-O-poly(ethylene glycol) graft copolymers. European Polymer Journal. 2004;**40**(4):685–691. doi:10.1016/j.eurpolymj.2003.12.005

[56] De Abreu F.R., Campana-Filho S.P. Preparation and characterization of carboxymethyl-chitosan. Polímeros. 2005;**15**(2):79–83. doi:10.1590/S0104-14282005000200004

[57] Badawy M.E., Rabea E.I., Rogge T.M., Stevens C.V., Smagghe G., Steurbaut W., Höfte M. Synthesis and fungicidal activity of new N,O-Acyl chitosan derivatives. Biomacromolecules. 2004;**5**(2):589–595. doi:10.1021/bm0344295

[58] Badawy M.E., Rabea E.I., Rogge T.M., Stevens C.V., Steurbaut W., Höfte M., Smagghe G. Fungicidal and insecticidal activity of O-Acyl chitosan derivatives. Polymer Bulletin. 2005;**54**(4):279–289. doi:10.1007/s00289-005-0396-z

[59] Jayakumar R., Selvamurugan N., Nair S.V., Tokura S., Tamura H. Preparative methods of phosphorylated chitin and chitosan–an overview. International Journal of Biological Macromolecules. 2008;**43**(3):221–225. doi:10.1016/j.ijbiomac.2008.07.004

[60] Mori T., Okumura M., Matsuura M., Ueno K., Tokura S., Okamoto Y., Minami S., Fujinaga T. Effects of chitin and its derivatives on the proliferation and cytokine production of fibroblasts in vitro. Biomaterials. 1997;**18**(13):947–951. doi:10.1016/S0142-9612(97)00017-3

[61] Costa-Júnior E.D.S., Pereira M.M., Mansur H.S. Properties and biocompatibility of chitosan films modified by blending with PVA and chemically crosslinked. Journal of Materials Science: Materials in Medicine. 2009;**20**(2):553–561. doi:10.1007/s10856-008-3627-7

[62] Boot R.G., Renkema G.H., Strijland A., Van Zonneveld A.J., Aerts J.M. Cloning of a cDNA encoding chitotriosidase, a human chitinase produced by macrophages. The Journal of Biological Chemistry. 1995;**270**(44):26252–26256. doi:10.1074/jbc.270.44.26252

[63] Wang Y.C., Lin M.C., Wang D.M., Hsieh H.J. Fabrication of a novel porous PGA-chitosan hybrid matrix for tissue engineering. Biomaterials. 2003;**24**(6):1047–1057. doi:10.1016/S0142-9612(02)00434-9

[64] Zhu A., Zhang M., Wu J., Shen J. Covalent immobilization of chitosan/heparin complex with a photosensitive hetero-bifunctional crosslinking reagent on PLA surface. Biomaterials. 2002;**23**(23):4657–4665. doi:10.1016/S0142-9612(02)00215-6

[65] Jarry C., Chaput C., Chenite A., Renaud M.A., Buschmann M., Leroux J.C. Effects of steam sterilization on thermogelling chitosan-based gels. Journal of Biomedical Materials Research. 2001;**58**(1):127–135. doi:10.1002/1097-4636(2001)58:1<127::AID-JBM190>3.0.CO;2-G

[66] Jayakumar R., Prabaharan M., Nair S.V., Tamura H. Novel chitin and chitosan nanofibers in biomedical applications. Biotechnology Advances. 2010;**28**(1):142–150. doi:10.1016/j.biotechadv.2009.11.001

[67] LogithKumar R., KeshavNarayan A., Dhivya S., Chawla A., Saravanan S., Selvamurugan N. A review of chitosan and its derivatives in bone tissue engineering. Carbohydrate Polymers. 2016;**151**:172–188. doi:10.1016/j.carbpol.2016.05.049

[68] Shimono N., Takatori T., Ueda M., Mori M., Higashi Y., Nakamura Y. Chitosan dispersed system for colon-specific drug delivery. International Journal of Pharmaceutics. 2002;**245**(1–2):45–54. doi:10.1016/S0378-5173(02)00344-7

[69] Lee K.Y., Kwon I.C., Kim Y.H., Jo W.H., Jeong S.Y. Preparation of chitosan self-aggregates as a gene delivery system. Journal of Controlled Release. 1998;**51**(2–3):213–220. doi:10.1016/S0168-3659(97)00173-9

[70] Fang N., Chan V., Mao H.Q., Leong K.W. Interactions of phospholipid bilayer with chitosan: effect of molecular weight and pH. Biomacromolecules. 2001;**2**(4):1161–1168. doi:10.1021/bm015548s

[71] Cui Z., Mumper R.J. Chitosan-based nanoparticles for topical genetic immunization. Journal of Controlled Release. 2001;**75**(3):409–419. doi:10.1016/S0168-3659(01)00407-2

[72] Mansouri S., Cuie Y., Winnik F., Shi Q., Lavigne P., Benderdour M., Beaumont E., Fernandes J.C. Characterization of folate-chitosan-DNA nanoparticles for gene therapy. Biomaterials. 2006;**27**(9):2060–2065. doi:10.1016/j.biomaterials.2005.09.020

[73] You J.O., Liu Y.C., Peng C.A. Efficient gene transfection using chitosan–alginate core-shell nanoparticles. International Journal of Nanomedicine. 2006;**1**(2):173–180.

[74] Jayakumar R., Chennazhi K.P., Muzzarelli R.A., Tamura H., Nair S.V., Selvamurugan N. Chitosan conjugated DNA nanoparticles in gene therapy. Carbohydrate Polymers. 2010;**79**(1):1–8. doi:10.1016/j.carbpol.2009.08.026

[75] Azuma K., Izumi R., Osaki T., Ifuku S., Morimoto M., Saimoto H., et al. Chitin, chitosan, and its derivatives for wound healing: old and new materials. Journal of Functional Biomaterials. 2015;**6**(1):104–142. doi:10.3390/jfb6010104

[76] Silva D.J. B., Zuluaga F., Carlos H. Evaluation of Biocompatibility of Chitosan Films from the Mycelium of Aspergillus niger in Connective Tissue of Rattus norvegicus. Journal of Molecular and Genetic Medicine. 2015;9(3):1-8. DOI: 10.4172/1747-0862.1000174

[77] Jayakumar R., Menon D., Manzoor K., Nair S.V., Tamura H. Biomedical applications of chitin and chitosan based nanomaterials – a short review. Carbohydrate Polymers. 2010;**82**(2):227–232. doi:10.1016/j.carbpol.2010.04.074

[78] Choi B.K., Kim K.Y., Yoo Y.J., Oh S.J., Choi J.H., Kim C.Y. In vitro antimicrobial activity of a chitooligosaccharide mixture against Actinobacillusactinomycetemcomitans and Streptococcus mutans. International Journal of Antimicrobial Agents. 2001;**18**(6):553–557. doi:10.1016/S0924-8579(01)00434-4

[79] Fernandez-Saiz P., Ocio M.J., Lagaron J.M. The use chitosan in microbial films for food protection. Reviews: Perspectives in Agriculture, Veterinary Science, Nutrition and Natural Resources. 2010;**5**(24):1–11. doi:10.1079/PAVSNNR20105024

[80] Xie W., Xu P., Liu Q. Antioxidant activity of water-soluble chitosan derivatives. Bioorganic and Medicinal Chemistry Letters. 2001;**11**(13):1699–1701. doi:10.1016/S0960-894X(01)00285-2

[81] Wang L., Dong Y., Men H., Tong J., Zhou J. Preparation and characterization of active films based on chitosan incorporated tea polyphenols. Food Hydrocolloids. 2013;**32**(1):35–41. doi:10.1016/j.foodhyd.2012.11.034

[82] Yuan G., Lv H., Yang B., Chen X., Sun H. Physical properties, antioxidant and antimicrobial activity of chitosan films containing carvacrol and pomegranate peel extract. Molecules. 2015;**20**(6):11034–11045. doi:10.3390/molecules200611034

[83] Moreno-Vásquez M.J., Valenzuela-Buitimea E.L., Plascencia-Jatomea M., Encinas-Encinas J.C., Rodríguez-Félix F., Sánchez-Valdes S., Rosas-Burgos E.C., Ocaño-Higuera V.M., Graciano-Verdugo A.Z. Functionalization of chitosan by a free radical reaction: characterization, antioxidant and antibacterial potential. Carbohydrate Polymers. 2017;**155**(2):117–127. doi:10.1016/j.carbpol.2016.08.056

[84] Liu J., Meng C.G., Liu S., Kan J., Jin C.H. Preparation and characterization of protocatechuic acid grafted chitosan films with antioxidant activity. Food Hydrocolloids. 2017;**63**:457–466. doi:10.1016/j.foodhyd.2016.09.035

[85] Khan I., Ullah S., Oh D.H. Chitosan grafted monomethyl fumaric acid as a potential food preservative. Carbohydrate Polymers. 2016;**152**(5):87–96. doi:10.1016/j.carbpol.2016.06.073

[86] Geisberger G., Gyenge E.B., Hinger D., Käch A., Maake C., Patzke G.R. Chitosan-thioglycolic acid as a versatile antimicrobial agent. Biomacromolecules. 2013;**14**(4):1010–1017. doi:10.1021/bm3018593

[87] Ma G., Zhang X., Han J., Song G., Nie J. Photo-polymeriable chitosan derivative prepared by Michael reaction of chitosan and polyethylene glycol diacrylate (PEGDA). International Journal of Biological Macromolecules. 2009;**45**(5):499–503. doi:10.1016/j.ijbiomac.2009.08.007

Chapter 6

Chitosan: Strategies to Increase and Modulate Drug Release Rate

David Lucio and María Cristina Martínez-Ohárriz

Additional information is available at the end of the chapter

http://dx.doi.org/10.5772/65714

Abstract

Chitin is the second most abundant polysaccharide present in nature; however, chitin has more applications when transformed into chitosan (CS). It is biocompatible, biodegradable, mucoadhesive, soluble in acidic-solutions, nontoxic and nonallergenic. The main drawback of chitosan in pharmaceutical procedures is its low solubility in physiological medium. Chitosan shows physicochemical characteristics that allow it to interact with a wide variety of molecules. This is of particular interest when increasing the solubility of poor water-soluble drugs. For this purpose, chitosan can be used in oral, nasal and ocular routes. In order to modulate drug release rate and achieve a proper drug delivery in physiological medium, some parameters can be modified when solid dispersions or nanoparticles (NPs) based on chitosan are being designed. In case of nanoparticles, chitosan can be used as the main component or as a modifying agent. In order to optimize drug loading and drug delivery, response surface methodology (RSM) is an interesting tool usually underestimated in the pharmaceutical field, which allows us to optimize the parameters involved in the process simultaneously and not by different steps, which usually lead to mistakes.

Keywords: chitosan, carboxymethyl chitosan, drug delivery, nanoparticles, solid dispersions, porous microstructure, response surface methodology

1. Introduction

Chitin is the second most abundant polysaccharide and the only cationic one present in nature; these characteristics make it suitable and interesting for industrial uses [1]. However, chitin has more applications when transformed into chitosan (CS) by partial deacetylation under alkaline conditions [2].

Chitosan [(1–4)-2-amino-2-deoxy-β-D-glucan] is a linear cationic polysaccharide comprising by N-acetyl glucosamine and glucosamine units (**Figure 1**). Although research is more advanced in the case of cellulose, amino groups provide an important advantage to chitosan due to its ease to be chemically modified, including functional groups that would give new properties for more specific uses [3, 4].

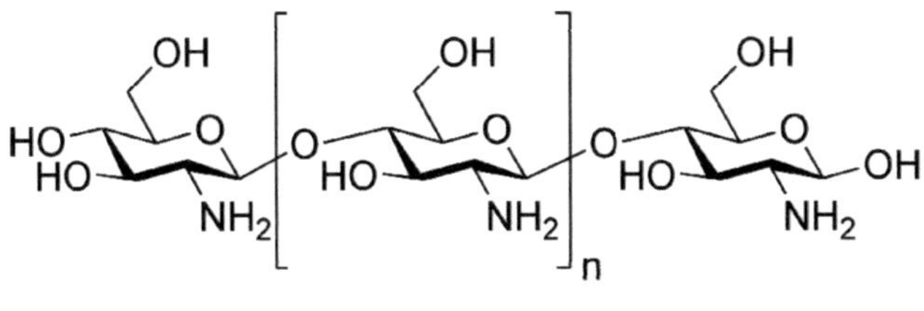

Figure 1. Representation of chitosan molecule.

Chitosan are receiving increasing attention in the industrial domain as a consequence of their special properties. It is a biocompatible, biodegradable, mucoadhesive, and transmucosal penetration enhancer, soluble in acidic solutions, nontoxic and nonallergenic [5]. Moreover, because it is generally recognized as safe (GRAS), which together with its abundance and low cost make chitosan a particularly interesting material for its use in the pharmaceutical field [6, 7]. In addition, its antacid and antiulcer activities made chitosan and its derivatives an interesting alternative if gastric irritation wants to be avoided when anti-inflammatory drug are administered [8, 9]. Finally, chitosan is an excellent direct compression adjuvant, an important characteristic for potential pharmaceutical formulations [10].

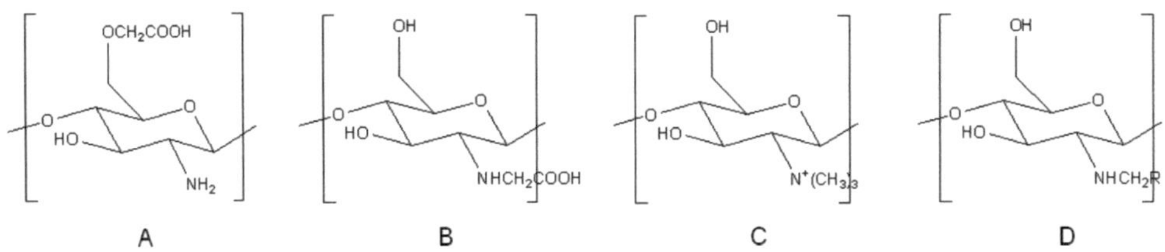

Figure 2. Representation of molecules of O-carboxymethyl chitosan (A), N-carboxymethyl chitosan (B), trimethyl chitosan (C), and N-alkyl chitosan (D).

The main drawback of chitosan in pharmaceutical procedures is its low solubility in aqueous solutions and physiological medium [11, 12]; acetylation degree and molecular weight are the most relevant variables which condition chitosan solubility [13]. However, this problem can be solved, thanks to the easy modification of its primary aliphatic amine group. This fact allows us to obtain water-soluble derivatives such as carboxymethyl chitosan, trimethyl chitosan, and N-alkyl chitosans (**Figure 2**) [14]. Both chitosan and its derivatives have physicochemical characteristics that allow them to interact with a wide variety of drugs. This is of particular interest when trying to increase the solubility of poor water-soluble drugs [15, 16], which is a fundamental strategy to increase low bioavailability of drugs belonging to Groups II (low solubility, high permeability) and IV (low solubility, low permeability) of Biopharmaceutical Classification System (BCS). In this sense, different strategies have been developed to increase the solubility of these drugs, such as particle size reduction, drug fusion and preparation of solid dispersions, nanosuspensions, amorphous states, inclusion complexes, or nanoparticles (NPs).

For this purpose, chitosans can be used in oral, nasal, and ocular routes, being the first one considered as the safest and most advisable way. Drug-controlled release by oral route has been achieved using chitosan in several pharmaceutical dosage forms such as solid dispersions, cross-linked hydrogels or nanoparticles [17, 18]. In case of nanoparticles, chitosan can be used both as modifying agent and as the main component by different methods such as cross-linking, ionic gelation, or solvent evaporation.

As it is well known, drug delivery from polymeric systems is influenced by several factors: polymer molecular weight, deacetylation and substitution degree of polysaccharides, nanoparticle size, porous structure, or compression force are some of the factors which directly determine the release rate from a solid matrix. The length and conformation of polysaccharide chains influence the accessibility of chitosan functional groups by drug molecules, which is crucial to establish electrostatic interactions and the subsequent incorporation of drug molecules into these systems.

Furthermore, different preparation methods can lead to different interactions and porous structures when nanoparticles are formed or tablets are prepared by compression [19]. Porous microstructures (pore size distribution, porosity, tortuosity, etc.) strongly affect the penetration of solvent into matrixes, dissolution of drug molecules, and subsequent release of these molecules to the medium. This fact can be observed when the release rate studies were conducted for similar systems but different porous structures as consequence of preparation method.

In these pharmaceutical processes, there are some parameters which can be modified with the typical aims to incorporate the maximum amount of drug into the systems or to achieve a proper drug delivery in physiological medium. In order to optimize these characteristics, design of experiments is an interesting tool usually underestimated in the pharmaceutical field; however, recently, it is being considered a point of interest to maximize the benefits of some processes. Simplifying the process, reducing the economic cost, and improving the characteristics of the final formulation are key features that can make a formulation cross the barrier from laboratory to market. For this purpose, response surface methodology (RSM) has proven to be a suitable tool when trying to optimize a procedure [20]. This strategy is especially interesting when the response to be optimized is influenced by two or more parameters simultaneously or when two different responses must be optimized at the same time.

It is worthy to note that one of the most common mistakes when a response is being optimized is trying to modify each parameter independently. All individual parameters may produce a determined variation in the response depending on the specific value taken by the other influential parameters. For this reason, this parameter modification should be done simultaneously and not individually.

In the past, our group has explored the use of different polymers such as PEG [21], PVP [22], poly(anhydride) [23], chitosans [18], or cyclodextrins [24, 25] to obtain systems, which can increase the bioavailability of different drugs. To improve the properties of these delivery systems, along with the drug of interest, various polymers can be combined to provide improvements in several properties, such as the drug solubility increase and the achievement of a controlled release system at the same time.

The aim of the current chapter was to investigate the possibilities and advantages of chitosan-based system for the oral administration of drugs, which show low bioavailability as consequence of its low solubility in different dosage forms: tablets and nanoparticles. Diflunisal (DF) has been chosen as model molecule (**Figure 3**). It is a drug with analgesic and anti-inflammatory activity, and it has been shown to interact with a high variety of polymers [26].

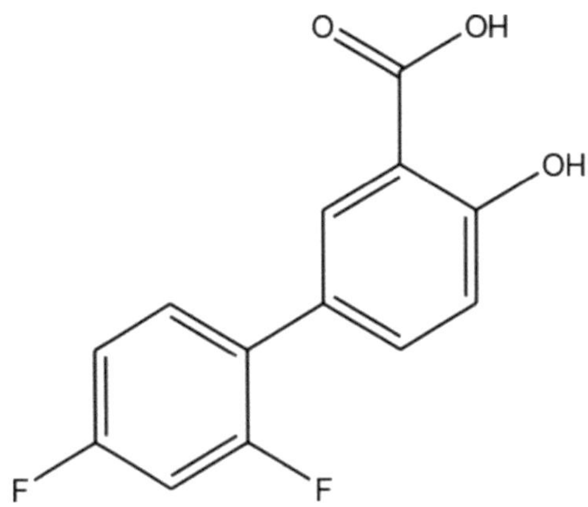

Figure 3. Representation of diflunisal molecule.

2. Materials and methods

2.1. Materials

Diflunisal (DF) was kindly supplied by Merck Sharp and Dohme (Spain). Chitosan (>375 kDa) was supplied by Sigma-Aldrich with a degree of deacetylation of 75–85%. 2-(Hydroxypropyl)-β-cyclodextrin (HPβCD) was purchased from Cyclolab (Hungary). Hydroxypropyl methylcellulose (HPMC) K4M was obtained from Colorcon (Orpington, UK), and sodium alginate (Alg) was supplied by Sigma-Aldrich. The reagents ethanol (Scharlau, PA), ammonium hydroxide, and hydrochloric acid (Panreac, analytical grade) were used as received. All aqueous solutions were prepared with deionized water obtained from a commercial Millipore Elix 3 system (0.1 μS/cm conductivity).

2.2. Preparation of solid dispersions

Diflunisal-chitosan solid dispersions were prepared by kneading at 20:80 drug/polymer ratio. Chitosan was wetted in a mortar with a minimum volume of an ethanol/water solution (50%, v/v) and kneaded with a pestle, while diflunisal was added until a dense paste was formed.

2.3. Preparation of nanoparticles

Chitosan-based nanoparticles can be easily formed via ionic gelation due to the interaction between the positive charges of chitosan amine groups and negative charges of sulfate ions. In order to incorporate the maximum amount of drug into nanoparticles, inclusion complexes with cyclodextrins were previously formed by a coevaporation method.

Chitosan- and drug-cyclodextrin complexes were dissolved in water and added over an aqueous solution containing sodium sulfate (SS). After the immediate formation by ionic gelation, nanoparticles were centrifugated at 20,000 × *g*, frozen, and then freeze-dryed.

2.4. Optimization of nanoparticles: response surface methodology

The effective parameters in the NP formation process, including CS, sodium sulfate, and DF-HPβCD complex concentrations were optimized using response surface methodology. Statgraphics Centurion XVI.I software was used in order to find the best combination of factors which minimize nanoparticle diameter and polydispersity index (PDI) simultaneously. Central composite design (CCD) was chosen due to its orthogonal and rotable properties. It is also an appropriate design of experiments when a process is influenced by more than two factors at the same time because of the reduction of the experiment number:

$$\text{Number of experiments} = 2^n + 2 \cdot n + \text{central point replicates} \tag{1}$$

where n is the number of factors involved in the process optimization.

2.5. Characterization techniques

2.5.1. X-ray diffractometry (XRD)

X-ray powder diffraction patterns were collected on a X Bruker axs D8 Advance diffractometer (Karlsruhe, Germany) using a CuK_α radiation.

2.5.2. Scanning electron microscopy

The shape and surface morphology were examined by scanning electron microscopy (SEM) (Zeiss DSM 940 A apparatus) provided with a digital image capture system (DISS de Point Electronic GmBh).

2.5.3. Transmission electron microscopy (TEM)

The morphology and size of nanoparticles in water solution were analyzed using a transmission electron microscope (TEM). Samples were analyzed using a LIBRA 120 energy-filtering TEM (Zeiss) operated at 80 kV.

2.5.4. Dynamic light scattering

The size distribution of nanoparticles was determined by dynamic light scattering (DLS) using a DynaPro photon correlation spectrometer at 25.0 ± 0.1°C. The intensity of size distributions, expressed in terms of the hydrodynamic radius (R_h), was calculated by the Stokes-Einstein equation:

$$R_h = \frac{kT}{6\pi\eta_0 D_0} \quad (2)$$

2.5.5. Zeta potential

The zeta potential of nanoparticles was measured by electrophoretic laser Doppler anemometry using a Zetamaster analyzer system (Malvern Instruments, UK).

2.6. Dissolution rate studies

The in vitro dissolution studies of pure drug and tablets including solid dispersions were performed using tablets containing 150 mg of diflunisal-chitosan solid dispersions. These dispersions were pressed on an instrumented single-punch tablet press (Kilian SP300, IMA, Germany) using round flat-faced punches (7 mm diameter and 3.15 mm thickness). HPMC and sodium alginate were added directly before the compression process.

The dissolution rate assays were performed in a SOTAX AT 7smart (SOTAX, Basel, Switzerland) tablets, or NPs were placed in 900 mL of PBS (Phosphate Buffer Solution) 7.4 medium at 37.0°C and stirred at 50 rpm.

3. Results and discussion

3.1. Optimization of nanoparticle formation process

Nanoparticles were easily formed via ionic gelation due to the interactions between positive charges of chitosan and negative charges of sodium sulfate.

Optimization of particle size was carried out using response surface methodology and central composite design of experiments. In previous experiments, several characteristics of the process were studied. Nanoparticles were formed using chitosan of high, medium, and low molecular weight; high molecular weight was the most suitable material because the chain length is longer, and this fact facilitates folding and therefore the nanoparticle formation when contact to the negative charges of sodium sulfate. The volume of solutions employed and stirring rate were shown to be non-influential factors in the process. **Table 1** shows factors to optimize particle size and polydispersity index responses.

Factors			Responses	
Chitosan (%)	**HPβCD-DF (%)**	**Sodium sulfate (%)**	**NP size (nm)**	**PDI**
0.3	0.3	3	821	0.201
0.3	0.3	3	813	0.175
0.3	0.3	3	788	0.234
0.3	0.3	3	801	0.197
0.3	0.3	3	801	0.185
0.5	0.5	5	5217	0.721
0.3	0.3	6.4	2415	0.485
0.3	0.0	3	1851	0.258
0.3	0.3	3	783	0.414
0.3	0.3	3	797	0.294
0.3	0.3	3	846	0.215
0.1	0.5	5	697	0.109
0.3	0.3	0	–	–
0.3	0.3	3	825	0.253
0.5	0.5	1	3655	0.449
0.5	0.1	1	504	0.263
0.1	0.1	5	1290	0.503
0.1	0.5	1	505	0.432
0.64	0.3	3	1002	0.218
0.5	0.1	5	1454	0.298
0.3	0.64	3	3811	0.516
0.1	0.1	1	488	0.218
0.0	0.3	3	–	–

Table 1. Experiments corresponding to the central composite design study.

Analysis of variance (ANOVA) for particle size and polydispersity index allows us to identify which factors and interactions between factors influence the responses. Pareto chart indicates the statistically significant factors in order to build the response surface mathematical model

(**Figure 4**). Subsequently, the mathematical model can be built to predict the response depending on the value taken by the factors.

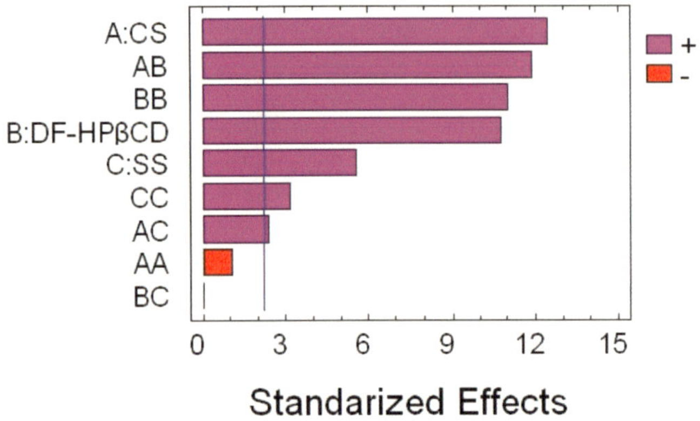

Figure 4. Pareto chart (Factor A = chitosan concentration, Factor B = complex concentration, and Factor C = sodium sulfate concentration).

The model obtained for the optimization of particle diameter explains 98.3% of the variability in the experimental data, whereas for the optimization of polydispersity index, the model obtained explains only 88% of the variability of experimental data. **Figure 5** shows a schematic representation of the response surface for a pair of factors influencing the particle size.

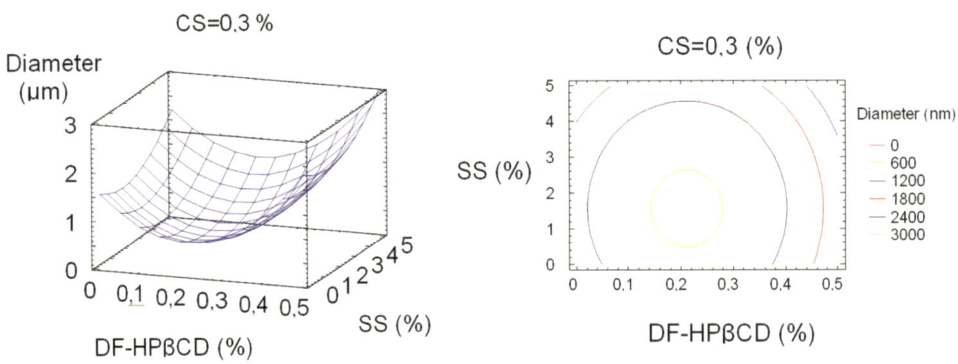

Figure 5. Schematic representation of response surface for particle diameter as a function of DF-HPβCD and sodium sulfate concentration (variation in the third factor cannot be shown in the same figure; for this reason, chitosan concentration has been set on 0.3%).

The three-dimensional representation of the response as a function of two factors (**Figure 4**) does not allow us to visualize the minimum value for the selected surface. **Figure 6** allows us to carry out a preliminary analysis of the influence of each individual factor and its interactions in the response.

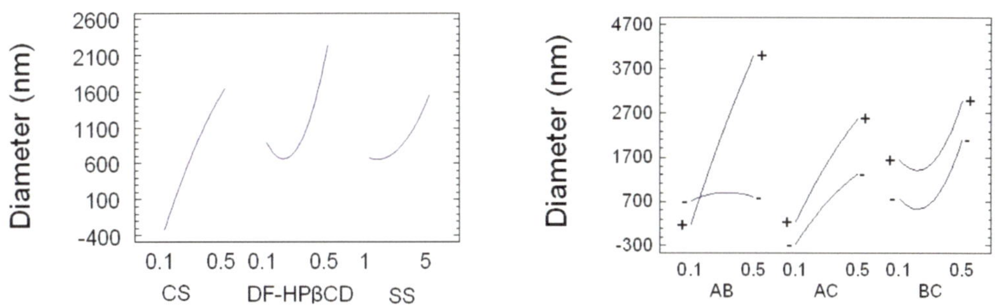

Figure 6. Schematic representation of the influence of each factor and the interactions between factors in the particle diameter (Factor A = chitosan concentration, Factor B = complex concentration, and Factor C = sodium sulfate concentration).

The minimum particle diameter was obtained for concentrations of chitosan 0.12%, DF-HPβCD 0.34%, and sodium sulfate 2.2% according to the prediction of the mathematical model. Moreover, the minimum polydispersity index was obtained for concentrations of chitosan 0.61% and sodium sulfate 1.2% (DF-HPβCD concentration is not an influence factor for PDI).

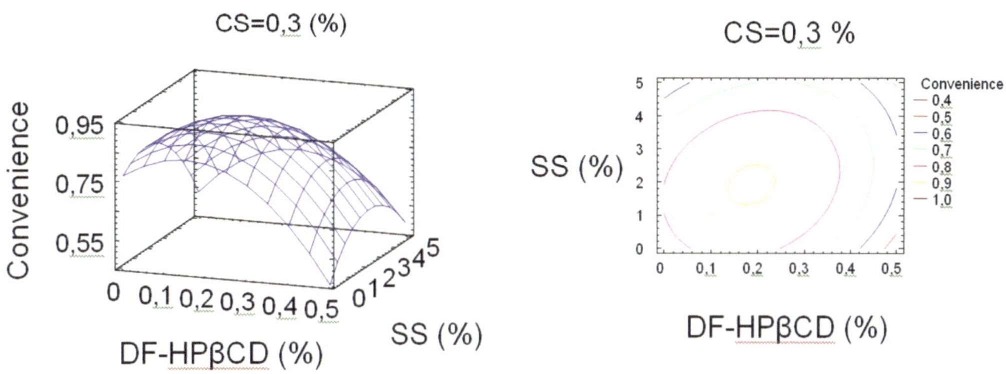

Figure 7. Schematic representation of response surface for convenience of responses as a function of DF-HPβCD and sodium sulfate concentration (variation in the third factor cannot be shown in the same figure; for this reason, chitosan concentration has been set on 0.3%).

As can be observed, the optimization of both responses leads to different values for the three factors. It is therefore necessary to define a new function that optimizes the convenience of both responses simultaneously. For that purpose, a weight three times greater was conferred for the particle diameter than for polydispersity index. The representation of the response surface desirability function is shown in **Figure 7**. The values of factors for obtaining the best nanoparticles in terms of particle diameter and polydispersity index simultaneously are chitosan 0.14%, DF-HPβCD 0.28%, and sodium sulfate 1.9%.

Once the model that describes the function convenience was built, the nanoparticles were performed using the optimal conditions calculated showing a particle diameter of 515 nm and a polydispersity index of 0.14. The optimized nanoparticles showed a formation yield of 72.6%, an encapsulation efficiency of 34% and the amount of drug encapsulated was 42.3 μg/mg.

3.2. Characterization of nanoparticles

After the optimization from based on response surface methodology, physicochemical characterization of the optimized formulation was carried out.

It is important to ensure that all drug molecules are contained inside the nanoparticle and not distributed on the surface in order to prevent releases in two stages when it is intended to obtain controlled release systems [27]. With the aim to verify the distribution of drug in these systems, X-ray powder diffractometry was carried out (**Figure 8**). The absence of drug crystals on nanoparticle surface was corroborated, so a release in one single mode could be expected.

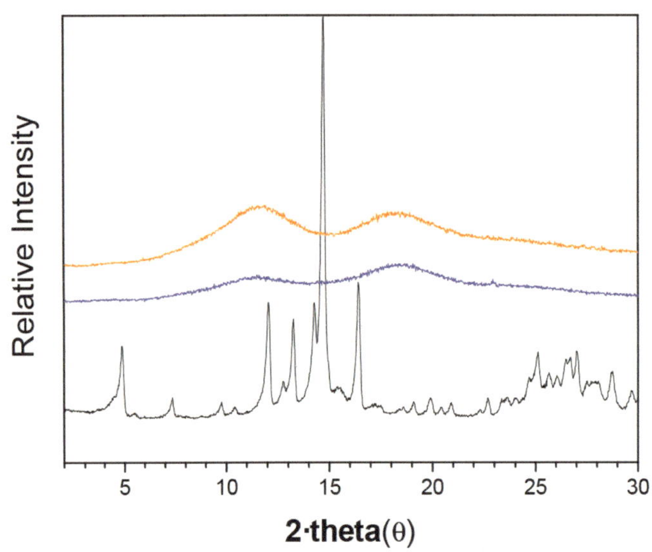

Figure 8. X-ray diffraction patterns of diflunisal (—), chitosan empty nanoparticles (—), and optimized nanoparticle formulation CS-(DF-HPβCD) (—).

The study of nanoparticle morphology was carried out by transmission electron microscopy. Round shape of nanoparticles can be visualized in **Figure 9**, confirming the previously measured particle size by dynamic light scattering (DLS), which reported a particle diameter distribution between 500 and 550 nm. The size measured by DLS is slightly higher than shown in the TEM images. This fact is due to the measurements of hydrodynamic radius obtained by DLS (which include solvent molecules displaced together with solid nanoparticles) and is always higher than the real size of nanoparticles. The zeta potential of the systems was 33.2 mV ensuring the stability of the nanoparticles in solution.

http://dx.doi.org/10.5772/65714

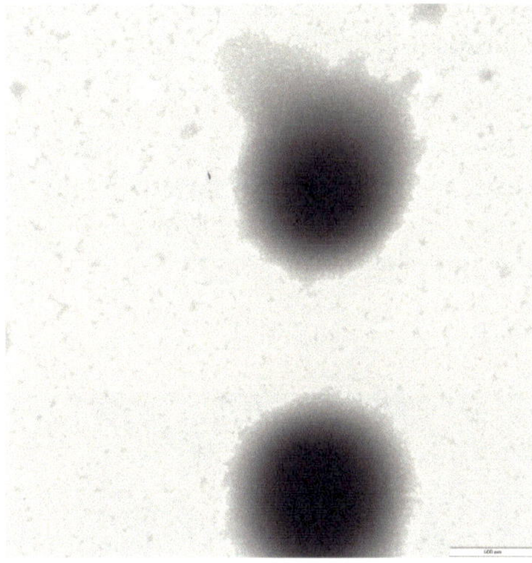

Figure 9. TEM image of the optimized nanoparticle formulation CS-(DF-HPβCD).

3.3. Characterization of solid dispersion systems

Diflunisal and chitosan solid dispersions were analyzed by X-ray diffractometry (XRD) and scanning electron microscopy (SEM). It is important to characterize the crystalline state of drug molecules if it will be administered by oral route. Drugs in crystalline state usually show a lower dissolution rates, because it is necessary to use a certain amount of energy to break the crystal lattice formed by drug molecules.

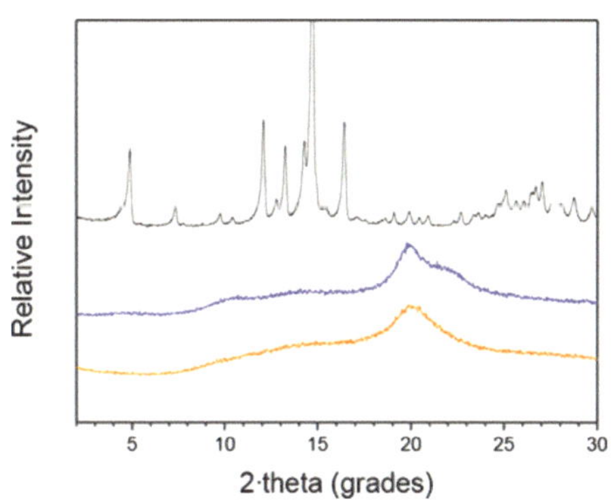

Figure 10. XRD patterns of diflunisal (–), DF-CS kneaded system (—), and chitosan (—).

The X-ray diffraction patterns of pure components, diflunisal and chitosan, and DF-CS-kneaded systems are shown in **Figure 10**. Chitosan shows the typical profile of amorphous materials. Diflunisal diffraction pattern showed the peaks of polymorph II [28, 29]. The kneading of diflunisal and chitosan gave rise to amorphous systems, and the X-ray diffraction pattern showed the absence of reflections as a consequence of the breaking of crystal lattice formed by molecules of diflunisal, which is molecularly dispersed within the amorphous polymer.

SEM micrographs show diflunisal as long prismatic, rodlike crystals typical of crystalline form II [19]. Chitosan particles showed rough surface with very different particle sizes. The disappearance of diflunisal needle agglomerates when the complex was observed corroborates the interaction between drug and polymer (**Figure 11**).

Figure 11. SEM micrographs of chitosan (A), diflunisal (B), and DF-CS-kneaded system (C).

3.4. Dissolution rate studies

3.4.1. Dissolution from tablets

Drug solubility enhancement caused by interaction with chitosan has been described previously in the literature [19]. The interaction by hydrogen bonds between diflunisal and chitosan is the main reason by which a faster release of diflunisal can be expected when it is compressed in the presence of chitosan. Different agents commonly used in the pharmaceutical industry have been employed to modulate diflunisal release rate from solid dispersions formed with chitosan by kneading. The dissolution rates from the different matrices prepared are shown in **Figure 12**.

Diflunisal release from tablets formed by kneading with chitosan showed a very fast release of diflunisal; in this case, the maximum amount of diflunisal was released to the medium in less than 1 h. In order to modulate drug release rate, it is therefore necessary to employ agents able to delay drug delivery. Sodium alginate and hydroxypropyl methylcellulose are widely used in the pharmaceutical domain for this purpose. These agents cause matrix gelification in aqueous solution promoting the slowly release of drug by diffusion from inside the matrix. For both concentrations of alginate, the plateau was reached at 15 h; in case of the addition of HPMC, the complete drug release was not achieved until 30 h, thereby achieving a much slower release.

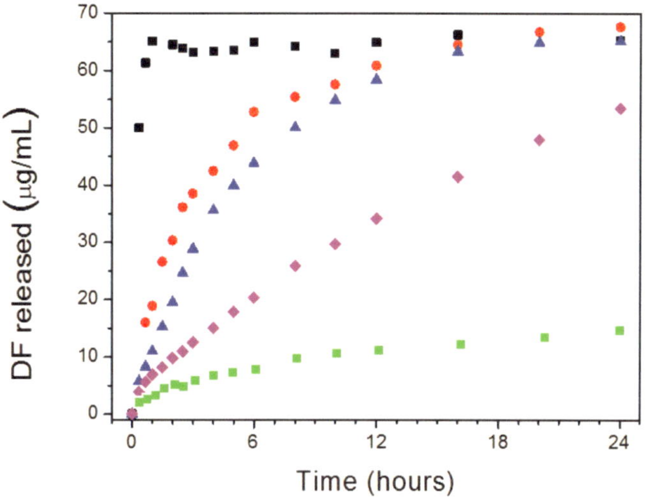

Figure 12. Dissolution rate of diflunisal from tablets compressed DF-CS systems alone (■) or in combination with 20% alginate (●), 40% alginate (▲), and 40% HPMC (♦). Tablets compressed with pure diflunisal (■).

Kinetic parameters were calculated with the aim to facilitate a rapid and easy comparison of release profiles. Dissolution efficiency (DE) is defined as the percentage of area under the dissolution curve at a fixed time. It is expressed as a percentage of the rectangular area described by 100% dissolution in the same time [30]:

$$DE = \frac{\int_0^t y \times dt}{y_{100} \times t} \times 100 \tag{3}$$

$$MDT = \frac{\sum_{j=1}^{n} \hat{t}_j \times \Delta M_j}{\sum_{j=1}^{n} M_j} \tag{4}$$

For calculating mean dissolution time (MDT), j is the sample number, $\hat{t}_j$ is the time at the midpoint between t_j and t_{j-1}, and ΔM_j is the additional amount of drug dissolved between t_j and t_{j-1}. M_{12h} indicates the amount of drug released at a fixed time (12 h). $t_{5\%}$ indicates the time needed to release a fixed amount of drug (5%). The trends exhibited by different profiles according to the agents used can be quantified from the kinetic parameters shown in **Table 2**.

Formulation tablets	DE (%)	MDT (h)	M_{12h} (mg/mL)	$t_{5\%}$ (h)
DF-CS	94,2	0.17	64.90	<0.3
DF-CS-Alg (20%)	68.8	2.03	60.85	0.3
DF-CS-Alg (40%)	60.4	2.96	58.41	0.3
DF-CS-HPMC (40%)	31.7	8.83	34.16	0.7

Table 2. Dissolution parameters for diflunisal release from different tablets.

As can be observed, DE value closes to 100% and low mean dissolution time indicates a fast release of diflunisal from chitosan matrix causing this release that could be considered as burst-type profile. When alginate or HPMC was added to DF-CS system, a significant decrease in the dissolution rate was found, being more effective in case of HPMC as was evidenced by the low DE value and highs MDT and $t_{5\%}$ values.

To test whether the addition of a higher amount of alginate results in a significant change in the rate of drug release, the comparison between different profiles can also be performed by pair-wise procedures, including difference and similarity factors (**Table 3**). The different factor (f_1) measures the percent error between two curves over all time points, where n is the number of sampling points and R_j and T_j are the percent dissolved of the reference and test products at each time point j. Similar profiles yield f_1 values lower than 15. Similarity factors (f_2) higher than 50 indicate similar release profiles. These parameters can be calculated by the following equations:

$$f_1 = \frac{\sum_{j=1}^{n} \left| R_j - T_j \right|}{\sum_{j=1}^{n} R_j} \times 100 \tag{5}$$

$$f_2 = 50 \times \log \left\{ \left[1 + \left(\frac{1}{n} \right) \sum_{j=1}^{n} \left| R_j - T_j \right|^2 \right]^{-1/2} \times 100 \right\} \tag{6}$$

Profiles compared	f_1	f_2
DF-CS vs DF-CS-Alg (20%)	71	26
DF-CS-Alg (20%) vs DF-CS-Alg (40%)	13	62
DF-CS-Alg (40%) vs DF-CS-HPMC (40%)	78	19

Table 3. Comparative parameters of diflunisal release profiles.

From the data obtained for difference and similarity factors, it can be concluded that the addition of HPMC or alginate leads to different dissolution rates compared to DF-CS systems.

Different amounts of alginate (20% vs 40%) do not involve significantly changes in diflunisal release ($f_1 < 15$ and $f_2 > 50$). However, the addition of alginate and HPMC clearly proves to be significant in the dissolution rate of the drug. The different dissolution rate reduction between alginate and HPMC is also evident by comparison through f_1 and f_2.

In the light of the different dissolution profiles shown in **Figure 12**, the study of the mechanisms involved in drug release from the different matrices is important for a better understanding of the phenomena involved in this process.

The *Korsmeyer-Peppas* equation is a simple semiempirical model which relates drug release with the elapsed time [31]. n Values indicate release exponent; for our systems, when it is close to 0.43 means that drug release occurs by pure Fickian diffusion process, when relaxation phenomena and erosion of matrix are involved in drug release, the n-value is approaching to 1. k_{KP} is a constant incorporating the structural and geometrical characteristics of the matrix system:

$$\frac{M_t}{M_\infty} = k_{KP} t^n \tag{7}$$

The *Peppas-Sahlin* model fitted to the experimental data provides values of both diffusion constant (k_D) for the Fickian contribution and erosion constant (k_E) for the erosional/relaxational contribution, the m exponent is 0.43 for our systems [32]:

$$\frac{M_t}{M_\infty} = k_D t^m + k_E t^{2m} \tag{8}$$

Kinetic fitting to aforementioned mathematical models was carried out in order to analyze the phenomena involving in drug release (**Table 4**).

Tablets	Korsmeyer-Peppas			Peppas-Sahlin (m = 0.43)		
	k_{KP} (h^{-n})	n	R^2	k_D (h^{-1})	k_E (h^{-1})	R^2
DF-CS	–	–	–	–	–	–
DF-CS-Alg (20%)	0.315	0.50	0.9794	0.275	0.039	0.9801
DF-CS-Alg (40%)	0.188	0.76	0.9946	0.062	0.125	0.9941
DF-CS-HPMC (40%)	0.110	0.63	0.9978	0.067	0.048	0.9986

Table 4. Kinetic constants and correlation coefficients after data fitting to Korsmeyer-Peppas and Peppas-Sahlin equations (DF-CS systems could not be fitted to kinetic equations because of the fast drug release rate corresponding to burst type).

As can be observed in the values shown in **Table 4**, a mixture of phenomena that influence drug release to the dissolution medium occurs for the systems studied (except DF-CS). In order to assess the contribution of erosion and diffusion phenomena, erosion (E) and diffusion (D) relative contributions have been calculated [33] and represented as function of time in **Figure 13**:

$$D = \frac{1}{1 + \left(k_E / k_D \right) t^m} \tag{9}$$

$$1 = D + E \tag{10}$$

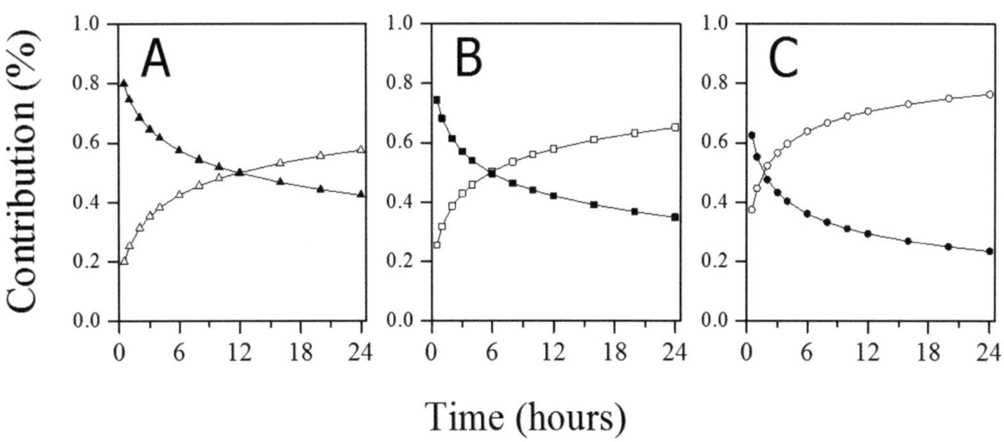

Figure 13. Erosion (empty points) and diffusion (filled points) relative contribution for DF-CS-Alg (20%) (A), DF-CS-Alg (40%) (B), and DF-CS-HPMC (40%) (C).

In all systems, the erosion contribution begins as less important than the diffusion contribution. However, this trend was inverted for the three systems. For each system a time in which the erosion is becoming more important than diffusion can be calculated. For Alg-20% system, this change occurs around 12 h while occurring at 6 h for Alg-40% system and in less than 2 h for HPMC-40% system. This contribution ratio explains why drug release from system containing HPMC is slower; in fact, the amount of drug release is practically independent of time since the erosion phenomenon reaches a greater importance than the contribution of diffusion phenomena. The same trend, but at different time, can be observed for alginate systems where the diffusion phenomenon remains longer.

3.4.2. Dissolution from nanoparticles

Diflunisal release from the previously optimized nanoparticle formulation has been studied. The same amount of diflunisal contained in nanoparticles was tested for comparison purposes. Drug dissolution profiles from nanoparticle system and pure diflunisal are shown in **Figure 14**.

http://dx.doi.org/10.5772/65714

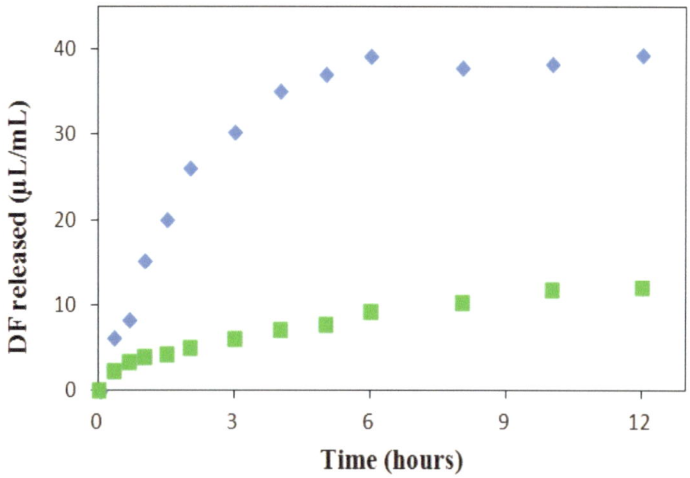

Figure 14. Dissolution rate of diflunisal from diflunisal powder (■) and CS-(DF-HPβCD) nanoparticles (♦).

Diflunisal release from nanoparticles has proven to be faster than directly from diflunisal powder. The total amount of drug encapsulated into nanoparticles was released within 6 h. However, the amount of diflunisal released in the same time is less than 10 mg/L when drug release occurs from diflunisal powder directly. This fact is due on one hand to the increase in drug solubility produced by the presence of small amounts of chitosan and HPβCD in solution; on the other hand, diflunisal molecules in nanoparticles are presented in the amorphous state, which allow to a faster release than in the case of pure diflunisal, where an additional energy is needed to break the crystal lattice.

The dissolution efficiency was 69.5% for nanoparticle system and 18.2% for pure diflunisal. This difference shows clearly a faster dissolution rate and a greater amount of drug released.

Mathematical models cannot be applied to these systems, which do not have a controlled and constant geometry that allows ascertain what proportion involved diffusion and erosion/relaxation phenomena.

4. Conclusions

Chitosan can interact with a high variety of drugs due to its physicochemical properties. The abundance and low cost associated with the production of chitosan make that it can be considered as a promising material in pharmaceutical industry. Chitosan is a versatile product able to be chemically modified and the ease to control the length of its polymer chains makes chitosan an ideal product for use in controlled release formulations.

Response surface methodology is a good tool to optimize fabrication process of pharmaceutical formulations when it depends on two or more factors. It is a good strategy to minimize the number of experiments and to avoid mistakes usually made when the factors involved in the process are optimized individually.

The use of chitosan alone or in the presence of other excipients is made possible to obtain delivery systems on demand in solid oral dosage forms as tablets or nanoparticles. It allows us to modulate drug release from close to zero-order to type-burst kinetics.

Acknowledgements

The authors thank the University of Navarra (PIUNA project) for the financial support and the ADA for the grant of David Lucio.

Abbreviations

CS	Chitosan
DF	Diflunisal
HPβCD	2-Hydroxypropyl-β-cyclodextrin
SS	Sodium sulfate
Alg	Sodium alginate
HPMC	Hydroxypropyl methylcellulose
NPs	Nanoparticles
RSM	Response surface methodology
CCD	Central composite design
DTA	Differential thermal analysis
SEM	Scanning electron microscopy
TEM	Transmission electron microscopy
DLS	Dynamic light scattering
PDI	Polydispersity index

Author details

David Lucio* and María Cristina Martínez-Ohárriz

*Address all correspondence to: dlucio@alumni.unav.es

Department of Chemistry, Faculty of Sciences, University of Navarra, Navarra, Spain

References

[1] Bernkop-Schnürch, A., Dünnhaupt, S. Chitosan-based drug delivery systems. Eur. J. Pharm. Biopharm. 81, 463–469. 2012.

[2] Kurita, K. Chemical modifications of chitin and chitosan. In R. Muzzarelli, C. Jeuniaux & G.W. Gooday (Eds.), Chitin in nature and technology. (pp. 287–295) New York: Plenum. 1986.

[3] Rinaudo, M. Chitin and chitosan: properties and applications. Prog. Polym. Sci. 31, 603–632. 2006.

[4] Kato, Y., Onishi H., Machida, Y. Application of chitin and chitosan derivatives in the pharmaceutical field. Curr. Pharm. Biotechnol. 4, 303–309. 2003.

[5] Domínguez-Delgado, C.L., Rodríguez-Cruz, I.M., Fuentes-Prado, E., Escobar-Chávez, J.J., Vidal-Romero, G., García-González, L., Puente-Lee, R.I. Drug carrier systems using chitosan for non-parenteral routes. Pharmacology and Therapeutics. 273–325. 2014

[6] He, W., Guo, X., Xiao, L., Feng, M. Study on the mechanisms of chitosan and its derivatives used as transdermal penetration enhancers. Int. J. Pharm. 457, 158–167. 2009.

[7] Sinha, V.R., Singla, A.K., Wadhawan, S., Kaushik, R., Kumria, R., Bansal, K., Dhawan, S. Chitosan microspheres as a potential carrier for drugs. Int. J. Pharm. 274, 1–33. 2004.

[8] Hillyard, I.W., Doczi, J., Kiernan, P.B. Antacid and antiulcer properties of the polysaccharide Chitosan in the rat. Exp. Biol. Med. 115, 1108–1112. 1964.

[9] Acikgoz, M., Kas, H.S., Hascelik, Z., Milli, Û., Hincal, A.A. Chitosan microspheres of diclofenac sodium, II: in vivo and in vitro evaluation. Pharmazie. 117, 41–48. 1995.

[10] García Mir, V., Heinämäki, J., Antikainen, O., Sandler, N., Bilbao Revoredo, O., Iraizoz Colarte, A., Nieto, O.M., Yliruusi, J. Application of crustacean chitin as a co-diluent in direct compression of tablets. Pharm. Sci. Technol. 11, 409–415. 2010.

[11] Prabaharan, M. Chitosan derivatives as promising materials for controlled drug delivery. J. Biomater. Appl. 23, 5–36. 2008.

[12] Zhou, X., Hu, Y., Tian, Y., Hu, X. Effect of N-trimethyl chitosan enhancing the dissolution properties of the lipophilic drug cyclosporine A. Carbohyd. Polym. 76, 285–290. 2009

[13] Uragami, T., Tokura, S. Material Science of chitin and Chitosan. Tokyo: Kodansha. 2006.

[14] Riva, R., Ragelle, H., des Rieux, A., Duhem, N., Jérôme, C., Préat, V. Chitosan and chitosan derivatives in drug delivery and tissue engineering. In Chitosan for biomaterials II (pp. 19–44). Springer, Berlin Heidelberg. 2011.

[15] Sugimoto, M., Morimoto, M., Sashiwa, H., Saimoto, H., Shigemasa, Y. Preparation and characterization of water-soluble chitin and chitosan derivatives. Carbohyd. Polym. 36, 49–59. 1998.

[16] Lu, G., Kong, L., Sheng, B., Wang, B., Gong, Y., Zhang, X. Degradation of covalently cross-linked carboxymethyl chitosan and its potential application for peripheral nerve regeneration. Eur. Polym. J. 43, 3807–3818. 2007.

[17] Aiping, Z., Wenjie, J., Lanhua, Y., Gongjun, Y., Hui, Y., Hao, W. O-Carboxymethylchitosan-based novel gatifloxacin delivery system. Carbohyd. Polym. 68, 693–700. 2007

[18] Lucio, D., Irache, J.M., Martínez-Ohárriz, M.C., Response surface methodology for optimization of chitosan coated nanoparticles. 19th International Symposium on Microencapsulation, Pamplona, September 2013. pp 64.

[19] Lucio, D., Zornoza, A., Martínez-Ohárriz, M.C. Influence of chitosan and carboxymethylchitosan on the polymorphism and solubilisation of diflunisal. Int. J. Pharm. 467(1), 19–26. 2014.

[20] Sarabia, L.A., Ortiz, M.C. Response surface methodology. Compr. Chemom. 1, 345–390. 2009.

[21] Martínez-Ohárriz, M.C., Martín, C., Goñi, M.M., Rodriguez-Espinosa, C., Tros-Ilarduya, M.C., Zornoza, A. Influence of polyethylene glycol 4000 on the polymorphic forms of diflunisal. Eur. J. Pharm. Sci. 8, 127–132. 1999.

[22] Martínez-Ohárriz, M.C., Rodriguez-Espinosa, C., Martín, C., Goñi, M.M., Tros-Ilarduya, M.C., Sánchez, M. Solid dispersions of diflunisal-PVP: polymorphic and amorphous states of the drug. Drug Dev. Ind. Pharm. 28, 717–725. 2002.

[23] Agüeros, M., Zabaleta, V., Espuelas, S., Campanero, M.A., Irache, J.M. Increased oral bioavailability of paclitaxel by its encapsulation through complex formation with cyclodextrins in poly (anhydride) nanoparticles. J. Controlled Release, 145(1), 2–8. 2010.

[24] Zornoza, A., Sánchez, M., Vélaz, I., Fernández, L. Diflunisal and its complexation with cyclodextrins. A fluorimetric study. Biomed. Chromatogr. 13, 111–112. 1999.

[25] Zugasti, M.E., Zornoza, A., Goñi, M.M., Isasi, J.R., Vélaz, I., Martín, C., Sánchez, M., Martínez-Ohárriz, M.C. Influence of soluble and insoluble cyclodextrin polymers on drug release from hydroxypropyl methylcellulose tablets. Drug Dev. Ind. Pharm. 35, 1264–1270. 2009.

[26] Tempero, K.F., Cirillo, V.J., Steelman, S.L. Diflunisal: a review of pharmacokinetic and pharmacodynamic properties, drug interactions, and special tolerability studies in humans. Br. J. Clin. Pharmacol. 4(S1), 31S–36S. 1977.

[27] Agnihotri, S.A., Mallikarjuna, N.N., Aminabhavi, T.M. Recent advances on chitosan-based micro-and nanoparticles in drug delivery. J. Controlled Release, 100(1), 5–28. 2004.

[28] Cotton, M., Hux, R. Diflunisal. In K. Florey (Ed.), Analytical profiles of drug substances. (pp. 491–526) London: Academic Press Inc. 1985.

[29] Martínez-Ohárriz, M.C., Martín, C., Goñi, M.M., Rodriguez-Espinosa, C., Tros-Ilarduya, M.C., Zornoza, A. Polymorphism of diflunisal: isolation and solid-state characteristics of a new crystal form. J. Pharm. Sci. 83, 174–177. 1994.

[30] Costa, P., Sousa Lobo, J.M. Modeling and comparison of dissolution profiles. Eur. J. Pharm. Sci. 13, 123–133. 2001.

[31] Korsmeyer, R.W., Gurny, R., Doelker, E., Buri, P., Peppas, N.A. Mechanism of solute release from porous hydrophilic polymers. Int. J. Pharm. 15, 25–35. 1983.

[32] Peppas, N.A., Sahlin, J.J. A simple equation for the description of solute release. III Coupling of diffusion and relaxation. Int. J. Pharm. 57, 169–172. 1989.

[33] Sujja-areevath, J., Munday, D.L., Cox, P.J., Khan, K.A. Relationship between swelling, erosion and drug release in hydrophilic natural gum mini-matrix formulations. Eur. J. Pharm. Sci. 6, 207–217. 1998.

Chapter 7

Marine Polysaccharides as Multifunctional Pharmaceutical Excipients

G. Thirumurugan and M.D. Dhanaraju

Additional information is available at the end of the chapter

http://dx.doi.org/10.5772/66191

Abstract

This chapter is presented to depict the chance of marine polysaccharides satisfying the properties of good pharmaceutical excipients and the potential of that could be utilized as multifunctional pharmaceutical excipients in the pharmaceutical solid dosage form manufacturing as fillers, diluents/vehicles, binders, glidants, binders and disintegrants, and so on. In addition, this chapter discusses the use of marine polysaccharide holding a specific pharmacological activity in the formulation/dosage form, which is used for the same aliment/disease.

Keywords: marine polysaccharide, pharmaceutical excipients, bioavailability, drug delivery, dissolution, disintegrants, pharmacological activity

1. Introduction

Drugs, which are obtained from plants, animals, marine source or minerals, are not very often administered in their pure form. Before releasing into the market, these drugs are combined with a variety of inert substances (excipients/adjuvants) and modified into a dosage form, which is convenient to be administered by a specified route [1]. Similar to the drugs, these excipients are also obtained from many sources, namely natural, synthetic sources. Nowadays, the naturally obtained excipients have shown more interest due to the various advantages like low cost, biocompatibility, biodegradability, non-toxicity, easy availability, eco-friendly processing, superior patient tolerance and acceptance [2]. Before, it was thought that the therapeutic reaction to a drug is a characteristic of its inherent pharmacological action. However, nowadays, it is much implicit that the dose-response connection obtained after drug administration by various routes—for example, oral and parenteral—is not identical. The difference is also observed when the same medicine is administered as dissimilar dosage forms or similar formulations produced by dissimilar manufacturers, which in turn depends upon the physical, chemical properties of the drug, the excipients present in the formulation, the technique of formulation and the way of administration [1].

Natural marine origins are among the most widely used excipients in the formulation of different dosage forms. A number of marine-based polysaccharides, such as agar, alginate, carrageenan, fucoidan, chitosan and hyaluronan are utilized in pharmaceutical dosage form as binders, vehicles, disintegrating agents, gelling agents and drug release sustaining agents [3].

The purpose of this review is to discuss potential pharmaceutical formulation development applications of marine polysaccharides, with a special emphasis in multifunctional excipients development for enhancing bioavailability, drug delivery applications. In addition, this review discusses the use of marine polysaccharide holding a specific pharmacological activity in the formulation/dosage form, which is used for the same aliment/disease, for example, incorporating the marine polysaccharide having anti-cancer activity in anti-cancer drug formulation as excipients. In fact, data has added that such excipients operate on well-distinct biological pathways and receptors in order to implement their valuable properties. If additives that work on several cancer pharmacological pathways or receptors targets are jointed together in concentrations that are at and/or more than their recommended levels, then there is a possibility exists that such a formulated dosage form may offer "standalone" control of cancer.

2. Marine polysaccharides, as multifunctional pharmaceutical excipients

A drug injected intravascularly (I.V/I.A) directly enters the blood and produces its pharmacological effects. The majority of drugs are administered extravascularly, usually through oral route. If anticipated to act systemically, such drugs can produce their pharmacological actions only when they come into the blood circulation from their site of application. In order to reach the blood circulation, orally administered formulation must disintegrate, deaggregate and dissolution of the drug in the aqueous fluid at the absorption site must occur [4]. If the drugs are not hydrophilic in nature, the absorption process of drugs like these is usually dissolution rate-limited (**Figure 1**).

Better absorption can be achieved by altering the characteristics of the dosage form using pharmaceutical excipients. Moreover, pharmaceutical excipients help in the manufacturing process by serving as a binder, diluents, wetting agent, filling agent, disintegrating agents, dissolution enhancers, and so on. [3]. Now their increasing demands and expectations

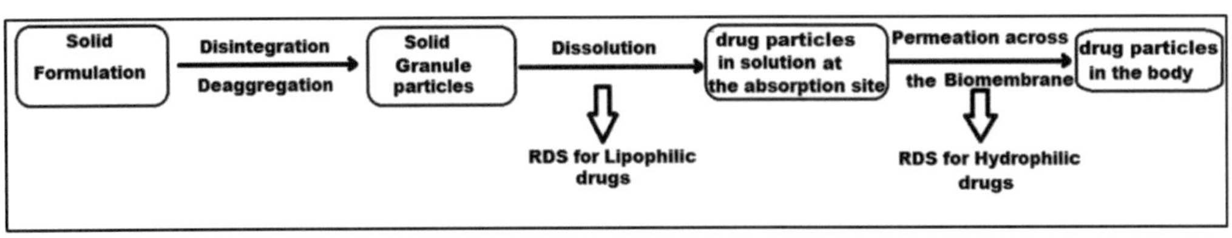

Figure 1. The two rate-determining steps in the absorption of orally administered dosage form.

http://dx.doi.org/10.5772/66191

with regard to quality have stimulated the development of new additives characterized by advanced assay and lesser content of impurities. Quality of formulation does depend on the quality of excipient [4]. The purpose of this study is to give a scope for the researchers to try and develop a formulation that is made up of just the drug and natural marine polysaccharides. This review is done to depict the chance of some marine polysaccharides satisfying the properties of good pharmaceutical excipients if the stability of them has been explored sufficiently.

2.1. As binder

Marine polysaccharides can serve as a good binder in the production of tablets by the wet granulation method of manufacture. In this role, binders are either added as a solution or as a solid into the powder mix (following which the granulating fluid, typically water, is added). Due to the high concentration of hydroxyl groups in the polysaccharide, generally have a high water-binding capacity that makes wet granulation easier [5]. The responsible bio-adhesives marine polysaccharides have extraordinarily high cohesive strength and binding strength to the solid surfaces, enabling the API to remain attached under tensional conditions.

2.2. As diluents

Marine polysaccharides can be employed as diluents/fillers in the formulation of tablets (by all methods) to increase the mass of the solid dosage forms that hold a low concentration of therapeutic agent and thereby render the manufacturing process more reliable and reproducible [6]. For example, chitin and chitosan are used as a diluent or filler and as a binder in direct compression of tablet processing, as a disintegrant, and so on. Chitin and chitosan have a lowest bulk and tapped density that cause good flow as well as compaction during filling and tablet compression processing. **Figure 2** shows the structure of cellulose chitin and chitosan.

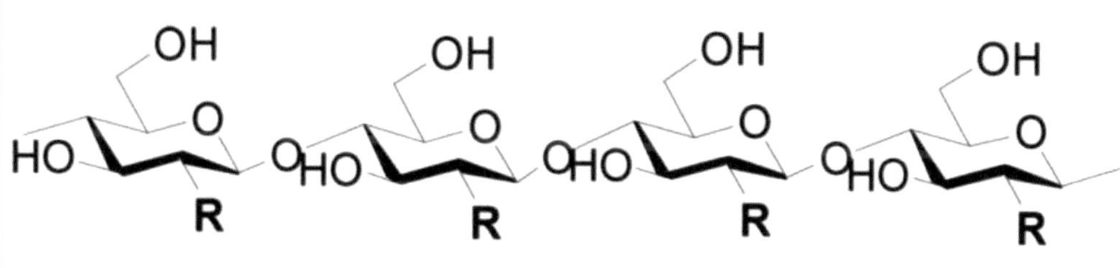

Chemical structures of cellulose (R = OH), chitin (R = NHCOCH3), and chitosan (R = NH2).

Figure 2. Structure of cellulose, chitin and chitosan.

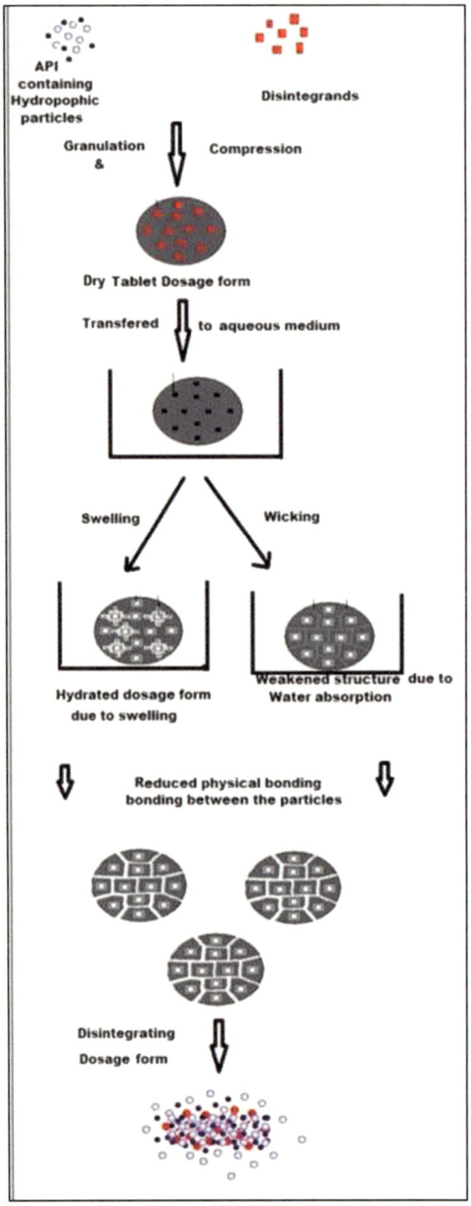

Figure 3. Disintegrants facilitating the breakdown of the tablet into granules.

2.3. As disintegrants

Disintegrants are materials added to the dosage forms that enhance the breakup or disintegration of tablet formulations into smaller particles that dissolve faster than in the absence of disintegrants. These materials have the main role to oppose the efficiency of tablet binder and physical forces that behave under compression to form the tablets. Tablet disintegrant usually considered as the rate determining step (RDS) in a faster drug release. Marine polysaccharides may behave as a good disintegrating agent by increasing the porosity, wettability and wicking or capillary action and operate by swelling in the presence of aqueous fluids in tablet formulations to facilitate the breakdown of the tablet into granules upon entry into the stomach [7] (**Figure 3**).

2.4. As dissolution enhancers

Marine polysaccharides may serve as a dissolution enhancer for the poor soluble drug. These powders can reduce cohesive forces holding a tablet dosage form together and induce the breakup into smaller granules, thus increasing the effective surface area for dissolution (**Figure 4**).

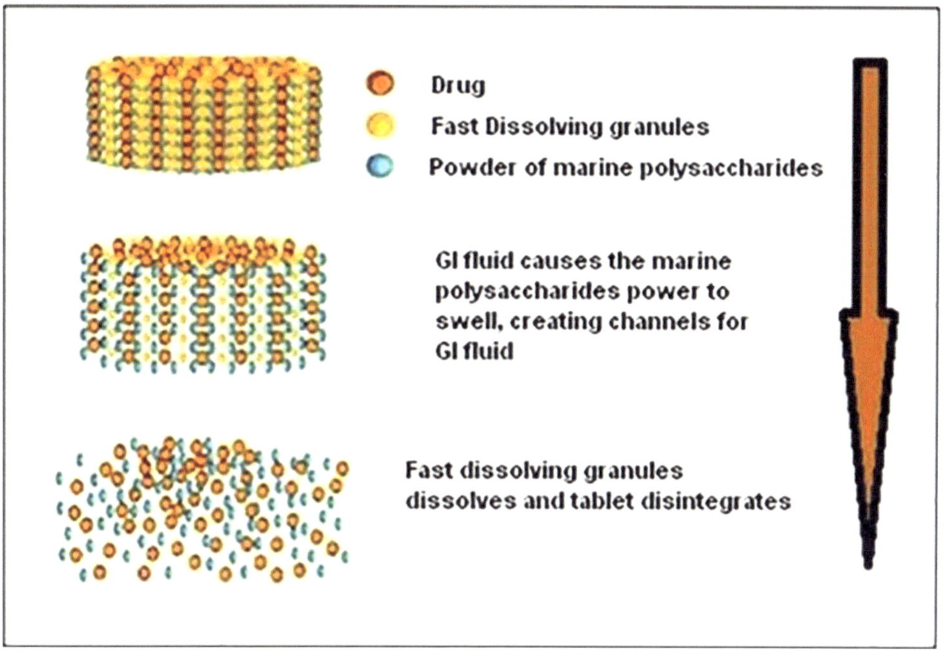

Figure 4. Dissolution enhancer facilitating dissolution for the poor soluble drug.

2.5. As drug delivery carriers

Marine polysaccharides have been widely used to synthesize drug delivery carriers. They are bio-compatible, non-toxic and bio-degradable and stimuli-responsive makes marine polysaccharides appropriate sources for the building of complex loading devices with a release that can be effectively controlled [4]. **Table 1** shows the compilation of marine polysaccharides utilized as drug delivery carriers, and **Figure 5** illustrates the mechanisms of various marine polysaccharide carriers and its method of preparation.

Drug delivery devices can be constructed using various methods and can be synthesized in a variety of shapes, such as hydro gels, micro or nanoparticles and capsules, capable of protecting a variety of bioactive agents such as proteins and nucleic acids.

S. no	Name of marine polysaccharide	Source	Drug delivery use
1.	Alginate 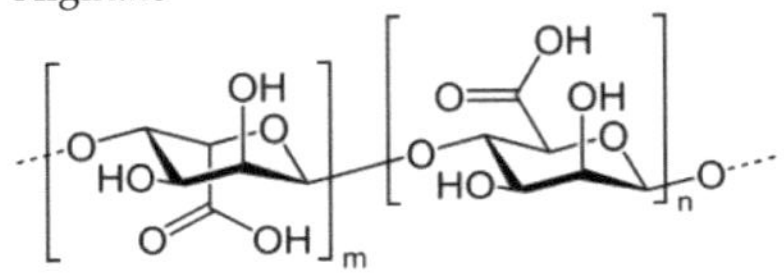	Brown sea weed	Excipient, stabilizer [8], hydrogel matrices, beads, particles [9–12], micro particles [13, 14]
2.	Carrageenans 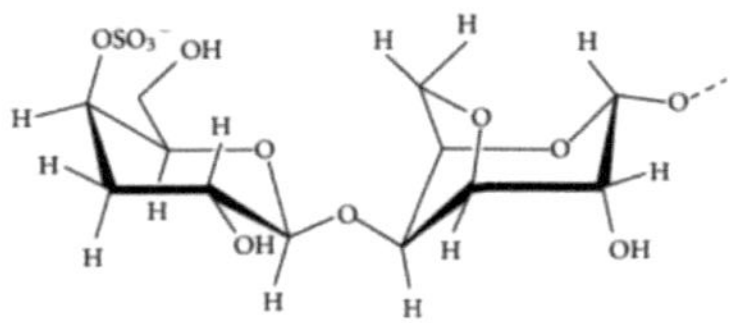	Red algae	Tablet compressor [15], controlled release (temperature sensitive) [16], fast release [17], sustained release [18] (Ph sensitive)
3.	Fucoidans 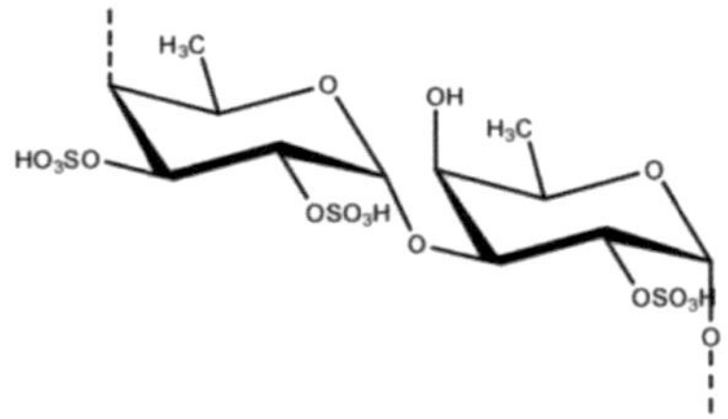	Brown algae	Microspores, fucospheres [19], insulin controlled release [20], nano-particles [21]
4.	Ulvans A 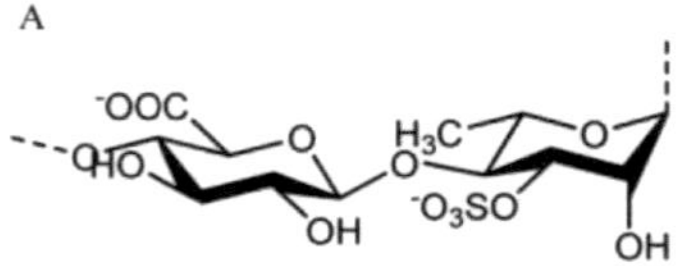	Green algae	Cosmetic delivery [22], nano-fibres (tissue engineering and regenerative medicine) [23], hydrogel [24]

S. no	Name of marine polysaccharide	Source	Drug delivery use
5.	Chitosans 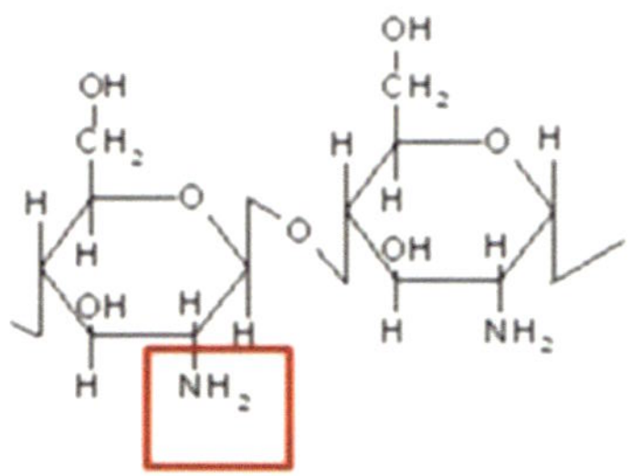	Marine animal (chitin)	Nano particles, beads of capsules for controlled release, membranes, films and scaffolds for tissue engineering and regenerative medicine [25], stabilization, acceleration (release), sustained release [26]
6.	Hyaluronan 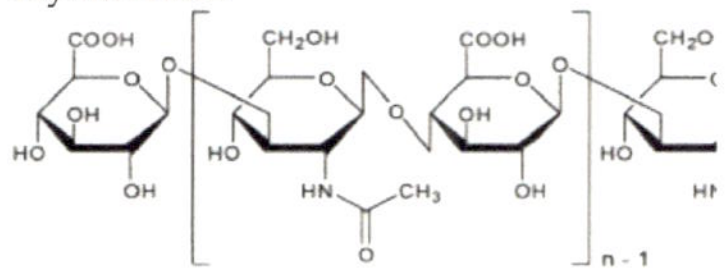	Marine animal (glycosaminoglycans)	Hydrogels [27], micro [28], nano particles [29], coating material [30], liposomes [31]
7.	Chondroitin sulfate 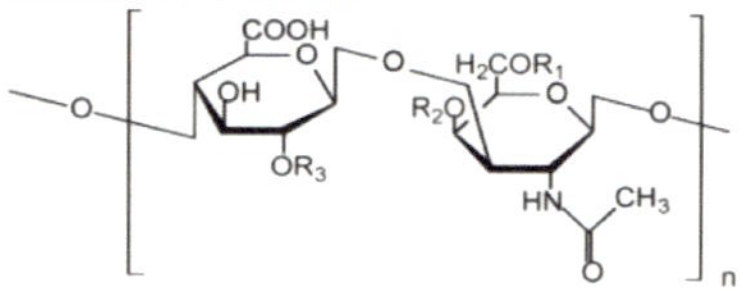	Marine animals (whale and shark)	Controlled release [32], anti-cancer drug delivery [33]
8.	Dermatan sulfate 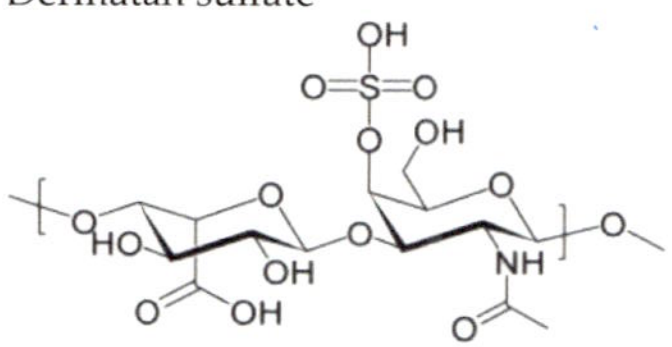	Ray skin (glycosaminoglycans)	Stabilizer for growth factor, cytokines [34]

S. no	Name of marine polysaccharide	Source	Drug delivery use
9.	Heparan sulfate 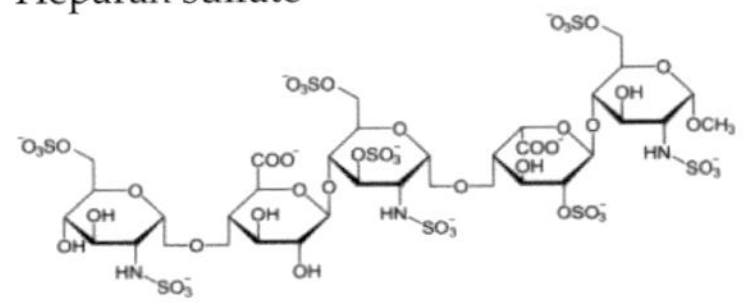	Ray skin (glycosaminoglycans)	Cancer treatment [35]
10.	Keratan sulfate 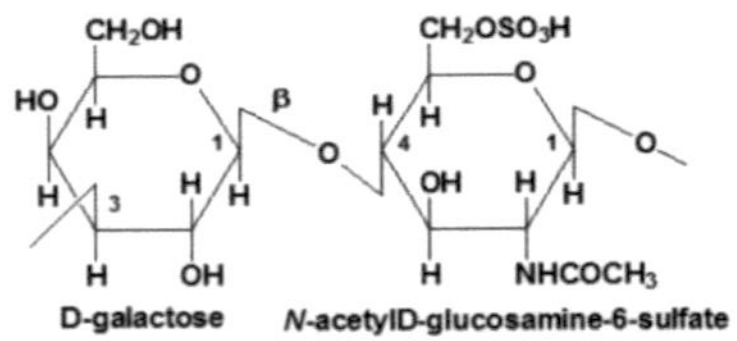	Ray skin (glycosaminoglycans)	
11.	Agarose 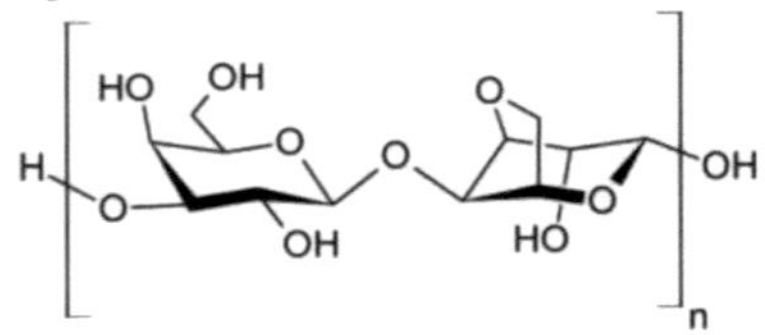	Red algae	Hydrogels [36]

Table 1. Compilation of marine polysaccharides utilized as drug delivery carriers.

http://dx.doi.org/10.5772/66191

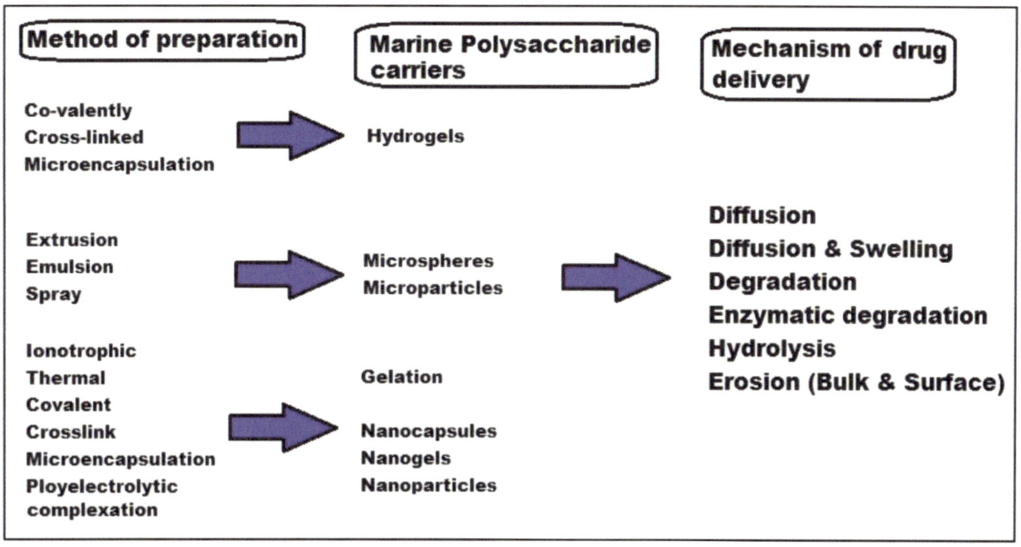

Figure 5. Illustrating the mechanism of various marine polysaccharide carriers and their method of preparation.

3. Pharmacological activity of marine polysaccharides and potential possibilities of standalone effects when incorporating as excipients

Marine polysaccharides stand for a number of abundant bio-active substances in marine organisms. In fact, numerous marine macro- and microorganisms are good quality possessions of carbohydrates with miscellaneous applications due to their bio-functional efficacies. By involving on cell propagation and cycle, and by modulating metabolic pathways, marine polysaccharides have numerous pharmacological efficacies, such as antioxidative, antibacterial, antiviral, immuno-stimulatory, anticoagulant and anticancer effects. Besides the polysaccharides, monosaccharides are useful for humans and can cure many diseases, mainly those linked to metabolism deficiency such as diabetes.

There has been growing facts in recent years that pharmaceutical additives may not be harmless, "inert" components of a formulation, but may hold either "stand alone" pharmacological activity or may act to modify the pharmacological efficacy of the API. Because most excipients have historically derived from food or food products, they have been assigned—and in most cases, experimentally confirmed—to have widespread pharmacological activity based on the ingestion of those products. Marine polysaccharides holding a specific pharmacological activity can be incorporated as an excipient in the formulation/dosage form, which is used for the same aliment/disease. **Table 2** shows the pharmacological activity of marine polysaccharides, and **Figure 6** illustrates the mechanism of action involved in the pharmacological activity of marine polysaccharides.

S. no	Marine polysaccharide	Source	Pharmacological activity
1.	Alginate	Brown sea weed	
2.	Carrageenans	Red algae	Anticoagulant [37], Anti-tumour [38] , Immuno-modulator [39], anti-hyperlipidaemic [40], Antioxidant [41], Antibacterial, viral [42], bird flu, dengue, Hepatitis A, HIV [43]
3.	Fucoidans	Brown algae	Anti-tumour [44, 45], anti-thrombin [19], anti-coagulant, anti-inflammatory, anti-adhesive and anti-viral [46], burn treatment [47]
4.	Ulvans	Green algae	Anti-viral, anti-oxidant, anti-coagulant, anti-tumour, anti-hyperlipidaemic and immune system enhancer [48]
5.	Chitosans	Marine animal (chitin)	Antimicrobial [49], anti-tumour and inflammatory [50]
6.	Hyaluronans	Marine animal (glycosaminoglycans)	Wound treatment [51], supplement for arthritis [52]
7.	Chondroitin sulfate	Marine animals (whale and shark)	Anti-coagulant [53], supplement for arthritis [53]
8.	Dermatan sulfate	Ray skin (glycosaminoglycans)	Anti-coagulant [54]
9.	Heparan sulfate	Ray skin (glycosaminoglycans)	
10.	Keratan sulfate	Ray skin (glycosaminoglycans)	Anti-adhesive, osteoarthritis [54]
11.	Agarose	Red algae	Glucose intolerance in type 2 diabetes mellitus, weight loss, anti-bacterial [40]

Table 2. Pharmacological activity of marine polysaccharides.

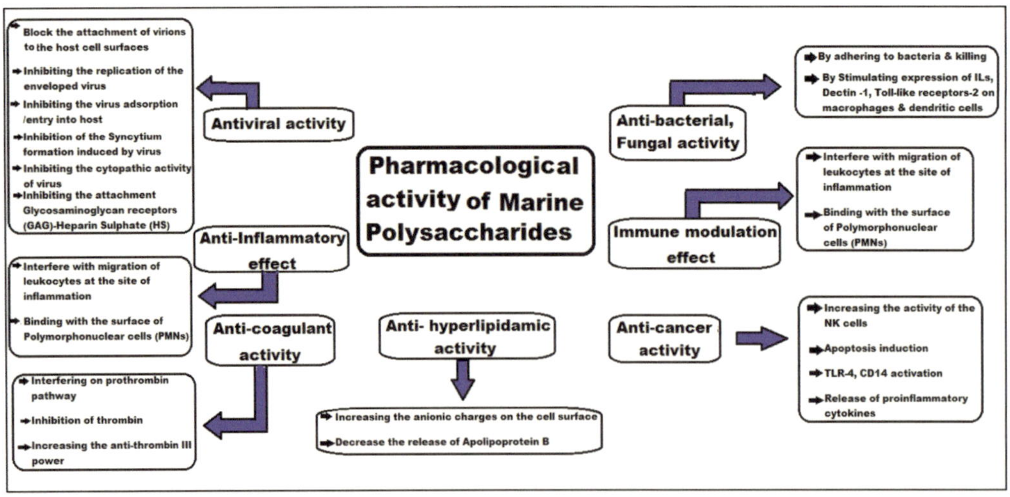

Figure 6. Illustrating the mechanism of action involved in the pharmacological activity of marine polysaccharides.

However, the majority of the pathways or modes of action has not been revealed in humans. In addition, fairly only some have been investigated in-vitro. These in-vitro findings are unable to extrapolate into their actions in-vivo in humans. There should be well-controlled human clinical investigations using such a mixture of additives are essential prior to any rationalization for effectiveness can be asserted.

Ideation of using "marine polysaccharide bio-active excipients" is possible for diseases that are chronic in nature and for whom gradual and lasting beneficial efficacy may be observed from the modulation of signalling pathways that direct to incremental but visible improvements in the quality of life. Such "marine polysaccharide bio-active excipients" may not be efficient for acute diseases for which radical interventions are necessary.

4. Conclusion

Marine polysaccharides have a huge potential as multifunctional pharmaceutical excipients for commercial needs. However, there is a need to characterize, check compatibility of these polysaccharides to produce pharmaceutical grade additives that could find the utility in drug dosage form. In addition, research is warranted in pharmacodynamic interaction of these marine polysaccharides and particular drug when using for same aliments.

Author details

G. Thirumurugan* and M.D. Dhanaraju

*Address all correspondence to: thiru.mpharm@gmail.com

Research Lab, GIET School of Pharmacy, Rajahmundry, Andhra Pradesh, India

References

[1] Brahmankar DM, Jaiswal SB. Biopharmaceutics and Pharmacokinetics: A Treatise. 1st ed. Delhi: Vallabh Prakashan Publishers; 1995, pp. 296–297.

[2] Ramachandran S, Shaheedha SM, Thirumurugan G and Dhanaraju MD. Floating controlled drug delivery system of famotidine loaded hollow microspheres (microballoons) in the stomach. Current Drug Delivery. 2010;**7**(1):93–97. doi: 10.2174/156720110790396436

[3] Saravanan M, Bhaskar K, Srinivasa Rao G, Dhanaraju MD. Ibuprofen-loaded ethylcellulose/polystyrene microspheres: An approach to get prolonged drug release with reduced burst effect and low ethylcellulose content. Journal of Microencapsulation: Micro and Nano Carriers. 2003;**20**(3)289–302. doi:10.1080/0265204031000093087

[4] Dhanaraju MD, Elisabeth S, Thirumurugan G. Triamcinolone-loaded glutaraldehyde cross-linked chitosan microspheres: Prolonged release approach for the treatment of rheumatoid arthritis. Drug Delivery. 2011;**18**(3):198–207. doi:10.3109/10717544.2010.528069

[5] Paolucci M, Fasulo G, Volpe MG. Employment of marine polysaccharides to manufacture functional biocomposites for aquaculture feeding applications. Marine Drugs. 2015;**13**(5):2680–2693. doi:10.3390/md13052680.

[6] De Jesus Raposo MF, de Morais AMB, de Morais RMSC. Marine polysaccharides from algae with potential biomedical applications. Marine Drugs. 2015;**13**(5):2967–3028. doi:10.3390/md13052967.

[7] Thomas NV, Kim S-K. Beneficial effects of marine algal compounds in cosmeceuticals. Marine Drugs. 2013;**11**(1):146–164. doi:10.3390/md11010146.

[8] Mano J.F. Stimuli-responsive polymeric systems for biomedical applications. Advanced Engineering Materials. 2008;**10**:515–527. doi: 10.1002/adem.200700355.

[9] Laurienzo P. Marine polysaccharides in pharmaceutical applications: An Overview. Marine Drugs. 2010;**8**:2435–2465. doi: 10.3390/md8092435.

[10] Gombotz WR, Wee SF. Protein release from alginate matrices. Advanced Drug Delivery Review. 2012;**64**:194–205. doi: 10.1016/j.addr.2012.09.007.

[11] Beneke CE, Viljoen AM, Hamman JH. Polymeric plant-derived excipients in drug delivery. Molecules. 2009;**14**:2602–2620. doi: 10.3390/molecules14072602.

[12] Sudhakar Y, Kuotsu K, Bandyopadhyay AK. Buccal bioadhesive drug delivery: A promising option for orally less efficient drugs. Journal of Controlled Release. 2006;**114**:15–40. doi: 10.1016/j.jconrel.2006.04.012.

[13] Lima AC, Sher P, Mano JF. Production methodologies of polymeric and hydrogel particles for drug delivery applications. Expert Opinion on Drug Delivery. 2012;**9**:231–248. doi: 10.1517/17425247.2012.652614.

[14] Soppimath KS, Aminabhavi TM, Kulkarni AR, Rudzinski WE. Biodegradable polymeric nanoparticles as drug delivery devices. Journal of Controlled Release. 2001;**70**:1–20. doi: 10.1016/S0168-3659(00)00339-4.

[15] Picker KM. Matrix tablets of carrageenans. I. A compaction study. Drug Development and Industrial Pharmacy. 1999;**25**:329–337. doi: 10.1081/DDC-100102178.

[16] Ganta S, Devalapally H, Shahiwala A, Amiji M. A review of stimuli-responsive nanocarriers for drug and gene delivery. Journal of Controlled Release 2008;**126**:187–204. doi: 10.1016/j.jconrel.2007.12.017.

[17] Bornhoft M, Thommes M, Kleinebudde P. Preliminary assessment of carrageenan as excipient for extrusion/spheronisation. European Journal of Pharmaceutics and Biopharmaceutics. 2005;**59**:127–131. doi: 10.1016/j.ejpb.2004.05.007.

[18] Bonferoni MC, Rossi S, Ferrari F, Bertoni M, Bolhuis GK, Caramella C. On the employment of *lambda carrageenan* in a matrix system. III. Optimization of a lambda carrageenan-HPMC hydrophilic matrix. Journal of Controlled Release. 1998;**51**:231–239. doi: 10.1016/S0168-3659(97)00175-2.

[19] Sezer AD, Akbuga J. Fucosphere: New microsphere carriers for peptide and protein delivery: Preparation and *in vitro* characterization. Journal of Microencapsulation. 2006;**23**:513–522. doi: 10.1080/02652040600687563.

[20] Sato K, Takahashi S, Anzai J. Layer-by-layer thin films and microcapsules for biosensors and controlled release. Analytical Sciences. 2012;**28**:929–938. doi: 10.2116/analsci. 28.929.

[21] Pinheiro AC, Bourbon AI, Cerqueira MA, Maricato E, Nunes C, Coimbra MA, Vicente AA. Chitosan/fucoidan multilayer nanocapsules as a vehicle for controlled release of bioactive compounds. Carbohydrate Polymer. 2015;**115**:1–9. doi: 10.1016/j.carbpol.2014.07.016.

[22] Ahmed OM, Ahmed RR. Anti-proliferative and apoptotic efficacies of ulvan polysaccharides against different types of carcinoma cells *in vitro* and *in vivo*. Journal of Cancer Science & Therapy. 2014;**6**:202–208. doi: 10.4172/1948-5956.1000272.

[23] Alves A, Sousa RA, Reis RL. A practical perspective on ulvan extracted from green algae. Journal of Applied Phycology. 2013;**25**:407–424. doi: 10.1007/s10811-012-9875-4.

[24] Morelli A, Chiellini F. Ulvan as a new type of biomaterial from renewable resources: Functionalization and hydrogel preparation. Macromolecular Chemistry and Physics. 2010;**211**:821–832. doi: 10.1002/macp.200900562.

[25] Prabaharan M, Mano JF. Chitosan-based particles as controlled drug delivery systems. Drug Delivery. 2005;**12**:41–57. doi: 10.1080/10717540590889781.

[26] Felt O, Buri P, Gurny R. Chitosan: A unique polysaccharide for drug delivery. Drug Development and Industrial Pharmacy. 1998;**24**:979–993. doi: 10.3109/03639049809089942.

[27] Santos JR, Alves NM, Mano JF. New thermo-responsive hydrogels based on poly (*N*-isopropylacrylamide)/hyaluronic acid semi-interpenetrated polymer networks: Swelling properties and drug release studies. Journal of Bioactive and Compatible Polymers. 2010;**25**:169–184. doi: 10.1177/0883911509357863.

[28] Lim ST, Martin GP, Berry DJ, Brown MB. Preparation and evaluation of the *in vitro* drug release properties and mucoadhesion of novel microspheres of hyaluronic acid and chitosan. Journal of Controlled Release. 2000;**66**:281–292. doi: 10.1016/S0168-3659(99)00285-0.

[29] Oyarzun-Ampuero FA, Brea J, Loza MI, Torres D, Alonso MJ. Chitosan-hyaluronic acid nanoparticles loaded with heparin for the treatment of asthma. International Journal of Pharmaceutics. 2009;**381**:122–129. doi: 10.1016/j.ijpharm.2009.04.009.

[30] Grech JMR, Mano JF, Reis RL. Chitosan beads as templates for layer-by-layer assembly and their application in the sustained release of bioactive agents. Journal of Bioactive and Compatible Polymers. 2008;**23**:367–380. doi: 10.1177/0883911508093389.

[31] Simoes S, Moreira JN, Fonseca C, Duzgunes N, de Lima MC. On the formulation of pH-sensitive liposomes with long circulation times. Advanced Drug Delivery Review. 2004;**56**:947–965. doi: 10.1016/j.addr.2003.10.038.

[32] Santo VE, Gomes ME, Mano JF, Reis RL. Chitosan-chondroitin sulphate nanoparticles for controlled delivery of platelet lysates in bone regenerative medicine. Journal of Tissue Engineering and Regenerative Medicine. 2012;**6**:S47–S59. doi: 10.1002/term.1519.

[33] Guo YM, Shi XM, Fang QL, Zhang J, Fang H, Jia WL, Yang G, Yang L. Facile preparation of hydroxyapatite-chondroitin sulfate hybrid mesoporous microrods for controlled and sustained release of antitumor drugs. Materials Letters. 2014;**125**:111–115. doi: 10.1016/j.matlet.2014.03.084.

[34] Vitale C, Berutti S, Bagnis C, Soragna G, Gabella P, Fruttero C, Marangella M. Dermatan sulfate: An alternative to unfractionated heparin for anticoagulation in hemodialysis patients. Journal of Nephrology. 2013;**26**:158–163. doi: 10.5301/jn.5000105.

[35] Knelson EH, Nee JC, Blobe GC. Heparan sulfate signaling in cancer. Trends in Biochemical Sciences. 2014;**39**:277–288. doi: 10.1016/j.tibs.2014.03.001.

[36] Hoare TR, Kohane DS. Hydrogels in drug delivery: Progress and challenges. Polymer. 2008;**49**:1993–2007. doi: 10.1016/j.polymer.2008.01.027.

[37] Silva FRF, Dore CMPG, Marques CT, Nascimento MS, Benevides NMB, Rocha HAO, Chavante SF, Leite EL. Anticoagulant activity, paw edema and pleurisy induced carrageenan: Action of major types of commercial carrageenans. Carbohydrate Polymers. 2010;**79**:26–33. doi: 10.1016/j.carbpol.2009.07.010.

[38] Zhou G, Sun Y, Xin H, Zhang Y, Li Z, Xu Z. *In vivo* antitumor and immunomodulation activities of different molecular weight lambda-carrageenans from *Chondrus ocellatus*. Pharmacological Research. 2004;**50**:47–53. doi: 10.1016/j.phrs.2003.12.002.

[39] Panlasigui LN, Baello OQ, Dimatangal JM, Dumelod BD. Blood cholesterol and lipid-lowering effects of carrageenan on human volunteers. Asia Pacific Journal of Clinical Nutrition. 2003;**12**:209–214.

[40] De Souza MCR, Marques CT, Dore CMG, da Silva FRF, Rocha HAO, Leite EL. Antioxidant activities of sulfated polysaccharides from brown and red seaweeds. Journal of Applied Phycology. 2007;**19**:153–160. doi: 10.1007/s10811-006-9121-z.

[41] Carlucci MJ, Ciancia M, Matulewicz MC, Cerezo AS, Damonte EB. Antiherpetic activity and mode of action of natural carrageenans of diverse structural types. Antiviral Research. 1999;**43**:93–102. doi: 10.1016/S0166-3542(99)00038-8.

[42] Schaeffer DJ, Krylov VS. Anti-HIV activity of extracts and compounds from algae and cyanobacteria. Ecotoxicology Environment Safety. 2000;**45**:208–227. doi: 10.1006/eesa.1999.1862.

[43] Ermakova S, Sokolova R, Kim SM, Um BH, Isakov V, Zvyagintseva T. Fucoidans from brown seaweeds *sargassum hornery, eclonia cava, costaria costata*: Structural characteristics and anticancer activity. Applied Biochemistry and Biotechnology. 2011;**164**:841–850. doi: 10.1007/s12010-011-9178-2.

[44] Anastyuk SD, Shevchenko NM, Ermakova SP, Vishchuk OS, Nazarenko EL, Dmitrenok PS, Zvyagintseva TN. Anticancer activity *in vitro* of a fucoidan from the brown alga *fucus evanescens* and its low-molecular fragments, structurally characterized by tandem mass-spectrometry. Carbohydrate Polymers. 2012;**87**:186–194. doi: 10.1016/j.carbpol.2011.07.036.

[45] Kim SK, Ravichandran YD, Khan SB, Kim YT. Prospective of the cosmeceuticals derived from marine organisms. Biotechnology and Bioprocess Engineering. 2008;**13**:511–523. doi: 10.1007/s12257-008-0113-5.

[46] Alves A, Sousa RA, Reis RL. *In vitro* cytotoxicity assessment of ulvan, a polysaccharide extracted from green algae. Phytotherapy Research. 2013;**27**:1143–1148. doi: 10.1002/ptr.4843.

[47] Rabea EI, Badawy ME, Stevens CV, Smagghe G, Steurbaut W. Chitosan as antimicrobial agent: Applications and mode of action. Biomacromolecules. 2003;**4**:1457–1465. doi: 10.1021/bm034130m.

[48] Qin CQ, Du YM, Xiao L, Li Z, Gao XH. Enzymic preparation of water-soluble chitosan and their antitumor activity. International Journal of Biological Macromolecules. 2002;**31**:111–117. doi: 10.1016/S0141-8130(02)00064-8.

[49] Chen WY, Abatangelo G. Functions of hyaluronan in wound repair. Wound Repair Regen.1999;**7**:79–89. doi: 10.1046/j.1524-475X.1999.00079.x.

[50] Jiang GB, Quan D, Liao K, Wang H. Novel polymer micelles prepared from chitosan grafted hydrophobic palmitoyl groups for drug delivery. Molecular Pharmaceutics. 2006;**3**:152–160. doi: 10.1021/mp050010c.

[51] Teien AN, Abildgaard U, Hook M. The anticoagulant effect of heparan sulfate and dermatan sulfate. Thrombosis Research. 1976;**8**:859–867. doi: 10.1016/0049-3848(76)90014-1.

[52] Wang DA, Varghese S, Sharma B, Strehin I, Fermanian S, Gorham J, Fairbrother DH, Cascio B, Elisseeff JH. Multifunctional chondroitin sulphate for cartilage tissue-biomaterial integration. Nature Materials. 2007;**6**:385–392. doi: 10.1038/nmat1890.

[53] Mourao PA, Pereira MS. Searching for alternatives to heparin: Sulfated fucans from marine invertebrates. Trends in Cardiovascular Medicine. 1999;**9**:225–232. doi: 10.1016/S1050-1738(00)00032-3.

[54] Funderburgh JL. Keratan sulfate biosynthesis. IUBMB Life. 2002;**54**:187–194. doi: 10.1080/15216540214932.

Chapter 8

Speculations on the Use of Marine Polysaccharides as Scaffolds for Artificial Nerve 'Side-'Grafts

Sherif M. Amr and Basma Ekram

Additional information is available at the end of the chapter

http://dx.doi.org/10.5772/66460

Abstract

Marine polysaccharides (e.g., chitosan, alginate and agarose) meet the requirements of artificial nerve grafting. These include the following: (1) prerequisites of biopolymers used as scaffolds; (2) conditions required by nerve autografts; (3) macroengineering requirements (form, design); (4) microengineering requirements (microgrooves, inclusion filaments); (5) mechanical conditions required by nerve autografts; (6) molecular aspects of peripheral nerve regeneration; space and adherence for: (i) chondroitin sulfate proteoglycans, (ii) neurite outgrowth promoting factors, (iii) neurotrophic factors, (iv) cells; (7) artificial side grafts should be compatible with autologous nerve grafts; (8) spatial distribution of neurotrophic factor gradients; (9) modulation of fibrosis; (10) renewal of luminal fillers. The mechanical stability of chitosan should be increased by adding other polymeric chains and cross-linking. As a result of deacetylation of chitin, chitosan scaffolds have got numerous positively charged free amino groups that provide adherence for growth factors and cells. To modulate fibrosis, heparin cross-linked chitosan microspheres have proved effective for delivering/transplanting cells, heparin and growth factors. To renew luminal fillers, chitosan microspheres may be injected through an indwelling catheter incorporated into the nerve conduit. Agarose and alginate have gained more acceptance as hydrogel lumen fillers. Interacting with positively charged chitosan, negatively charged alginate may form versatile chitosan-calcium-alginate microspheres.

Keywords: nerve grafting, side grafting, brachial plexus lesions, artificial nerve grafts, chitosan, agarose, alginate

1. Introduction

After brachial plexus traction injuries, the nerve gap may be too large to be bridged by autogenous nerve grafts alone. This has prompted neuroscientists to investigate the use of artificial nerve grafts. Materials used as scaffolds for these grafts are summarized in **Table 1** [1–4]. Biodegradable materials have displaced nondegradable silicone nerve conduits because these have been associated with a chronic tissue response and poor regeneration [4–6]. Several tube and sheet forms are readily available on the market for use in peripheral nerve repair [e.g., Neurotube [polyglycolic acid (PGA), Synovis], Neurolac (poly-DL-lactide-caprolactone, PLCL, Polyganics BV), NeuraGen (collagen type I, Integra NeuroSciences) and Neuro-Matrix/ Neuroflex (collagen type I, Collagen Matrix Inc) [7–10]. The degradation rate of these conduits typically ranges from months (Neurotube in 3 months and NeuroMatrix in 7 months to years; Neurolac in 16 months and NeuraGen in 4 years) [4, 11]. A few of these resorbable conduits have been tested clinically in short nerve defects [2, 12–15]. Comparable or even superior results have been reported compared to nerve autografts [13, 15, 16]. In reconstructing long nerve defects, Neurolac and NeuraGen nerve conduits have induced functional recovery similar to nerve autografts [4, 17, 18]. Nevertheless, use of nerve conduits has been limited to nerve gaps less than 3 cm in length [2, 19]. Because of this, their clinical use has been limited, especially in brachial plexus lesions.

1. Autogenous biological conduits
1.1. Artery
1.2. Vein
1.3. Muscle
1.4. Tendon
2. Nonautogenous biological conduits
2.1. Nerve allografts
2.2. Decellularized biomatrices
2.3. Conduits consisting of basal laminae
2.4. Natural polymers; in vivo extracellular matrix polymers
2.4.1. Collagen
2.4.2. The glycosaminoglycan hyaluronic acid. Hyaluronic acid has been developed into an extremely useful copolymer gel with methylcellulose
2.5. Natural polymers; polymers derived from blood
2.5.1. Plasma derived polymers
2.5.2. Fibronectin
2.5.3. Fibrin
2.6. Natural polymers; polymers from marine life

http://dx.doi.org/10.5772/66460

2.6.1. Agarose is a linear polysaccharide derived from seaweed and cross-linked by temperature gradients through hydrogen bonding

2.6.2. Alginate is obtained from algae and the polymer solution is cross-linked by calcium into a sponge-like structure

2.6.3. Chitosan is a glycosaminoglycan (GAG) carbohydrate polymer derived by chemical deacetylation of chitin, the major structural polysaccharide found in crustacean, shellfish and insect shells

2.6.4. Carrageenan

2.6.5. Fucoidans

2.6.6. Ulvans

2.6.7. Hyaluronans, chondroitin sulfates, dermatan sulfates, heparan sulfates, keratin sulfates

3. Nonbiological conduits, nonabsorbable nerve conduits

3.1. Silicone

3.2. Polytetrafluoroethylene (ePTFE)

3.3. Gore-tex

3.4. Longitudinally oriented suture material (silk)

4. Nonbiological conduits, absorbable nerve conduits

4.1. Aliphatic polyesters

Members of this family include polyglycolic acid PGA, poly(lactic acid) (PLA), polycaprolactone (PCL) and their copolymers like PLGA

4.2. Poly(phosphor)esters (PPE)

Such polymers contain a characteristic phosphoester linkage in the backbone, which is more easily cleaved via hydrolysis in physiologically relevant conditions than the ester linkage. One type is poly(bis(hydroxyethyl) terephthalate-ethylphosphoester). Another one is poly(caprolactone-co-ethyl ethylene phosphate) (PCLEEP)

4.3. Polyurethanes

DegraPol is an example. It is synthesized by coupling poly(R-3-hydroxybutyric acid-co-R-3-hydroxyvaleric acid)-diol and poly(glycolide-co-caprolactone)-diol (PGCL) with 2, 2, 4-trimethylhexamethylene diisocyanate

4.4. Hydrogel-based conduits

These are based on polyethyleneglycol, a biodegradable synthetic polymer of ethylene oxide units. Hydrogel tubes are fabricated from poly(2- hydroxyethyl methacrylate) (PHEMA). PHEMA homopolymer may be copolymerized with methyl methacrylate (MMA) to provide greater rigidity. These PHEMA-MMA tubes have a biphasic wall structure, with an outer gel-like layer and inner porous layer

4.5. Piezoelectric polymers

Poled poly(vinylidene fluoride) (PVDF) conduits are fabricated by stretching PVDF conduits (to mechanically align dipoles) and then subjecting them to an electric field nA copolymer of PVDF and trifluoroethylene (PVDF–TFE) does not require mechanical stretching to align the dipoles. Positively poled PVDF–TFE conduits have shown the best results

Table 1. Scaffolds serving as nerve conduits.

The clinical use of artificial nerve grafts has gained new impetus, however. First, advances in conduit lumen filler technology have made it possible to use nerve scaffolds to bridge large peripheral nerve defects [20]. Second, after conventional end-to-end grafting, side grafting has made it possible to increase the incidence of nerve regeneration by applying additional (artificial) grafts extending from the side of the donor end to the side of the recipient end, so-called co-grafting [21]. Third, the axonal growth cone has been found to be sensitive to spatial molecular concentration gradients of nerve growth factors [22, 23]. Thus, artificial nerve side grafts supplied with spatial molecular concentration gradients of nerve growth factors might not only act as additional grafts to autografts but might provide additional stimulation to axonal growth cone progression within autografts as well. This might provide a solution for limited axonal progression in long-standing lesions (of more than one-year duration) or for partially regenerated lesions (obstetric brachial plexus lesions after 3 years of age). In light of these findings and expectations, the requirements of an artificial nerve graft have to be redefined. Marine polysaccharides (chitosan, alginate and agarose) may be modified or combined with other polymers, thus offering wide versatility to meet these requirements.

2. Requirements of an artificial nerve graft

Biopolymer scaffolds used as artificial nerve grafts should be biocompatible; their absorption and degradation kinetics should match the degree of in vivo cell/tissue growth; they should have adequate surface for cell access, proliferation and cell differentiation and lastly, they should not be toxic or carcinogenic; the same applies to their degradation products [24]. In addition, they should meet specific requirements

2.1. Artificial nerve grafts should fulfill the same conditions required by nerve autografts

In 1943, Seddon recognized three nerve injury types as follows: neuropraxia, axonotmesis and neurotmesis [25]. In 1951, Sunderland [25, 26] elaborated on this classification recognizing five degrees of nerve injury. A sixth degree has recently been added by Mackinnon [25]. Millesi [26] has subclassified Sunderland's classification according to the degree of fibrosis (**Table 1**). Autogenous nerve grafting is the standard for repair of irreducible nerve gaps (Millesi Subtypes 3C, 4N, 4S; Sunderland Type 5, Mackinnon Type 6) [26]. Autogenous grafts act as immunogenically inert scaffolds for axonal cone progression, providing simultaneously neurotrophic factors and Schwann cells for axonal regeneration [26].

2.1.1. *Artificial nerve grafts should meet macroengineering requirements*

To act as interposition nerve grafts, biomaterial polymer nerve scaffolds should meet macro-engineering requirements, that is, they should be of proper form, design (shape) and size (diameter). In addition, they should be supplied with macrogrooves and have reasonable wall thickness. As to *form*, scaffolds may be delivered as gels, as sponges, or as solid scaffolds [10, 27]. Comparing various *designs* of porous poly(ε-caprolactone) synthetic polymer scaffolds (cylinder, tube, multichannel and open-path design with and without a central core), Wong et al. [28] have concluded that open path designs are superior in terms of regenerative capacity and biocompatibility (**Figure 1a** and **b**). The relationship between scaffold channel *diameter* and the number of axons regenerating has been studied. Larger areas of fibrosis and a reduced axonal numbers are associated with larger tube diameters [29]. PLGA scaffolds with multiple longitudinally aligned *macrogroove channels* of 450 and 660 μm constructed using injection molding and seeded with Schwann cells support robust axonal growth [1, 30]. A close relationship between the formation of neuromas in regenerated tissues and the *thickness of the conduit tube wall* [31] has been observed. Tube wall thickness greater than 0.81 mm significantly attenuates axon growth [32], probably by decreasing nutrient diffusion and wall porosity. Kokai et al. have demonstrated that a wall thickness of 0.6 mm, a porosity of 80% and a pore size range of 10–40 μm are optimal for peripheral nerve repair [33, 34].

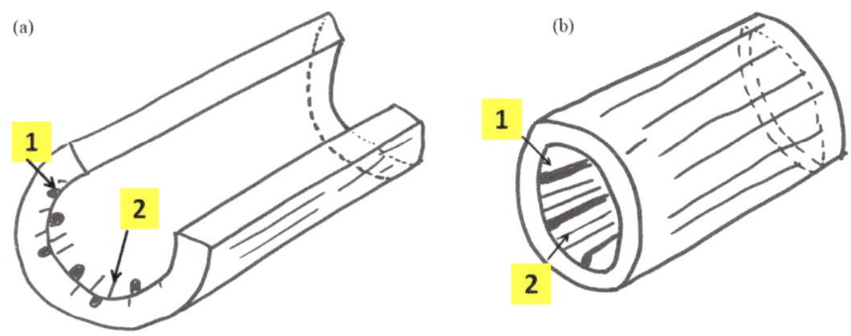

Figure 1. Nerve conduit design. According to Wong et al. [28], open path designs are preferable. It follows, that an open ¾ cylinder (a) is superior to a closed cylinder (b). However, both should be supplied with longitudinal macrogrooves (1) for fascicular regeneration and microgrooves (2) for axonal regeneration.

2.1.2. Artificial nerve grafts should meet microengineering requirements

Microengineering refers primarily to microgrooves directing axonal growth [35–37]. Microgrooves can be placed in the polymer surface by laser etching. A minimum groove depth of 2 μm [38], associated with narrow ridges (5 versus 10 μm) have improved neurite outgrowth and promoted cell adhesion [39]. Further improvements in neurite outgrowth can be obtained with coating the grooved surface with collagen or laminin peptides [40–42].

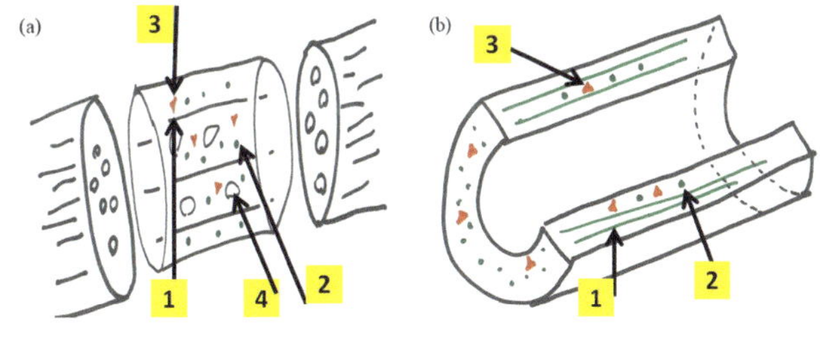

Figure 2. To enhance axonal progression, a closed cylinder nerve conduit should be filled with filaments (1), growth factors (2) and cells (3). In addition, it should be porous (4) (a). In a ¾ cylinder, to increase surface area to volume ratio, the wall of the cylinder should be hollow and filled with filaments, growth factors and cells (b).

Another microengineering aspect refers to axonal preference to grow along the length of micro and nanofibers of polymers [1]. The inclusion of filaments into the lumen of large tube diameter silicone tubes has been suggested by Lundborg et al. [43]. The aim has been to increase the overall cross-sectional area of the regenerated nerve fibers. Itoh et al. [44] have used eight collagen or polyester filaments. Synthetic or biological filament lumen fillers have been used ever since to enhance nerve regeneration (**Figure 2a** and **b**). Increasing the whole filament surface area by increasing their number and reducing their diameter (increased surface area-to-volume ratio) is a serious consideration [4, 43–47].

2.1.3. Artificial nerve grafts should fulfill the same mechanical conditions required by nerve autografts

A nerve conduit should possess sufficient toughness to resist compression or collapse, yet still be flexible and suturable [4, 48]. Ideally, a nerve conduit should have an elastic modulus comparable with that of a nerve autograft. In end-to-end grafting, nerve end/artificial graft elastic or shear modulus inequality might produce tension, compression or shear at the repair site, ending up with fibrosis and hampering progression of regeneration. Such inequality might also lead to rupture of the small caliber sutures (nylon or prolene 9/0 or 10/0 sutures) used for fascicular grafting. On the contrary, during side grafting, cografting with an artificial graft of slightly higher modulus of elasticity (e.g., 64.3–69.8 MPa PLLA versus 11.7 MPa peripheral nerve tensile strength [48, 49]) aids in absorbing tension off the nerve ends and off the autograft. This may improve regeneration [50, 51] (**Figures 3a–c** and **4a–c**).

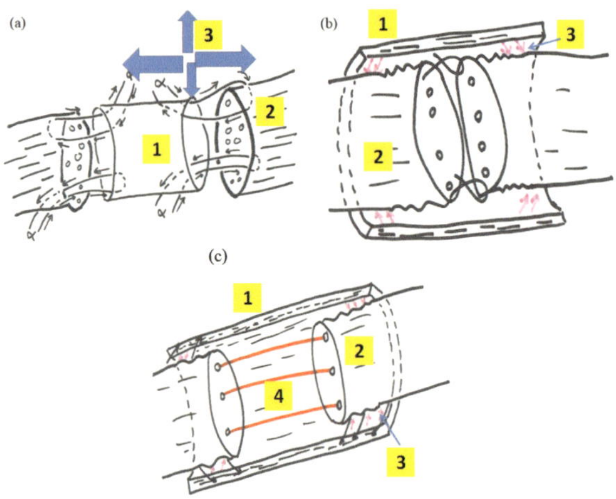

Figure 3. Suturing a nerve conduit (high tensile, compression and shear modulus of elasticity) (1) in an end-to-end fashion to a nerve end (low tensile, compression and shear modulus of elasticity) (2) produces high point contact stresses at the suture lines (3), leading to their rupture (a). Also, high point contact stresses lead to an inflammatory reaction and secondary fibrosis hampering further regeneration. After successful end-to-end suturing, if a ¾ nerve conduit is applied as a side graft to both ends (b), it transfers 'absorbs' stresses from the sutured ends and transfers them to a whole proximal or distal segment. Unloading the suture line promotes axonal sprouting; transferring stresses to a broad segmental area decreases stresses to a point that produces some damage which is necessary for side axonal sprouting but that will not hamper axonal regeneration. The same effect is produced, when a ¾ nerve conduit is side grafted to an end-to-end autografted area (4) (c).

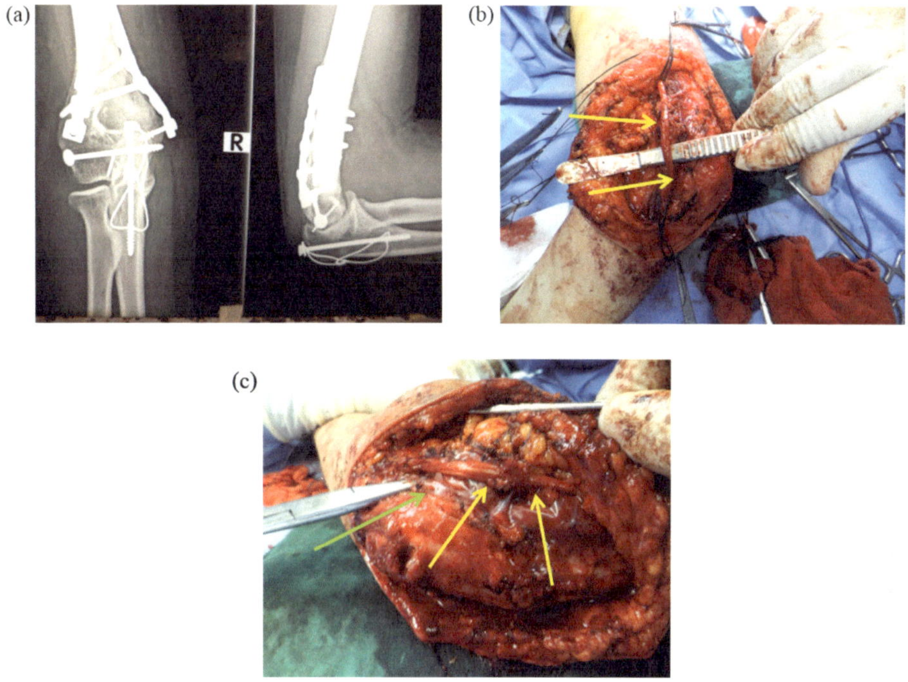

Figure 4. Stress unloading effect of side conduits; illustrative clinical case. During a road traffic accident, a 24 year lady had sustained an intercondylar fracture of the lower end of the humerus associated with a partial injury of the ulnar nerve. The intercondylar fracture had been fixed successfully (a). The patient presented to us with partial claw hand deformity. (b) The zone of injury on surgical exploration of the ulnar nerve (area between yellow arrows). The ulnar nerve was shortened, so that the injury zone formed a U-shaped loop. The healthy proximal and distal nerve ends were next sutured to an underlying silicone sheet (green arrow) to unload the U-shaped segment (c). The patient recovered full hand function in 9 months.

2.2. Requirements based on molecular aspects of peripheral nerve regeneration

Artificial nerve scaffolds should provide adequate space for the interplay and manipulation of the different molecular pathways for axonal (peripheral nerve) regeneration [35, 36, 52, 53] (**Figure 5**).

Biomaterial polymer nerve scaffolds should provide adequate space and adherence for the following: (1) *Fibrous tissue and chondroitin sulfate proteoglycans* necessary for basal lamina synthesis; (2) *Neurite outgrowth promoting factors* [e.g., laminin (LN), fibronectin (FN), heparin sulfate proteoglycans (HSP) and tenascin]; (3) *Neurotrophic factors* which include (a) the neurotrophins [nerve growth factor (NGF), brain-derived neurotrophic factor (BDNF), neurotrophin-3 (NT-3) and neurotrophin-4/5 (NT-4/5)]; (b) the neurokines [ciliary neurotrophic factor (CNTF) and leukemia inhibitory factor (LIF)]; (c) the transforming growth factor (TGF)-b family [TGF-b1, TGF-b2, TGF-b3, glial cell line derived neurotrophic factor (GDNF)]; (4) *Schwann cells and other cell types*. In a systematic review comparing functional results of different biomaterial polymer nerve scaffolds in rat experiments, Sinis et al. [54] have noted more favorable outcome of conduits coated with Schwann cells compared to plain synthetics.

To provide adequate space and adherence for cells and molecules, biomaterial polymer nerve scaffolds should be porous [1]. Currently, ideal scaffolding should have 80–90% porosity with a pore size of 50–250 μm. Its pores should be interconnected so as to allow for the introduction of lumen fillers [24, 55]. Lumen filler technology allows for incorporation of cells and molecular factors into nerve conduits using intermediary supporting matrices; its methods have been reviewed elsewhere (**Figure 2a** and **b**) [56]. One method involves incorporating *growth factors* into supporting matrices placed inside the lumen (e.g., hydrogel-forming collagen, fibrin, laminin, alginate, heparin and heparin sulfate). In general, relatively weak hydrogel-forming matrices are preferred to denser matrices because the latters can obstruct the path of axonal growth through the lumen [57, 58]. Growth factors interact with matrix materials in a number of important ways (ligand-receptor binding, ionic, electrostatic, hydrophobic, or covalent interactions). Ligand-receptor binding, for instance, may slow down the release of growth factors and protect against enzymatic degradation during nerve regeneration. A *second* method of lumen filling is incorporating *accessory cells* (e.g., *Schwann cells*) into a supporting matrix. Different matrices have been used successfully with Schwann cells for this purpose, such as alginate-fibronectin, collagen, gelatin and matrigel. Also, ingrowing Schwann cells and axons can be guided by such matrices for enhanced nerve regeneration. A *third* method of lumen filling is *impregnating the neural conduit wall* with *cells or neurotrophins* via cross-linking or immobilization in multichannel nerve conduits [56].

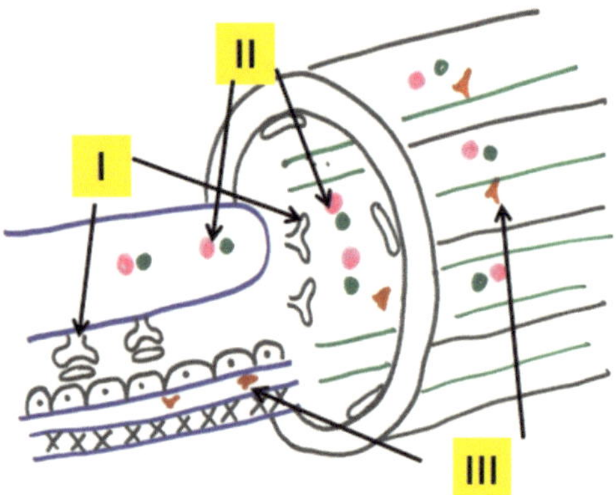

Figure 5. Molecular aspects of axonal regeneration. I. *Neurite outgrowth promoting factors* [laminin (LN), fibronectin (FN), heparin sulfate proteoglycans (HSP) and tenascin] pave the proper path by supplying orientation and adhesiveness for axons. II. *Neurotrophic factors* include a. neurotrophins (NGF, BDNF, NT-4/5); b. neurokines (CNTF, LIF); c. (TGF)-β family TGF-β1, TGF-β2, TGF-β3, GDNF. III. *Schwann cells*. All of these factors combined should be available in a nerve conduit. *The presence of single or few items (e.g., laminin, nerve growth factor, Schwann cells, mesenchymal stem cells) does not suffice for peripheral nerve regeneration.*

2.3. Requirements based on recent findings of side grafting (cografting autologous nerve grafts with biomaterial polymer nerve scaffolds)

In side grafting, the donor side is grafted to the recipient end; donor side collateral sprouting is stimulated by mechanical trauma or axotomy [25]. Similarly, the incidence of nerve regeneration after conventional end-to-end grafting can be increased by applying biomaterial polymer nerve side grafts extending from the side of the donor end to the side of the recipient end [21] (**Figure 3c**). Cografting autologous nerve grafts with biomaterial polymer nerve scaffolds presupposes biocompatibility between both.

2.4. Requirements based on spatial distribution of neurotrophic factor gradients

The axonal growth cone is sensitive to spatial molecular concentration gradients [22]. In a study by Rosoff et al. [22], axonal growth has been shown to be enhanced by a steep nerve growth factor (NGF) spatial concentration gradient. Axonal growth cones also preferentially advance up gradients of laminin [59]. For the spinal cord, gradients have been made in natural and synthetic polymers with laminin [1, 60]. Utilizing biomaterial polymer nerve scaffolds, axonal growth can be hypothetically made to bridge the whole length of the neural gap by seeding the scaffolds with multiple NGF spatial concentration gradients [23]. By diffusion, these NGF spatial concentration gradients might also enhance axonal growth within the adjacent natural nerve side and end-to-end autografts (**Figure 6**).

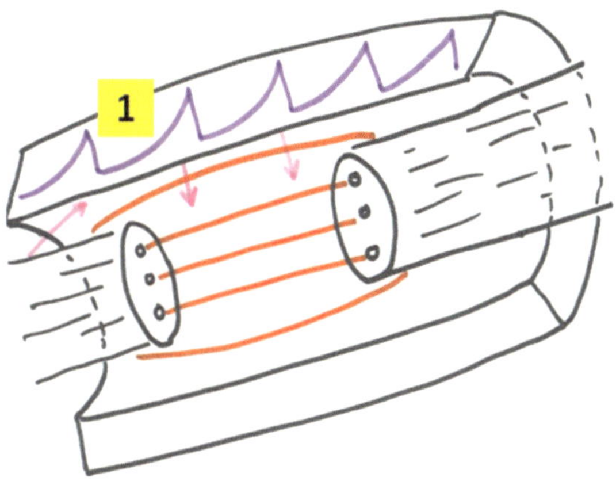

Figure 6. Spatial distribution of molecular factor gradients (1). Utilizing biomaterial polymer nerve scaffolds, axonal growth can be hypothetically made to bridge the whole length of the neural gap by seeding the scaffolds with multiple NGF spatial concentration gradients [23]. By diffusion, these NGF spatial concentration gradients might also enhance axonal growth within the adjacent natural nerve side and within the end-to-end autografts.

2.5. Requirements based on modulation of fibrosis

Scar formation prevents progression of axonal regeneration. Lysing the fibrosis/gliosis in the injury zone to an extent that allows settling of the basal lamina preventing meanwhile collapse of the neural tissue matrix is therefore essential. Recognizing this, Millesi [26] has modified Sunderland's classification of nerve injuries based on the degree of epineurial, perineurial and endoneurial fibrosis (**Table 2**). Clinically, surgical neurolysis of brachial plexus lesions has proven to be an effective procedure [50, 51, 61]. Heparin has been used for a long time in plastic surgery to lyse scar tissue [62]. Its perineural application has improved functional recovery

[63]. Among other actions, heparin influences fibroblast growth factor responsible for cell proliferation, differentiation, signal transduction and angiogenesis [64–66]. In the central nervous system, injured axons encounter a series of inhibitory factors [67]. Chondroitinase ABC lyses chondroitin sulfate proteoglycans allowing regenerating axons to progress to distal targets [68–70].

Seddon type	Sunderland type-millesi subtype
Neuropraxia	Sunderland type 1—there is transient conduction block without disruption of the anatomical structure of the nerve. Wallerian degeneration does not occur distal to the injury. Full nerve function is expected without intervention within 12 weeks
	Millesi subtype 1A—there is fibrosis of the epifascicular epineurium, necessitating an epineurotomy to enhance regeneration
	Millesi subtype 1B—there is fibrosis of the interfascicular epineurium, necessitating a partial epineurectomy to enhance regeneration
Axonotmesis: the axons are disrupted and Wallerian degeneration occurs	Sunderland type 2—the endoneurim and perineurium are intact and complete regeneration and functional recovery occurs.
	Millesi subtype 2A—there is fibrosis of the epifascicular epineurium, necessitating an epineurotomy to enhance regeneration
	Millesi subtype 2B. There is fibrosis of the interfascicular epineurium, necessitating a partial epineurectomy to enhance regeneration
	Sunderland type 3—the axon and endoneurium are damaged, but then perineurium and epineurium are intact. This results in disorganized regeneration which can result in obstruction or diversion of axons from their correct paths
	Millesi subtype 3A—there is fibrosis of epifascicular epineurium, necessitating an epineurotomy to enhance regeneration
	Millesi subtype 3B—there is fibrosis of the interfascicular epineurium, necessitating a partial epineurectomy to enhance regeneration
	Millesi subtype 3C—there is additional endoneural fibrosis. During surgery, electric stimulation of the affected nerve shows poor contraction of the muscle(s) supplied by it. Treatment is by excision of the fibrotic segment and nerve grafting
	Sunderland type 4—fourth degree injuries involve damage to all structures, however, continuity of the nerve trunk is bridged by a mass of connective tissue, Schwann cells and regenerating axons. Conduction down the nerve is not possible
	Millesi subtype 4N. The scar is invaded by neuroma. During surgery, electric stimulation of the affected nerve shows decreased or lost contraction of the muscle(s) supplied by it. Treatment is accordingly surgical by neurolysis or excision and nerve grafting
	Millesi subtype 4S—the scar consists of fibrosis only. During surgery, electric stimulation of the affected nerve shows lost contraction of the muscle(s) supplied by it. Treatment is surgical by excision and nerve grafting
Neurotmesis (complete division of	Sunderland type 5—the continuity of the nerve/brachial plexus roots, trunks, divisions or cords is completely interrupted. There may be brachial plexus root avulsions. Root avulsions are excluded by myelography or intraoperative foraminotomy. During surgery, electric stimulation of the interrupted

Seddon type	Sunderland type-millesi subtype
the nerve)	nerve shows lost contraction of the muscle(s) supplied by it. Treatment is surgical by excision and nerve grafting, or by nerve transfer in root avulsions
	MacKinnon type 6—a sixth degree of injury has been described by MacKinnon [25]. This refers to lesions with a mixed pattern of injury where there are varying degrees of injury in different sections of the nerve

Table 2. Treatment based classification of nerve injuries.

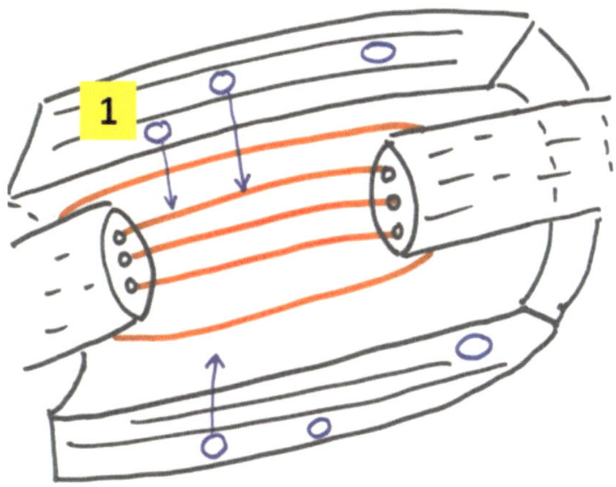

Figure 7. Modulation of fibrosis by heparin microspheres (1). Use can be made of both heparin as a neurolyzing agent as well as its ligand–receptor binding property. Knowing that heparin can bind FGF-2 and prevent its degradation, Han et al. have constructed a nerve conduit with a heparin-incorporated fibrin–fibronectin matrix to deliver this growth factor to the injury site [73].

Can we use artificial nerve side grafts supplied with spatial molecular concentration gradients of nerve growth factors for the following purposes simultaneously:

- to act as additional side grafts to end-to-end autografts as discussed under ' requirements based on recent findings of side grafting',

- to stimulate axonal growth cone progression within end-to-end autografts as discussed under 'requirements based on spatial distribution of neurotrophic factor gradients',

- to lyse postoperative fibrosis after brachial plexus surgery and direct cord implantation (using heparin and chondroitinase ABC respectively),

- as a result of the above, to regain full hand, elbow and shoulder functions in recent (less than 1 year) total brachial plexus palsy,

- as a result of the above, to provide a solution for limited axonal progression in long-standing brachial plexus lesions (of more than one-year duration) or for partially regenerated lesions (obstetric brachial plexus lesions after 3 years of age). This is an important consideration in view of the mounting evidence that motor endplates do not atrophy but are in a continuous regenerative process [71, 72].

Advances in lumen filling technology points to the possibility of all of the above (**Figures 2a, b** and **7**). Use can be made of both heparin as a neurolyzing agent as well as its ligand–receptor binding property. Heparin-incorporated fibrin–fibronectin matrices can enhance nerve regeneration by binding to FGF, GDNF and NGF [73–76], preventing their enzymatic cleavage, allowing them to bind to their respective growth factor receptors with greater affinity thus improving signal transduction [77, 78] and allowing for their release at a slower rate compared to when bound to a fibrin matrix alone [79, 80].

2.6. Requirements based on the necessity of renewal (recycling) of luminal fillers to allow for the replenishment of cells, molecular factors and concentration gradients

Closely linked to the above questions, is the following question. In global brachial plexus avulsions associated with extensive fibrosis and in which the whole brachial plexus has retracted to the axilla, can we use artificial side grafts to improve the results of direct cord implantation? In such a situation, the nerve gap is extensive and associated with excessive scarring. The axonal growth cone would thus take years to reach the target muscles. Consequently, the factors mentioned before have to be replenished continually.

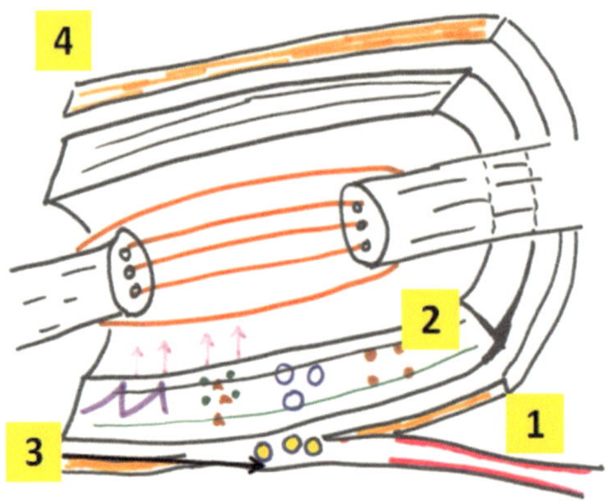

Figure 8. Catheter placement. Versatility is required to achieve the following three aims simultaneously: inserting and keeping the cell and drug delivery system (catheter (1)) patent; dissolving the fibrous tissue (2) that blocks pores after release of molecules and cells and replenishment of released molecules and cells associated with preservation of their spatial concentration gradients. To replenish cells and growth factors, chitosan microspheres (3) may be injected through an indwelling catheter incorporated into an external coat (4) of the nerve conduit. The internal surface of the catheter should be hydrophobic to inhibit adherence of cells and molecules to it, preventing its blockage. Although agarose and alginate have also served as nerve conduit scaffolds, they have gained more acceptance as hydrogel lumen fillers. Alginate, in particular, has been used both as a lumen filler and for encapsulation of cells and drugs. Interacting with positively charged chitosan, negatively charged alginate may form versatile chitosan-calcium-alginate microspheres. These may be useful tools for delivering cells and molecules through the catheter-conduit interface without blocking it.

This can take place through a continous cell and drug delivery system (catheter) [81, 82]. Versatility is required to achieve the following three aims simultaneously: keeping the cell and drug delivery system (catheter) patent; dissolving the fibrous tissue that blocks pores after release of molecules and cells and replenishment of released molecules and cells associated with preservation of their spatial concentration gradients. Microsphere, nanosphere and nanoshell technology is one method to achieve the previous aims [1, 83, 84]. Controlling surface tension as well as hydrophobic and hydrophilic properties of the conduit lumen and the microspheres may help us fulfill the simultaneous three aims described previously (**Figure 8**).

3. Chitosan as a biomaterial polymer nerve scaffold

3.1. Chitosan as a biopolymer; special properties

Chitosan [poly-(β-1/4)-2-amino-2-deoxy-D-glucopyranose] [1, 24] is a *biocompatible* biopolymer made of D-glucosamine and N-acetyl-D-glucosamine bonds and β bonds (1–4), in which glucosamine is the predominant repeating unit in its structure; it is a derivative of alkaline deacetylation of chitin.

The degree of deacetylation represents the rate of D-glucosamine units with respect to the total amount of N-acetyl-D-glucosamine that makes the chitosan molecule [24]. A deacetylated chitin over 60 or 70% is already considered to be chitosan. Deacetylation transforms the acetyl group into a primary amino group, which is more hydrophilic than the preceding molecule. *The degree of deacetylation influences biodegradation.* At a greater degree of deacetylation (between 84 and 90%), the degradation process is delayed. The free amino groups in the polymer chain exposed as a result of deacetylation confer a positive charge to chitosan. The positive charge in chitosan allows for many electrostatic interactions with negatively charged molecules including nervous tissue. Chitosan tubes with 5% acetylation have been found to be most supportive for peripheral nerve regeneration [85]. Chitosan as well as its biodegradation product, chitooligosaccharide, are neurosupportive, possibly by stimulating the proliferation of Schwann cells via the miRNA-27a/FOXO1 axis [20, 86–88].

3.2. Chitosan as an artificial nerve graft scaffold fulfills the same conditions required by nerve autografts

3.2.1. Chitosan as an artificial nerve graft scaffold meets macroengineering requirements

Chitosan can be easily processed and manufactured in a variety of forms including tubes, fibers, films, sponges and hydrogels. Although nonwoven chitosan mesh conduits [89] and cell-seeded chitosan films [90] have supported axonal growth, tubes are the most common in practice, either as hollow structures or combined with lumen fillers [2, 24]. Stenberg et al. [91] have used hollow chitosan tubes for reconstruction of rat sciatic nerve defects. Gonzalez-Perez et al. [92] and others [85] have proven the regenerative capability of chitosan tubes of low (~2%) and medium (~5%) degree of acetylation, to bridge critical nerve gaps (15 mm long) in the rat sciatic nerve. Meyer et al. [93] have used fine-tuned chitosan nerve tubes with a longitudinal chitosan film as a lumen filler to reconstruct long sciatic nerve defects in rats. Nerve scaffolds composed of a chitosan nerve guidance conduit filled with polyglycolic acid (PGA)/polylactic-co-glycolic acid (PLGA) filaments have been used to bridge large peripheral nerve defects both experimentally and clinically [20].

3.2.2. Chitosan as an artificial nerve graft scaffold meets microengineering requirements

Microgrooves, microfilaments as lumen fillers and surface treatment of chitosan are microengineering procedures. *Microgrooved* chitosan or poly(DL-lactide) (PLA) conduits have been demonstrated to enhance peripheral nerve regeneration as compared to smooth conduits [94]. Zhang et al. [95] have supplied omentum-wrapped collagen-chitosan scaffolds with longitudinally oriented micro-channels to promote axonal regeneration in rats. Hu [96] has devised a nerve-guiding collagen-chitosan scaffold with inner dimensions resembling the basal lamina microchannels of normal nerves.

As far as *lumen filling* is concerned, Hu et al. [97] have demonstrated that nerve conduits filled with longitudinal aligned polyglycolic acid filaments have a better regenerative outcome for bridging large peripheral nerve gaps than hollow nerve conduits. Patel et al. [98] have blended glial cell line-derived nerve growth factor (GDNF) and laminin with chitosan to fabricate GDNF-laminin blended chitosan (GLC) nerve guides.

Microengineering applies also to *surface treatment* of chitosan to make it ready to accept molecular factors of peripheral nerve regeneration. Huang et al. [99] have investigated the surface effects of laminin modified PLGA and chitosan membranes after chemical method and plasma treatment. Results have shown that laminin has been covalently bound onto the surface of both PLGA and chitosan membranes either by chemical method or by oxygen plasma treatment. Oxygen plasma has been a better method to incorporate laminin onto the surface of membrane.

3.2.3. Chitosan as an artificial nerve graft scaffold fulfills the same mechanical conditions required by nerve autografts

Chitosan is brittle, fracturing easily under low energy [24]. Several methods have been developed to improve its mechanical properties.

In the first and most reliable one, other polymeric chains are added (e.g., polyethyleneglycol, dialdehydes). Cross-linking agents possessing at least two functional reactive groups (e.g., formaldehyde, epoxides) are used to bridge all polymeric chains together (e.g., to form glutaraldehyde and glyoxal respectively) [24]. A bio-artificial polycaprolactone (PCL)/chitosan nanofibrous scaffold has been designed and evaluated [100]. Mechanical testing has shown that Young's modulus and strain at break of the electrospun PCL/chitosan nanofibers are better than those of the chitosan nanofibers. Huang et al. [101] have used electrospun collagen-chitosan-TPU nanofibrous scaffolds The scaffolds have been cross-linked by glutaraldehyde (GTA) vapor. Wang et al. [102] have used formaldehyde-cross-linked chitosan conduits containing PGA filaments [4].

The second method makes use of the high number of positively charged amine groups consequent upon deacetylation. They are made to react with the negatively charged factors of the molecular aspects of peripheral nerve regeneration. The resulting compounds improve both the mechanical and neurotrophic aspects of chitosan. Junka et al. [103] have used nanofibrous scaffolds electrospun from blends of poly(caprolactone) (PCL) and chitosan. Taking advantage of the amine groups on chitosan, the surface of the scaffolds have been functionalized with laminin by carbodiimide based cross-linking. Wang et al. [102] have compared chitosan nano/microfibrous tubes with a deacetylation rates of 78 and 93% to each other. Chitosan nano/microfiber mesh tubes with a deacetylation rate of 93% have had sufficient mechanical properties to preserve tube space and to provide a better scaffold for cell migration and attachment. Haastert-Talini et al. [85] have compared chitosan tubes of varying degrees of acetylation (low—2%, medium—5%, high—20%) for bridging peripheral nerve defects. Chitosan tubes with 5% acetylation have been found to be most supportive for peripheral nerve regeneration.

The third method consists of coating chitosan tubes with a film composed of a material of high mechanical toughness. Duda et al. [104] have reported on composite nerve grafts with an inner 3D multichannel porous chitosan core and an outer electrospun polycaprolactone shell. Also, cellulosic fibers in nano scale have been found to be an attractive reinforcement to chitosan, producing environmental friendly composite films with improved physical properties [105].

In the fourth method, the scaffold shape is additionally varied. Itoh et al. [106, 107] have developed nerve scaffolds using tendon chitosan tubes or hydroxyapatite-coated tendon chitosan tubes having either a circular or triangular cross-section, as well as triangular tubes combined with laminin. The mechanical strength of triangular tubes has been found to be higher than circular tubes and the inner volume of a triangular tube tends to be larger than in circular tubes. Nerve tissue regeneration along the tube wall has been present in both the laminin and laminin peptide groups.

In the fifth method, lumen fillers have been used. Xue et al. [108] have used a neural scaffold that consisted of a chitosan conduit inserted with poly(lactic-co-glycolic acid) (PLGA) fibers and mesenchymal stem cells to bridge an extra-large gap in dog sciatic nerve.

3.3. Chitosan as an artificial nerve graft scaffold fulfills requirements based on molecular aspects of peripheral nerve regeneration; chitosan fulfills requirements based on spatial distribution of neurotrophic factor gradients

The interaction of chitosan and molecular aspects of peripheral nerve regeneration obeys principles of luminal filling. The principles of lumen filling can also be applied to create spatial concentration gradients of molecular factors. Three main factors determine this: the degree of deacetylation of chitosan, its porosity and chitosan biodegradation products (chitooligosaccharides).

Impregnating the *neural conduit wall* with *cells* or *neurotrophins* via cross-linking or immobilization is one way of lumen filling [56]. *Cell adhesion* to the structure is determined by the extent of its positive charge, itself a function of the degree of alkaline deacetylation [1, 24]. This effect has been reported for a number of anchorage-dependent cells, such as dorsal root ganglion neurons, Schwann cells [86], neural stem cells [109] and mesenchymal stem cells [110–112]. Examples for impregnating the nerve conduit wall with *neurite outgrowth promoting factors and neurotrophins* include the use of laminin [113, 114], neurotrophin-3 [115] and transforming growth factor-β1 [116].

Combining *growth factors* with *a lumen growth-supporting matrix*; incorporating *accessory cells* into the *lumen matrix*; seeding (*genetically engineered*) *cells* that produce growth factors inside the lumen; and/or using *microspheres* to deliver growth factors or accessory cells to the NC lumen are four other methods of luminal filling [56]. The latter four methods require *pores*. Porous chitosan scaffolds contain pores with size ranges from 1 to 250 μm and that vary according to temperature and water content. The lower the temperature and the greater the water content, the smaller the pore size. The porous structure can be stabilized by adding glutaraldehyde, polyethyleneglycol, heparin, or collagen, allowing the structure to become more resistant and to maintain elasticity[24]. However, the need for cross-linking agents may be eliminated [20, 117, 118]. A natural and low-toxicity cross-linking agent, *genipin*, has been used to immobilize nerve growth factor (NGF) onto chitosan-based neural scaffolds [119].

Lumen matrix biomaterials include collagen [120], fibronectin or hydrogel [121]. All bind molecular factors and cells with high affinity and release them slowly over time [56].

The biodegradation products of chitosan are water-dissolvable chitooligosaccharides (COSs), which have been shown to support peripheral nerve regeneration. Jiang et al. [122] have demonstrated that the beneficial effect of *chitooligosaccharides* on cell behavior and function of primary Schwann cells is accompanied by up-regulation of adhesion proteins and neurotrophins.

3.4. Chitosan as an artificial nerve graft scaffold fulfills requirements based on recent findings of side grafting

Cografting autologous nerve grafts with biomaterial polymer nerve scaffolds presupposes biocompatibility between both. Evidence for neural cell/chitosan biocompatibility comes from an animal model of multiple sclerosis developed by Hoveizi et al. [123]. These authors have noted the role of polylactic acid/chitosan (PLA/CS) scaffold in increasing neural cell differentiation. The biocompatibility of functional Schwann cells induced from bone mesenchymal cells with a chitosan conduit membrane has been proven by Zhang et al. [124].

3.5. Chitosan as an artificial nerve graft scaffold fulfills requirements based on modulation of fibrosis

Firstly, through its cationic interactions, chitosan prevents secondary infection by many types of fungi, yeasts and bacteria [24]. Secondly, chitosan exerts an antifibrotic effect by binding to collagen via hydrogen bonding and polyanionic–polycationic interactions. Thus it prevents postoperative abdominal adhesions either singly, or as a polypropylene/chitosan mesh, or even as a chitosan-gelatin modified film. Experimentally, it has been used to prevent excessive fibrosis after intestinal resection anastomosis. The feasibility of using silk fibroin and chitosan blend scaffolds for preventing excessive fibrosis after ventral hernia repair has been investigated in guinea pigs [24].

Thirdly, heparin, both an anticoagulant and fibrotic drug, can be incorporated into chitosan microspheres or can be used as a luminal filling matrix molecule in chitosan conduits. Heparin cross-linked chitosan microspheres for the delivery of neural stem cells and growth factors for central nervous system repair have been used by Skop et al. [125]. This has proved effective for transplanting large numbers of neural stem cells, heparin and FGF-2. Using chitosan nerve conduits, Han et al. [73] have developed a basic fibroblast growth factor delivery system fabricated with heparin-incorporated fibrin-fibronectin matrices for repairing rat sciatic nerve disruptions.

3.6. Requirements based on the necessity of renewal (recycling) of luminal fillers to allow for the replenishment of cells, molecular factors and concentration gradients

Chitosan offers the versatility required in microsphere, nanosphere and nanoshell technology [1, 91] as one method to achieve three aims simultaneously: dissolving the fibrous tissue that blocks pores after release of molecules and cells as aforementioned, replenishment of released molecules and cells associated with preservation of their spatial concentration gradients and keeping the cell and drug delivery system (catheter) patent.

To replenish released molecules and cells, Zeng et al. [10] have incorporated chitosan microspheres into collagen-chitosan scaffolds for the controlled release of nerve growth factor. Wei et al. [126] have evaluated the potential of a chitosan/silk fibroin scaffold serving as a delivery vehicle for adipose-derived stem cells in neural tissue. Wang et al. [127] have used Schwann cells on oriented chitosan nanofiber mesh tubes as delivery vehicles. Maysinger et al. [128] have prepared microspheres containing ciliary neurotrophic factor (CNTF) or genetically engineered cells able to synthesize and release this cytokine. Raisi et al. [129] have resorted to mesenchymal stem cell-derived microvesicles to enhance sciatic nerve regeneration in the rat.

Keeping the cell and drug delivery system (catheter) patent may necessitate external coating of the nerve conduit [104]. Part of the external coating may be made of a hydrophobic material, preventing microsphere adherence to it and thus keeping it patent. Electrically and magnetically responsive microspheres may be exposed to repulsive fields in hydrophobic areas. Indeed these microspheres have been used to enhance nerve regeneration. Qi et al. [130] have electrically stimulated olfactory ensheathing cells (OECs) using a biodegradable conductive composite made of conductive polypyrrole (PPy, 2.5%) and biodegradable chitosan (97.5%). Similarly, Liu et al. [131] have activated Schwann cells in vitro using magnetically responsive magnetic nanoparticles (MNPs) and a biodegradable chitosan-glycerophosphate polymer.

4. Other biomaterial polymer nerve scaffolds

Although agarose and alginate have also served as nerve conduit scaffolds, they have gained more acceptance as hydrogel lumen fillers. Alginate, in particular has been used both as a lumen filler and for encapsulation of cells and drugs.

4.1. Agarose

4.1.1. Agarose as a biopolymer; general properties

Agarose consists of alternating units of b-D-galactopyranose and 3, 6-anhydro-a- L-galactopyranose; extracted from different sources it can have different chemical compositions depending on Sulphated L-hydroxyl group concentrations [1, 132]. Agarose is thermoresponsive, but poorly degradable

4.1.2. Agarose as a nerve scaffold

4.1.2.1. Macroengineering and microengineering requirements

Stokols and colleagues have developed agarose-based scaffolds with linear pores by using a freeze-drying process [133] as well as by templating [134]. These authors have used a freeze drying method to form agarose scaffolds containing linear guidance pores with a mean diameter of 120 μm [133, 135]. This process involves the formation of ice crystals whose size and direction of growth can be controlled by the temperature gradient [136]. Pore size in the scaffold can also be controlled by the freezing rate and pH; the faster rate the smaller the size [137]. Integrating BDNF into the scaffold material or using BDNF-secreting mesenchymal stem cells scaffold channels, has significantly improved the scaffold's capacity to promote regeneration [135]. Koffler et al. [138] have implanted linear-channeled agarose scaffolds in spinal cord injuries, providing growing axons with linear growth paths at the lesion site and minimizing regeneration distance. The linear channels have helped regenerating axons maintain the correct mediolateral and dorsoventral position in the spinal cord.

4.1.2.2. Pore size and mechanical properties

The average pore size of agarose hydrogels is influenced by the concentration of the polymer in solution and the settling temperature. An increase in concentration results in tightly packed helices that translate to a decrease in pore size. Decreasing the settling temperature results in small pore gels with higher elastic moduli [132].

4.1.2.3. Molecular aspects of peripheral nerve regeneration

The interaction of agarose and molecular aspects of peripheral nerve regeneration obeys principles of luminal filling. Agarose does not provide adhesion motifs to cells and does not allow interaction between adherent cells and the entrapping matrix. However, it can be supplemented with adhesion molecules of the extracellular matrix [56, 139, 140].

In repairing injuries in the CNS, agarose-based biomaterials have also had incredible success, at least in experimental settings [133, 134]. BDNF within lipid microtubules has been incorporated into agarose scaffolds, enhancing axonal growth for the length of the scaffold but not into the distal cord [70]. Living neurons and axonal tracts have been internalized within agarose hydrogel tubes by Cullen et al. [141]. In a study by Gros et al. [142], templated agarose scaffolds have substantially enhanced the organization and distance over which long-tract axons extend through a spinal cord lesion site in the presence of combinatorial therapies, but host-scaffold reactive interfaces have limited axon re-penetration of the host.

4.1.3. Agarose as a scaffold lumen filler

In order to promote sciatic nerve regeneration across a challenging 20 mm nerve gap in rats, Dodla and Bellamkonda [140] have used anisotropic scaffold lumen fillers of agarose hydrogels containing gradients of laminin-1 (LN-1) and nerve growth factor (NGF) molecules. Injectable agarose hydrogels have also been implemented [70, 143]. Using multi-photon chemistry, Wylie et al. [144] have created 3D agarose hydrogels that incorporate two bioactive molecules (a transcription factor and a growth factor).

4.2. Alginate

4.2.1. Alginate as a biopolymer; general properties

Composed of (1–4)-linked α-L-guluronate (G) and β-D-mannuronate (M) monomers, alginate has carboxyl groups which are charged at pH values higher than 3–4, making it soluble in neutral and alkaline media [145]. Alginate is biocompatible, has low toxicity and high bioavailability [132]. Alginates can form polyelectrolyte complexes in the presence of polycations such as poly-L-lysine or chitosan. Poly-L-lysine has been widely used to coat alginate beads [132].

4.1.2. Alginate as a nerve scaffold

4.1.2.1. Macroengineering and microengineering requirements

Anionic in nature, alginate can be assembled with polycations into polyelectrolyte free-standing films (i.e., films with a few micrometers in thickness) which are suitable drug reservoirs of biomolecules, such as growth factors and antibiotics [145]. Lu et al. [146] have reported on the use of chitin-alginate 3D microfibrous scaffolds to support efficient neuronal differentiation and maturation under chemically defined culture conditions. Ethylene diamine has been covalently cross-linked to further freeze-dried alginate, resulting in a porous foam. This has been subsequently implanted into a 7-mm sciatic nerve gap in the rat without suturing [147]. Alginate foams have also been tested alone or in combination with a protective PGA mesh in the treatment of a 50-mm sciatic nerve defect in the cat [148]. Alginate foams can promote nerve regeneration even without suturing to the nerve stumps, or enclosing within a tube structure [4].

4.1.2.2. Molecular aspects of peripheral nerve regeneration

Wnt proteins are bifunctional axon guidance molecules, several of which appear to mediate guidance of corticospinal tract axons along the spinal cord. Park et al. [149] have studied the effect on spinal cord regeneration by Wnt-containing alginate scaffolds. In another study

[56, 150], nerve growth factor (NGF) has been cross-linked to the wall of an alginate/chitosan scaffold.

4.1.2.3. Modulation of fibrosis

Heparin can be used both as a neurolyzing agent and as a matrix lumen filling ligand-binder. A heparin/alginate gel combined with basic fibroblast growth factor but without a tubular structure [151] has been used as an artificial nerve graft for digital nerve repair. A matrix consisting of heparin and alginate covalently cross-linked with ethylenediamine has been produced to stabilize and control the release of growth factors [152].

4.2.3. Alginate as a scaffold lumen filler

4.2.3.1. Hydrogel matrix as lumen filler

Alginate hydrogels have been used as lumen fillers in association with *growth factors* such as LIF, a pleiotrophic protein [74, 153], fibroblastic growth factor 2 (FGF2) [154] and glial derived growth factor (GGF) [155, 156]. Alginate hydrogels have been used as lumen fillers in association with *cells*, such as Schwann cells [58, 157], allogenic Schwann cells [158] and olfactory ensheathing cells (OECs), Schwann cells (SCs) and bone marrow stromal cells (BMSCs) [159].

4.2.3.2. Encapsulated particles as lumen fillers

Cells and growth factors may be encapsulated in protective alginate microcapsules, coated with multilayers of chitosan and alginate. Poly(lactic acid) microparticles may be co-encapsulated to provide anchorage points thus promoting cell survival [145]. Alginate encapsulation has involved, among other things, the encapsulation of olfactory ensheathing cells (OECs) [160], fibroblasts engineered to produce brain-derived neurotrophic factor (Fb/BDNF) [161] and NGF-expressing cells [162].

5. Conclusion

We have outlined the requirements of artificial nerve end-to-end or preferably side grafting. These include: general prerequisites of biopolymers used as scaffolds; conditions required by nerve autografting; macroengineering requirements; microengineering requirements; mechanical conditions required by nerve autografting; requirements based on molecular aspects of peripheral nerve regeneration; requirements of side grafting; requirements based on spatial distribution of neurotrophic factor gradients; requirements based on modulation of fibrosis; requirements based on the necessity of renewal of luminal fillers to allow the replenishment of cells, molecular factors and concentration gradients. We have also outlined that using nerve conduit as side grafts necessitates their design as open cylinders, so they can be easily placed around suture lines and nerve graft and so that to allow diffusion of nutrients to the autografts. Side-grafting unloads stresses from autografts and nerve ends. They allow of spreading nerve growth factor concentration gradients and dissolution of fibrosis. Chitosan satisfies all these requirements. Combined with other materials, it makes them less stiff, thus mechanically suitable as grafts. Produced as a result of deacetylation of chitin, chitosan nerve conduit scaffolds have got numerous positively charged free amino groups, which enable them to provide adherence for nerve growth factors and cells by lumen filling methods. To modulate fibrosis, heparin cross-linked chitosan microspheres have proved effective for delivering/ transplanting large numbers of neural stem cells, heparin (matrix molecule and neurolyzing agent) and FGF-2. To replenish cells and growth factors, chitosan microspheres may be injected through an indwelling catheter incorporated into an external coat of the nerve conduit. The internal surface of the catheter should be hydrophobic to inhibit adherence of cells and molecules to it, preventing its blockage. Polystyrene can be used to create superhydrophobic surfaces. The real challenge remains the catheter-conduit interface. This cannot be kept hydrophobic; nevertheless, it should be kept patent. Agarose and alginate have mainly served as hydrogel lumen fillers. Alginate, in particular, has been used both as a lumen filler and for encapsulation of cells and drug. Interacting with positively charged chitosan, negatively charged alginate may form versatile chitosan-calcium-alginate microspheres. Both cationic chitosan- and anionic alginate-coated poly(DL-lactide-co-glycolide) nanoparticles [163] and chitosan/alginate beads [164] have been produced and may be useful tools for delivering cells and molecules through the catheter-conduit interface without blocking it.

Author details

Sherif M. Amr[1*] and Basma Ekram[2]

*Address all correspondence to: sherifamrh@hotmail.com

1 The Department of Orthopaedics and Traumatology, Cairo University, Cairo, Egypt

2 The Department of Polymer Science, National Research Institute, Cairo, Egypt

References

[1] Madigan NN, McMahon S, O'Brien T, Yaszemski MJ, Windebank AJ. Current tissue engineering and novel therapeutic approaches to axonal regeneration following spinal cord injury using polymer scaffolds. Respir Physiol Neurobiol. 2009 Nov 30;169(2): 183–199.

[2] Konofaos P, Ver Halen JP. Nerve repair by means of tubulization: past, present, future. J Reconstr Microsurg. 2013 Mar;29(3):149–164. doi:10.1055/s-0032-1333316. Epub 2013 Jan 9. Review. PubMed PMID: 23303520.

[3] Lundborg G. Alternatives to autologous nerve grafts. Handchir Mikrochir Plast Chir. 2004;36:1–7.

[4] Jiang X, Lim SH, Mao HQ, Chew SY. Current applications and future perspectives of artificial nerve conduits. Exp Neurol. 2010 May;223(1):86–101. doi:10.1016/j.expneurol. 2009.09.009. Epub 2009 Sep 19. Review. PubMed PMID: 19769967.

[5] Panseri S, Cunha C, Lowery J, Del Carro U, Taraballi F, Amadio S, Vescovi A, Gelain F. Electrospun micro- and nanofiber tubes for functional nervous regeneration in sciatic nerve transections. BMC Biotechnol. 2008;8:39.

[6] Rangappa N, Romero A, Nelson KD, Eberhart RC, Smith GM. Laminincoated poly(L-lactide) filaments induce robust neurite growth while providing directional orientation. J Biomed Mater Res. 2000;51:625–634.

[7] Meek MF, Coert JH. US Food and Drug Administration/Conformit Europe approved absorbable nerve conduits for clinical repair of peripheral and cranial nerves. Ann Plast Surg. 2008;60(4):466–472.

[8] Mackinnon SE, Hudson AR, Bojanowski V, Hunter DA, Maraghi E. Peripheral nerve injection injury with purified bovine collagen—an experimental model in the rat. Ann Plast Surg. 1985;14:428–436.

[9] Meek MF, Dunnen WFD. Porosity of the wall of a neurolac nerve conduit hampers nerve regeneration. Microsurgery. 2009;29:473–478.

[10] Zeng W, Rong M, Hu X, Xiao W, Qi F, Huang J, Luo Z. Incorporation of chitosan microspheres into collagen-chitosan scaffolds for the controlled release of nerve growth factor. PLoS One. 2014 Jul 1;9(7):e101300. doi:10.1371/journal.pone.0101300. eCollection 2014. PubMed PMID: 24983464; PubMed Central PMCID: PMC4077743.

[11] Schlosshauer B, Dreesmann L, Schaller HE, Sinis N. Synthetic nerve guide implants in humans: a comprehensive survey. Neurosurgery. 2006;59:740–748.

[12] Mackinnon SE, Dellon AL. Clinical nerve reconstruction with a bioabsorbable polyglycolic acid tube. Plast Reconstr Surg. 1990;85:419–424.

[13] Weber RA, Breidenbach WC, Brown RE, Jabaley ME, Mass DP. A randomized prospective study of polyglycolic acid conduits for digital nerve reconstruction in humans. Plast Reconstr Surg. 2000;106:1036–1045, discussion 1046–1048.

[14] Kim J, Dellon AL. Reconstruction of a painful post-traumatic medial plantar neuroma with a bioabsorbable nerve conduit: a case report. J Foot Ankle Surg. 2001;40:318–323.

[15] Inada Y, Morimoto S, Takakura Y, Nakamura T. Regeneration of peripheral nerve gaps with a polyglycolic acid-collagen tube. Neurosurgery. 2004;55:640–646, discussion 646–648.

[16] Navissano M, Malan F, et al. Neurotube for facial nerve repair. Microsurgery. 2005;25(4): 268–271.

[17] Bertleff, M.J.O.E., Meek, M.F., et al.. A prospective clinical evaluation of biodegradable Neurolac nerve guides for sensory nerve repair in the hand. J Hand Surg. 2005;30(3): 513–518.

[18] Lohmeyer J, Zimmermann S, Sommer B, Machens HG, Lange T, Mailander P. Bridging peripheral nerve defects by means of nerve conduits. Chirurg. 2007;78:142–147.

[19] Pabari A, Yang SY, Seifalian AM, Mosahebi A. Modern surgical management of peripheral nerve gap. J Plast Reconstr Aesthet Surg. 2010;63:1941–1948.

[20] Gu X. Progress and perspectives of neural tissue engineering. Front Med. 2015;9(4):401–411. doi:10.1007/s11684-015-0415-x.

[21] Amr SM, Moharram AN. Repair of brachial plexus lesions by end-to-side side-to-side grafting neurorrhaphy: experience based on 11 cases. Microsurgery. 2005;25(2):126–146.

[22] Rosoff WJ, Urbach JS, Esrick MA, McAllister RG, Richards LJ, Goodhill GJ. A new chemotaxis assay shows the extreme sensitivity of axons to molecular gradients. Nat Neurosci. 2004 Jun;7(6):678–82. Epub 2004 May 25 (Erratum in: Nat Neurosci. 2004 Jul; 7(7):785).

[23] Tse TH, Chan BP, Chan CM, Lam J. Mathematical modeling of guided neurite extension in an engineered conduit with multiple concentration gradients of nerve growth factor (NGF). Ann Biomed Eng. 2007 Sep;35(9):1561–1572. Epub 2007 May 23.

[24] Rodríguez-Vázquez M, Vega-Ruiz B, Ramos-Zúñiga R, Saldaña-Koppel DA, Quiñones-Olvera LF. Chitosan and its potential use as a scaffold for tissue engineering in regenerative medicine. Biomed Res Int. 2015;2015:8212–8279.

[25] Chimutengwende-Gordon M, Khan W. Recent advances and developments in neural repair and regeneration for hand surgery. Open Orthopaedics J. 2012;6 (Suppl 1: M13): 103–107.

[26] Millesi H. Chirurgie der Peripheren Nerven. Munich: Urban &Schwarzenberg; 1992; pp. 14–15.

[27] Hiroshi N, Tator CH, Shoichet MS. Bioengineered strategies for spinal cord repair. J Neurotrauma. 2006;23:496–507 [PubMed: 16629632].

[28] Wong DY, Leveque JC, Brumblay H, Krebsbach PH, Hollister SJ, Lamarca F. Macro-architectures in spinal cord scaffold implants influence regeneration. J Neurotrauma. 2008;25:1027–1037 [PubMed: 18721107].

[29] Krych AJ, Rooney GE, Chen B, Schermerhorn TC, Ameenuddin S, Gross L, Moore MJ, Currier BL, Spinner RJ, Friedman JA, Yaszemski MJ, Windebank AJ. Relationship between scaffold channel diameter and number of regenerating axons in the transected rat spinal cord. Acta Biomater. 2009 Sep;5(7):2551–9. doi:10.1016/j.actbio.2009.03.021. PubMed PMID: 19409869; PubMed Central PMCID:PMC2731813.

[30] Moore MJ, Friedman JA, Lewellyn EB, Mantila SM, Krych AJ, Ameenuddin S, Knight AM, Lu L, Currier BL, Spinner RJ, Marsh RW, Windebank AJ, Yaszemski MJ. Multiple-channel scaffolds to promote spinal cord axon regeneration. Biomaterials. 2006b;27:419–429 [PubMed: 16137759].

[31] Ducker TB, Hayes GJ. Experimental improvements in the use of Silastic cuff for peripheral nerve repair. J Neurosurg. 1968;28:582.

[32] Rutkowski GE, Heath CA. Development of a bioartificial nerve graft. II. Nerve regeneration in vitro. Biotechnol Prog. 2002;18:373.

[33] Kokai LE, Lin YC, Oyster NM, Marra KG. Diffusion of soluble factors through degradable polymer nerve guides: controlling manufacturing parameters. Acta Biomater. 2009;5:2540.

[34] Nectow AR, Marra KG, Kaplan DL. Biomaterials for the development of peripheral nerve guidance conduits. Tissue Eng Part B Rev. 2012 Feb;18(1):40–50. doi:10.1089/ten.TEB.2011.0240. Epub 2011 Sep 23. Review. PubMed PMID: 21812591; PubMed Central PMCID: PMC3262974.

[35] Bellamkonda R. Peripheral nerve regeneration: an opinion on channels, scaffolds and anisotropy. Biomaterials. 2006;27:3515–3518 [PubMed: 16533522].

[36] Bellamkonda R, Ranieri JP, Bouche N, Aebischer P. Hydrogel-based three-dimensional matrix for neural cells. J Biomed Mater Res. 1995;29:663–671 [PubMed: 7622552].

[37] Khademhosseini A, Langer R. Microengineered hydrogels for tissue engineering. Biomaterials. 2007;28:5087–5092 [PubMed: 17707502].

[38] Clark P, Connolly P, Curtis A, Dow J, Wilkinson C. Cell guidance by ultafine topography in vitro. J Cell Sci. 1991;99:73–77 [PubMed: 1757503].

[39] Goldner JS, Bruder JM, Li G, Gazzola D. Neurite bridging across micropatterned grooves. Biomaterials. 2006;27:460–472 [PubMed: 16115675].

[40] Yao L, Damodaran G, Nikolskaya N, Gorman AM, Windebank AJ, Pandit A. The effect of laminin peptide gradient of enzymatically cross-linked collagen scaffold on neurite growth. J Biomed Mater Res A. 2009a;11 (Epub February 11, 2009).

[41] Yao L, O'Brien N, Windebank A, Pandit A. Orienting neurite growth in electrospun fibrous neural conduits. J Biomed Mater Res B Appl Biomater. 2009b;90(2):483–491 [PubMed: 19130615].

[42] Yao L, Wang S, Cui W, Sherlock R, O'Connell C, Damodaran G, Gorman A, Windebank A, Pandit A. Effect of functionalized micropatterned PLGA on guided neurite growth. Acta Biomater. 2009c;5:580–588 [PubMed: 18835227].

[43] Lundborg G, Dahlin L, Dohi D, Kanje M, Terada N. A new type of "bioartificial" nerve graft for bridging extended defects in nerves. J Hand Surg (Br.). 1997;22:299–303.

[44] Itoh S, Takakuda K, Ichinose S, Kikuchi M, Schinomiya K. A study of induction of nerve regeneration using bioabsorbable tubes. J Reconstr Microsurg. 2001;17:115–123.

[45] Chew SY, Mi R, Hoke A, Leong KW. Aligned protein–polymer composite fibers enhance nerve regeneration: a potential tissue-engineering platform. Adv Funct Mater. 2007;17:1288–1296.

[46] Zhang BG, Quigley AF, Myers DE, Wallace GG, Kapsa RM, Choong PF. Recent advances in nerve tissue engineering. Int J Artif Organs. 2014 Apr;37(4):277–91. doi:10.5301/ijao.5000317. Epub 2014 Apr 15. Review. PubMed PMID: 24811182.

[47] Hudson TW, Evans GR, Schmidt CE. Engineering strategies for peripheral nerve repair. Clin Plast Surg. 1999;26:617–628, ix.

[48] Weir NA, Buchanan FJ, Orr JF, Farrar DF, Boyd A. Processing, annealing and sterilisation of poly-L-lactide. Biomaterials. 2004;25:3939.

[49] Rydevik BL, Kwan MK, Myers RR, Brown RA, Triggs KJ, Woo SL, Garfin SR. An in vitro mechanical and histological study of acute stretching on rabbit tibial nerve. J Orthop Res. 1990;8:694.

[50] Millesi H. Factors affecting the outcome of peripheral nerve surgery. Microsurgery. 2006;26:295–302.

[51] Millesi H, Rath Th, Reihsner R, Zoch G. Microsurgical neurolysis: its anatomical basis and itsclassification. Microsurgery. 1993;14:430–439.

[52] Frostick SP, Yin Q, Kemp G. Schwann cells, neurotrophic factors and peripheral nerve regeneration. Microsurgery. 1998;18:397–405.

[53] Blottner D, Baumgarten HG. Neurotrophy and regeneration in vivo. Acta Anat. 1994;150:235–245.

[54] Sinis N, Kraus A, Tselis N, Haerle M, Werdin F, Schaller HE. Functional recovery after implantation of artificial nerve grafts in the rat—a systematic review. J Brachial Plex Peripher Nerve Inj. 2009 Oct 25;4:19. doi:10.1186/1749-7221-4-19. PubMed PMID: 19852862; PubMed Central PMCID: PMC2770034.

[55] Liu H, Yao F, Zhou Y et al. Porous poly (DL-lactic acid) modified chitosan-gelatin scaffolds for tissue engineering. J Biomater. Appl. 2005;19(4):303–322.

[56] Pabari A, Yang SY, Mosahebi A, Seifalian AM. Recent advances in artificial nerve conduit design: strategies for the delivery of luminal fillers. J Control Release. 2011 Nov 30;156(1):2–10. doi:10.1016/j.jconrel.2011.07.001. Epub 2011 Jul 6. Review. PubMed PMID: 21763371.

[57] Labrador RO, Buti M, Navarro IX. Influence of collagen and laminin gels concentration on nerve regeneration after resection and tube repair. Exp Neurol. 1998:149(1):243–252.

[58] Mosahebi A, Wiberg M, Terenghi G. Addition of fibronectin to alginate matrix improves peripheral nerve regeneration in tissue-engineered conduits. Tissue Eng. 2003;9(2): 209–218.

[59] Adams DN, Kao EY, Hypolite CL, Distefano MD, Hu WS, Letourneau PC. Growth cones turn and migrate up an immobilized gradient of the laminin IKVAV peptide. J Neurobiol 2005;62:134–147 [PubMed: 15452851].

[60] Dodla MC, Bellamkonda RV. Anisotropic scaffolds facilitate enhanced neurite extension in vitro. J Biomed Mater Res A. 2006;78A:213–221 [PubMed: 16892507].

[61] Mazal PR, Millesi H. Neurolysis: is it beneficial or harmful? Acta Neurochir. 2005;92(Suppl):3–6.

[62] Ho WS, Ying SY, Chan PC, Chan HH. Use of onion extract, heparin, allantoin gel in prevention of scarring in chinese patients having laser removal of tattoos: a prospective randomized controlled trial. Dermatol Surg. 2006 Jul;32(7):891–6. PubMed PMID: 16875470.

[63] Kahraman A, Kahveci R. Evaluating the effect of polytetrafluoroethylene and extractum cepae-heparin-allantoin gel in peripheral nerve injuries in a rat model. Plast Surg (Oakv). 2015 Spring;23(1):9–14. PubMed PMID: 25821766; PubMed Central PMCID: PMC4364148.

[64] Billings PC, Pacifici M. Interactions of signaling proteins, growth factors and other proteins with heparan sulfate: mechanisms and mysteries. Connect Tissue Res. 2015;56(4):272–80. doi:10.3109/03008207.2015.1045066.

[65] Olczyk P, Mencner Ł, Komosinska-Vassev K. Diverse roles of heparan sulfate and heparin in wound repair. Biomed Res Int. 2015;2015:549417. doi:10.1155/2015/549417. Epub 2015 Jul 7. Review. PubMed PMID: 26236728; PubMed Central PMCID: PMC4508384.

[66] Santos-Silva A, Fairless R, Frame MC, Montague P, Smith GM, Toft A, Riddell JS, Barnett SC. FGF/heparin differentially regulates Schwann cell and olfactory ensheathing cell interactions with astrocytes: a role in astrocytosis. J Neurosci. 2007 Jul 4;27(27): 7154–7167.

[67] Yiu G, He Z. Glial inhibition of CNS axon regeneration. Nat Rev Neurosci. 2006;7:617–627.

[68] Bradbury EJ, Moon LD, Popat RJ, King VR, Bennett GS, Patel PN, Fawcett JW, McMahon SB. Chondroitinase ABC promotes functional recovery after spinal cord injury. Nature. 2002;416:636–640 [PubMed: 11948352].

[69] Houle JD, Tom VJ, Mayes D, Wagoner G, Phillips N, Silver J. Combining an autologous peripheral nervous system "bridge" and matrix modification by chondroitinase allows robust, functional regeneration beyond a hemisection lesion of the adult rat spinal cord. J Neurosci. 2006 Jul 12;26(28):7405–7415.

[70] Jain A, Kim YT, McKeon RJ, Bellamkonda RV. In situ gelling hydrogels for conformal repair of spinal cord defects and local delivery of BDNF after spinal cord injury. Biomaterials. 2006;27:497–504 [PubMed: 16099038].

[71] Carraro U, Boncompagni S, Gobbo V, Rossini K, Zampieri S, Mosole S, Ravara B, Nori A, Stramare R, Ambrosio F, Piccione F, Masiero S, Vindigni V, Gargiulo P, Protasi F, Kern H, Pond A, Marcante A. Persistent muscle fiber regeneration in long term denervation. Past, present, future. Eur J Transl Myol. 2015 Mar 11;25(2):4832. doi: 10.4081/ejtm.2015.4832. eCollection 2015 Mar 11. Review. PubMed PMID: 26913148; PubMed Central PMCID: PMC4383182.

[72] Brenner HR, Lomo T, Williamson R. Control of end-plate channel properties by neurotrophic effects andby muscle activity in rat. J Physiol. 1987;388:367–381.

[73] Han H, Ao Q, Chen G, Wang S, Zuo H. A novel basic fibroblast growth factor delivery system fabricated with heparin-incorporated fibrin-fibronectin matrices for repairing rat sciatic nerve disruptions. Biotechnol Lett. 2010 Apr;32(4):585–91. doi:10.1007/s10529-009-0186-z. Epub 2009 Dec 24. PubMed PMID: 20033834.

[74] McKay Hart A, Wiberg M, Terenghi G. Exogenous leukaemia inhibitory factor enhances nerve regeneration after late secondary repair using a bioartif icial nerve conduit. Br J Plast Surg. 2003;56(5):444–450.

[75] Neufeld G, Gospodarowicz D, Dodge L, Fujii DK. Heparin modulation of the neurotropic effects of acidic and basic f ibroblast growth factors and nerve growth factor on PC12 cells. J Cell Physiol. 1987;131(1):131–140.

[76] Rider CC. Interaction between glial-cell-line-derived neurotrophic factor (GDNF) and 2-O-sulphated heparin-related glycosaminoglycans. Biochem Soc Trans. 2003;31(2): 337–339.

[77] Barnett MW, FisherCE, Perona-Wright G, Davies JA. Signalling by glial cell linederived neurotrophic factor (GDNF) requires heparan sulphate glycosaminoglycan. J Cell Sci. 2002;115(Pt 23):4495–4503.

[78] Tanaka M, Xiao H, Kiuchi K. Heparin facilitates glial cell line-derived neurotrophic factor signal transduction. Neuroreport. 2002;13(15):1913–1916.

[79] Sakiyama-Elbert SE, Hubbell JA. Controlled release of nerve growth factor from a heparin-containing fibrin-based cell ingrowth matrix. J Control Release. 2000;69(1):149–158.

[80] Wood MD, Hunter D, Mackinnon SE, Sakiyama–Elbert SE. Heparin–binding-affinity-based delivery systems releasing nerve growth factor enhance sciatic nerve regeneration. J Biomater Sci Polym Ed. 2010;21(6):771–787.

[81] Amr S, Amr H. Bridging defects in chronic spinal cord injury using peripheral nerve grafts; results of indwelling catheter implantation; a comparative clinical study. In:

Research on complication—current issues and technology. ISBN 978-1-922227-37-9. iConcept Press Limited, Brisbane, Australia; 2016.

[82] Amr S. Bridging defects in chronic spinal cord injury using peripheral nerve grafts: nerve grafting techniques and overcoming the gliosis; from basic science to clinical experience. In: Fuller H (editor) Recovery of motor function following spinal cord injury. Intechopen, Croatia, EU. 2016.

[83] Georgiou M, Bunting SCJ, Davies HD, Loughlin AJ, Golding JP, Phillips JB. Engineered neural tissue for peripheral nerve repair. Biomaterials. 2013;34:7335–7343.

[84] Benoit JP, Faisant N, Venier-Julienne MC, Menei P. Development of micro-spheres for neurological disorders: from basics to clinical applications. J Control Release. 2000;65:285–296 [PubMed:10699288].

[85] Haastert-Talini K, Geuna S, Dahlin LB, Meyer C, Stenberg L, Freier T, Heimann C, Barwig C, Pinto LF, Raimondo S, Gambarotta G, Samy SR, Sousa N, Salgado AJ, Ratzka A, Wrobel S, Grothe C. Chitosan tubes of varying degrees of acetylation for bridging peripheral nerve defects. Biomaterials. 2013 Dec;34(38):9886–904. doi:10.1016/j.biomaterials.2013.08.074. Epub 2013 Sep 17. PubMed PMID: 24050875.

[86] Yuan Y, Zhang P, Yang Y, Wang X, Gu X. The interaction of Schwann cells with chitosan membranes and fibers in vitro. Biomaterials. 2004;25(18):4273–4278 [PubMed: 15046917].

[87] Yang Y, Liu M, Gu Y, Lin S, Ding F, Gu X. Effect of chitooligosaccharide on neuronal differentiation of PC-12 cells. Cell Biol Int. 2009;33(3):352–356.

[88] Wang Y, Zhao Y, Sun C, Hu W, Zhao J, Li G, Zhang L, Liu M, Liu Y, Ding F, Yang Y, Gu X. Chitosan degradation products promote nerve regeneration by stimulating Schwann cell proliferation via miR-27a/FOXO1 axis. Mol Neurobiol. 2014 Nov 18 [Epub ahead of print] doi:10.1007/s12035-014-8968-2.

[89] Kusaba H, Terada-Nakaishi M, Wang W, Itoh S, Nozaki K, Nagai A, Ichinose S, Takakuda K. Comparison of nerve regenerative efficacy between decellularized nerve graft and nonwoven chitosan conduit. Biomed Mater Eng. 2016 May12;27(1):75–85. doi: 10.3233/BME-161571. PubMed PMID: 27175469.

[90] Wrobel S, Serra SC, Ribeiro-Samy S, Sousa N, Heimann C, Barwig C, Grothe C, Salgado AJ, Haastert-Talini K. In vitro evaluation of cell-seeded chitosan films for peripheral nerve tissue engineering. Tissue Eng Part A. 2014 Sep;20(17–18):2339–2349.

[91] Stenberg L, Kodama A, Lindwall-Blom C, Dahlin LB. Nerve regeneration in chitosan conduits and in autologous nerve grafts in healthy and in type 2 diabetic Goto-Kakizaki rats. Eur J Neurosci. 2016 Feb;43(3):463–73. doi:10.1111/ejn.13068. Epub 2015 Nov 2. PubMed PMID: 26355640.

[92] Gonzalez-Perez F, Cobianchi S, Geuna S, Barwig C, Freier T, Udina E, Navarro X. Tubulization with chitosan guides for the repair of long gap peripheral nerve injury in

the rat. Microsurgery. 2015 May;35(4):300–308. doi:10.1002/micr.22362.Epub 2014 Dec 4. PubMed PMID: 25471200.

[93] Meyer C, Stenberg L, Gonzalez-Perez F, Wrobel S, Ronchi G, Udina E, Suganuma S, Geuna S, Navarro X, Dahlin LB, Grothe C, Haastert-Talini K. Chitosan-film enhanced chitosan nerve guides for long-distance regeneration of peripheral nerves. Biomaterials. 2016 Jan;76:33–51. doi:10.1016/j.biomaterials.2015.10.040. Epub 2015 Oct 21. PubMed PMID: 26517563.

[94] Hsu SH, Lu PS, Ni HC, Su CH. Fabrication and evaluation of microgrooved polymers as peripheral nerve conduits. Biomed Microdevices. 2007 Oct;9(5):665–74. PubMed PMID: 17562182.

[95] Zhang YG, Huang JH, Hu XY, Sheng QS, Zhao W, Luo ZJ. Omentum-wrapped scaffold with longitudinally oriented micro-channels promotes axonal regeneration and motor functional recovery in rats. PLoS One. 2011;6(12):e29184. doi:10.1371/journal.pone.0029184.

[96] Hu X, Huang J, Ye Z, Xia L, Li M, Lv B, Shen X, Luo Z. A novel scaffold with longitudinally oriented microchannels promotes peripheral nerve regeneration. Tissue Eng Part A. 2009 Nov;15(11):3297–308. doi:10.1089/ten.TEA.2009.0017. PubMed PMID: 19382873.

[97] Hu W, Gu J, Deng A, Gu X. Polyglycolic acid filaments guide Schwann cell migration in vitro and in vivo. Biotechnol Lett. 2008 Nov;30(11):1937–42. doi:10.1007/s10529–008–9795–1. Epub 2008 Jul 9. PubMed PMID: 18612593.

[98] Patel M, Mao L, Wu B, Vandevord PJ. GDNF-chitosan blended nerve guides: a functional study. J Tissue Eng Regen Med. 2007 Sep-Oct;1(5):360–7. PubMed PMID: 18038430.

[99] Huang YC, Huang CC, Huang YY, Chen KS. Surface modification and characterization of chitosan or PLGA membrane with laminin by chemical and oxygen plasma treatment for neural regeneration. J Biomed Mater Res A. 2007 Sep15;82(4):842–51. PubMed PMID: 17335016.

[100] Prabhakaran MP, Venugopal JR, Chyan TT, Hai LB, Chan CK, Lim AY, Ramakrishna S. Electrospun biocomposite nanofibrous scaffolds for neural tissue engineering. Tissue Eng Part A. 2008 Nov;14(11):1787–97. doi:10.1089/ten.tea.2007.0393. PubMed PMID: 18657027.

[101] Huang C, Chen R, Ke Q, Morsi Y, Zhang K, Mo X. Electrospun collagen-chitosan-TPU nanofibrous scaffolds for tissue engineered tubular grafts. Colloids Surf B Biointerfaces. 2011 Feb 1;82(2):307–15. doi:10.1016/j.colsurfb.2010.09.002.

[102] Wang W, Itoh S, Matsuda A, Ichinose S, Shinomiya K, Hata Y, Tanaka J. Influences of mechanical properties and permeability on chitosan nano/microfiber mesh tubes as a scaffold for nerve regeneration. J Biomed Mater Res A. 2008

Feb;84(2):557–66 (Erratum in: J Biomed Mater Res A. 2008 Mar 1;84(3):846) PubMed PMID: 17941013.

[103] Junka R, Valmikinathan CM, Kalyon DM, Yu X. Laminin functionalized biomimetic nanofibers for nerve tissue engineering. J Biomater Tissue Eng. 2013 Aug 1;3(4):494–502. PubMed PMID: 24083073; PubMed Central PMCID: PMC3785102.

[104] Duda S, Dreyer L, Behrens P, Wienecke S, Chakradeo T, Glasmacher B, Haastert-Talini K. Outer electrospun polycaprolactone shell induces massive foreign body reaction and impairs axonal regeneration through 3D multichannel chitosan nerve guides. Biomed Res Int. 2014;2014:835269. doi:10.1155/2014/835269. Epub 2014 Apr 9. PubMed PMID: 24818158; PubMed Central PMCID: PMC4000981.

[105] H P S AK, Saurabh CK, A S A, Nurul Fazita MR, Syakir MI, Davoudpour Y, Rafatullah M, Abdullah CK, M Haafiz MK, Dungani R. A review on chitosan-cellulose blends and nanocellulose reinforced chitosan biocomposites: properties and their applications. Carbohydr Polym. 2016 Oct 5;150:216–26. doi:10.1016/j.carbpol.2016.05.028. Epub 2016 May 14. Review.

[106] Itoh S, Suzuki M, Yamaguchi I, Takakuda K, Kobayashi H, Shinomiya K, Tanaka J. Development of a nerve scaffold using a tendon chitosan tube. Artif Organs. 2003 Dec;27(12):1079–88. PubMed PMID: 14678421.

[107] Itoh S, Yamaguchi I, Suzuki M, Ichinose S, Takakuda K, Kobayashi H, Shinomiya K, Tanaka J. Hydroxyapatite-coated tendon chitosan tubes with adsorbed laminin peptides facilitate nerve regeneration in vivo. Brain Res. 2003 Dec 12;993(1–2):111–23. PubMed PMID: 14642836.

[108] Xue C, Hu N, Gu Y, Yang Y, Liu Y, Liu J, Ding F, Gu X. Joint use of a chitosan/PLGA scaffold and MSCs to bridge an extra large gap in dog sciatic nerve. Neurorehabil Neural Repair. 2012 Jan;26(1):96–106. doi: 10.1177/1545968311420444.

[109] Nomura H, Zahir T, Kim H, Katayama Y, Kulbatski I, Morshead CM, Shoichet MS, Tator CH. Extramedullary chitosan channels promote survival of transplanted neural stem and progenitor cells and create a tissue bridge after complete spinal cord transection. Tissue Eng A 2008b;14:649–665.

[110] Yue W, Yan F, Zhang YL, Liu SL, Hou SP, Mao GC, Liu N, Ji Y. Differentiation of rat bone marrow mesenchymal stem cells into neuron-like cells in vitro and co-cultured with biological scaffold as transplantation carrier. Med Sci Monit. 2016 May 26;22:1766–1772. PubMed PMID: 27225035.

[111] Zhu L, Liu T, Cai J, Ma J, Chen AM. Repair and regeneration of lumbosacral nerve defects in rats with chitosan conduits containing bone marrow mesenchymal stem cells. Injury. 2015 Nov;46(11):2156–63. doi:10.1016/j.injury.2015.08.035. Epub 2015 Sep 1. PubMed PMID: 26429103.

[112] Hu N, Wu H, Xue C, Gong Y, Wu J, Xiao Z, Yang Y, Ding F, Gu X. Long-term outcome of the repair of 50 mm long median nerve defects in rhesus monkeys with marrow mesenchymal stem cells-containing, chitosan-based tissue engineered nerve grafts. Biomaterials. 2013 Jan;34(1):100–11. doi:10.1016/j.biomaterials.2012.09.020. Epub 2012 Oct 11. PubMed PMID: 23063298.

[113] Itoh S, Matsuda A, Kobayashi H, Ichinose S, Shinomiya K and Tanaka J. Effects of a laminin peptide (YIGSR) immobilized on crab-tendon chitosan tubes on nerve regeneration. J Biomed Mater Res Part B Appl Biomater. 2005;73(2):375–382.

[114] Hsu SH, Kuo WC, Chen YT, Yen CT, Chen YF, Chen KS, Huang WC, Cheng H. New nerve regeneration strategy combining laminin-coated chitosan conduits and stem cell therapy. Acta Biomater. 2013 May;9(5):6606–15. doi:10.1016/j.actbio.2013.01.025. Epub 2013 Feb 1. PubMed PMID: 23376237.

[115] Wang X, Li Y, Gao Y, Chen X, Yao J, Lin W, Chen Y, Liu J, Yang Y, Wang X. Combined use of spinal cord-mimicking partition type scaffold architecture and neurotrophin-3 for surgical repair of completely transected spinal cord in rats. J Biomater Sci Polym Ed. 2013;24(8):927–39. doi:10.1080/09205063.2012.727267. Epub 2012 Oct 2. PubMed PMID: 23647249.

[116] Nie X, Deng M, Yang M, Liu L, Zhang Y, Wen X. Axonal regeneration and remyelination evaluation of chitosan/gelatin-based nerve guide combined with transforming growth factor-β1 and Schwann cells. Cell Biochem Biophys. 2014 Jan;68(1):163–72.

[117] Yang Y, Gu X, Tan R, Hu W, Wang X, Zhang P, Zhang T. Fabrication and properties of a porous chitin/chitosan conduit for nerve regeneration. Biotechnol Lett 2004; 26(23): 1793–1797

[118] Yang YM, Hu W, Wang XD, Gu XS. The controlling biodegradation of chitosan fibers by N-acetylation in vitro and in vivo. J Mater Sci Mater Med. 2007;18(11):2117–2121.

[119] Yang Y, Zhao W, He J, Zhao Y, Ding F, Gu X. Nerve conduits based on immobilization of nerve growth factor onto modified chitosan by using genipin as a crosslinking agent. Eur J Pharm Biopharm. 2011;79(3):519–525.

[120] Yan F, Yue W, Zhang YL, Mao GC, Gao K, Zuo ZX, Zhang YJ, Lu H. Chitosan-collagen porous scaffold and bone marrow mesenchymal stem cell transplantation for ischemic stroke. Neural Regen Res. 2015 Sep;10(9):1421–1426.

[121] Meyer C, Wrobel S, Raimondo S, Rochkind S, Heimann C, Shahar A, Ziv-Polat O, Geuna S, Grothe C, Haastert-Talini K. Peripheral nerve regeneration through hydrogel-enriched chitosan conduits containing engineered schwann cells for drug delivery. Cell Transplant. 2016;25(1):159–182. doi:10.3727/096368915X688010. Epub2015 Apr 14. PubMed PMID: 25876520.

[122] Jiang M, Cheng Q, Su W, Wang C, Yang Y, Cao Z, Ding F. The beneficial effect of chitooligosaccharides on cell behavior and function of primary Schwann cells is

accompanied by up-regulation of adhesion proteins and neurotrophins. Neurochem Res. 2014 Nov;39(11):2047–2057.

[123] Hoveizi E, Tavakol S, Ebrahimi-Barough S. Neuroprotective effect of transplanted neural precursors embedded on PLA/CS scaffold in an animal model of multiple sclerosis. Mol Neurobiol. 2015;51(3):1334–1342.

[124] Zhang P, Xu H, Zhang D, Fu Z, Zhang H, Jiang B. The biocompatibility research of functional Schwann cells induced from bone mesenchymal cells with chitosan conduit membrane. Artif Cells Blood Substit Immobil Biotechnol. 2006;34(1):89–97. PubMed PMID: 16519406.

[125] Skop NB, Calderon F, Levison SW, Gandhi CD, Cho CH. Heparin crosslinked chitosan microspheres for the delivery of neural stem cells and growth factors for central nervous system repair. Acta Biomater. 2013 Jun;9(6):6834–43. doi:10.1016/j.actbio.2013.02.043. Epub 2013 Mar 5. PubMed PMID: 23467042.

[126] Wei Y, Gong K, Zheng Z, Wang A, Ao Q, Gong Y, Zhang X. Chitosan/silk fibroin-based tissue-engineered graft seeded with adipose-derived stem cells enhances nerve regeneration in a rat model. J Mater Sci Mater Med. 2011 Aug;22(8):1947–64. doi: 10.1007/s10856-011-4370-z. Epub 2011 Jun 8. PubMed PMID: 21656031.

[127] Wang W, Itoh S, Konno K, Kikkawa T, Ichinose S, Sakai K, Ohkuma T, Watabe K. Effects of Schwann cell alignment along the oriented electrospun chitosan nanofibers on nerve regeneration. J Biomed Mater Res A. 2009 Dec 15;91(4):994–1005. doi:10.1002/jbm.a. 32329. PubMed PMID: 19097155.

[128] Maysinger D, Krieglstein K, Filipovic-Grcic J, Sendtner M, Unsicker K, Richardson P. Microencapsulated ciliary neurotrophic factor: physical properties and biological activities. Exp Neurol. 1996 Apr;138(2):177–88. PubMed PMID: 8620916.

[129] Raisi A, Azizi S, Delirezh N, Heshmatian B, Farshid AA, Amini K. The mesenchymal stem cell-derived microvesicles enhance sciatic nerve regeneration in rat: a novel approach in peripheral nerve cell therapy. J Trauma Acute Care Surg. 2014 Apr;76(4): 991–7. doi:10.1097/TA.0000000000000186. PubMed PMID: 24662862.

[130] Qi F, Wang Y, Ma T, Zhu S, Zeng W, Hu X, Liu Z, Huang J, Luo Z. Electricalregulation of olfactory ensheathing cells using conductive polypyrrole/chitosanpolymers. Biomaterials. 2013 Feb;34(7):1799–809. doi:10.1016/j.biomaterials.2012.11.042. Epub 2012 Dec 7. PubMed PMID: 23228424.

[131] Liu Z, Huang L, Liu L, Luo B, Liang M, Sun Z, Zhu S, Quan X, Yang Y, Ma T, Huang J, Luo Z. Activation of Schwann cells in vitro by magnetic nanocomposites via applied magnetic field. Int J Nanomedicine. 2014 Dec 17;10:43–61. doi:10.2147/IJN.S74332. eCollection 2015. PubMed PMID: 25565803; PubMed CentralPMCID: PMC4275057.

[132] Gasperini L, Mano JF, Reis RL. Natural polymers for the microencapsulation of cells. J R Soc Interface. 2014;11:0817. doi:10.1098/rsif.2014.0817

[133] Stokols S, Tuszynski MH. The fabrication and characterization of linearly oriented nerve guidance scaffolds for spinal cord injury. Biomaterials. 2004;25:5839–5846.

[134] Stokols S, Sakamoto J, Breckon C, Holt T, Weiss J, Tuszynski MH, Templated agarose scaffolds support linear axonal regeneration. Tissue Eng. 2006;12:2777–2787.

[135] Stokols S, Tuszynski MH. Freeze-dried agarose scaffolds with uniaxial channels stimulate and guide linear axonal growth following spinal cord injury. Biomaterials. 2006;27:443–451.

[136] Tabesh H, Amoabediny G, Nik NS, Heydari M, Yosefifard M, Siadat SO, Mottaghy K. The role of biodegradable engineered scaffolds seeded with Schwann cells for spinal cord regeneration. Neurochem Int. 2009;54:73–83 [PubMed: 19084565].

[137] Sachlos E, Czernuszka JT. Making tissue engineering scaffolds work. Review: the application of solid freeform fabrication technology to the production of tissue engineering scaffolds. Eur Cell Mater. 2003;5:29–39 [PubMed: 14562270].

[138] Koffler J, Samara RF, Rosenzweig ES. Using templated agarose scaffolds to promote axon regeneration through sites of spinal cord injury. Methods Mol Biol. 2014;1162:157–165. doi:10.1007/978-1-4939-0777-9_13. PubMed PMID: 24838966.

[139] Khaing ZZ, Schmidt CE. Advances in natural biomaterials for nerve tissue repair. Neurosci Lett. 2012 Jun 25;519(2):103–114.

[140] Dodla MC, Bellamkonda RV. Differences between the effect of anisotropic and isotropic laminin and nerve growth factor presenting scaffolds on nerve regeneration across long peripheral nerve gaps. Biomaterials. 2008 Jan;29(1):33–46.

[141] Cullen DK, Tang-Schomer MD, Struzyna LA, Patel AR, Johnson VE, Wolf JA, Smith DH. Microtissue engineered constructs with living axons for targeted nervous system reconstruction. Tissue Eng Part A. 2012 Nov;18(21–22):2280–9. doi:10.1089/ten.TEA.2011.0534

[142] Gros T, Sakamoto JS, Blesch A, Havton LA, Tuszynski MH. Regeneration of long-tract axons through sites of spinal cord injury using templated agarose scaffolds. Biomaterials. 2010 Sep;31(26):6719–6729. doi:10.1016/j.biomaterials.2010.04.035

[143] Chvatal SA, Kim YT, Bratt-Leal AM, Lee H, Bellamkonda RV. Spatial distribution and acute anti–inflammatory effects of Methylprednisolone after sustained local delivery to the contused spinal cord. Biomaterials 2008;29:1967–1975 [PubMed: 18255138].

[144] Wylie RG, Ahsan S, Aizawa Y, Maxwell KL, Morshead CM, Shoichet MS. Spatially controlled simultaneous patterning of multiple growth factors in three-dimensional hydrogels. Nat Mater. 2011;10:799–806.

[145] Cardoso MJ, Costa RR, Mano JF. Marine origin polysaccharides in drug delivery systems. Mar Drugs. 2016 Feb 5;14(2). pii: E34. doi:10.3390/md14020034. Review. PubMed PMID: 26861358; PubMed Central PMCID: PMC4771987.

[146] Lu HF, Lim SX, Leong MF, Narayanan K, Toh RP, Gao S, Wan AC. Efficient neuronal differentiation and maturation of human pluripotent stem cells encapsulated in 3D microfibrous scaffolds. Biomaterials. 2012 Dec;33(36):9179–9187.

[147] Suzuki K, Suzuki Y, Tanihara M, Ohnishi K, Hashimoto T, Endo K, Nishimura Y. Reconstruction of rat peripheral nerve gap without sutures using freezedried alginate gel. J Biomed Mater Res. 2000;49:528–533.

[148] Sufan W, Suzuki Y, Tanihara M, Ohnishi K, Suzuki K, Endo K, Nishimura Y. Sciatic nerve regeneration through alginate with tubulation or nontubulation repair in cat. J Neurotrauma. 2001;18:329–338.

[149] Park JH, Min J, Baek SR, Kim SW, Kwon IK, Jeon SR. Enhanced neuroregenerative effects by scaffold for the treatment of a rat spinal cord injury with Wnt3a-secreting fibroblasts. Acta Neurochir (Wien). 2013 May;155(5):809–16. doi:10.1007/s00701-013-1663-7

[150] Pfister LA, Alther E, Papaloizos M, Merkle HP, Gander B. Controlled nerve growth factor release from multi-ply alginate/chitosan-based nerve conduits. Eur J Pharm Biopharm. 2008;69(2):563–572.

[151] Suzuki Y, Ishikawa N, Tanihara M, Saito S. Nontubulation repair of peripheral nerve gap using heparin/alginate gel combined with b-FGF. Plast Reconstr Surg Glob Open. 2016 Feb 5;4(1):e600. doi:10.1097/GOX.0000000000000581. eCollection 2016 Jan. PubMed PMID: 27104099; PubMed Central PMCID: PMC4801089.

[152] Tanihara M, Suzuki Y, Yamamoto E, Noguchi A, Mizushima Y. Sustained release of basic fibroblast growth factor and angiogenesis in a novel covalently crosslinked gel of heparin and alginate. J Biomed Mater Res. 2001 Aug;56(2):216–21. PubMed PMID: 11340591.

[153] Szarek D, Marycz K, Bednarz P, Tabakow P, Jarmundowicz W, Laska J. Influence of calcium alginate on peripheral nerve regeneration: in vivo study. Biotechnol Appl Biochem. 2013 Nov-Dec;60(6):547–556. doi:10.1002/bab.1096. Epub 2013 Aug 3. PubMed PMID: 23909973.

[154] Ohta M, Suzuki Y, Chou H, Ishikawa N, Suzuki S, Tanihara M, Suzuki Y, Mizushima Y, Dezawa M, Ide C. Novel heparin/alginate gel combined with basic fibroblast growth factor promotes nerve regeneration in rat sciatic nerve. J Biomed Mater Res A. 2004;71(4):661–668.

[155] Mohanna PN, Terenghi G, Wiberg M. Composite PHB-GGF conduit for long nerve gap repair: a long-term evaluation. Scand J Plast Reconstr Surg Hand Surg. 2005;39(3):129–137. PubMed PMID: 16019744.

[156] Mohanna PN, Young RC, Wiberg M, Terenghi G. A composite poly-hydroxybutyrate-glial growth factor conduit for long nerve gap repairs. J Anat. 2003 Dec;203(6):553–565.

[157] Mosahebi A, Simon M, Wiberg M, Terenghi G. A novel use of alginate hydrogel as Schwann cell matrix. Tissue Eng. 2001;7(5):525–534.

[158] Mosahebi A, Fuller P, Wiberg M, Terenghi G. Effect of allogeneic Schwann celltransplantation on peripheral nerve regeneration. Exp Neurol. 2002 Feb;173(2):213–223. PubMed PMID: 11822885.

[159] Quigley AF, Bulluss KJ, Kyratzis IL, Gilmore K, Mysore T, Schirmer KS, Kennedy EL, O'Shea M, Truong YB, Edwards SL, Peeters G, Herwig P, Razal JM, Campbell TE, Lowes KN, Higgins MJ, Moulton SE, Murphy MA, Cook MJ, Clark GM, Wallace GG, Kapsa RM. Engineering a multimodal nerve conduit for repair of injured peripheral nerve. J Neural Eng. 2013 Feb;10(1):016008.

[160] Zhao H, Yang BL, Liu ZX, Yu Q, Zhang WJ, Yuan K, Zeng HH, Zhu GC, Liu DM, Li Q. Microencapsulation improves inhibitory effects of transplanted olfactory ensheathing cells on pain after sciatic nerve injury. Neural Regen Res. 2015Aug;10(8):1332–1337.

[161] Francis NL, Shanbhag MS, Fischer I, Wheatley MA. Influence of alginate cross-linking method on neurite response to microencapsulated neurotrophin-producing fibroblasts. J Microencapsul. 2011;28(5):353–362.

[162] Song M, Chen SZ, Han H, Xiong Y. An experimental study on repair of peripheral nerve injury by transplantation of microcapsulated NGF-expressing NIH 3T3 cells. Zhonghua Zheng Xing Wai Ke Za Zhi. 2005 Jan;21(1):53–7. Chinese.PubMed PMID: 15844601.

[163] Sanna V, Roggio AM, Siliani S, et al. Development of novel cationic chitosan-and anionic alginate–coated poly(d, l-lactide-co-glycolide) nanoparticles for controlled release and light protection of resveratrol. Int J Nanomed. 2012;7:5501–5516.

[164] Takka S, Gürel A. Evaluation of chitosan/alginate beads using experimental design: formulation and in vitro characterization. AAPS PharmSciTech. 2010;11(1):460–466. doi:10.1208/s12249–010–9406-z.

Chapter 9

Marine Polysaccharides in Medicine

Rajendra S. Dongre

Additional information is available at the end of the chapter

http://dx.doi.org/10.5772/65786

Abstract

About 70% of the Earth's surface is covered with seawater and 90% biosphere wraps maximum biodiversity that offers resourceful novel bio-molecules. Marine species are enriched with organic compounds viz. terpenoids, polyethers/ketides, lipo-glycoproteins, peptides and polysaccharides that act as cell surface receptors and involve in cell development/differentiation, besides being antimicrobial agents. Algae, sponge and fish have various defense mechanisms developed via specific/potent natural molecules to survive under hostile, extreme conditions such as various degrees of salinity, pressure, temperature, darkness, besides microbial and viral attacks. Marine seaweeds and algae enriched with polysaccharides such as glycosaminoglycans, agar, alginate and chitin/chitosan owing to their diversified significance have received growing attention among researchers. Currently, marine-derived biomolecules cater 20% market drug load while other natural products bear 30% share. Chitins exhibit various biological and physicochemical properties that can be exploited in biotechnology and medicine/drug, cosmetic, food and textile industries. This chapter focuses on chitin/chitosan production, its physicochemical characterization and biological activities and relationship between its chemical structure and bio-activity, including chemical modification reactions such as acylation, substitution, sulfonation and other cross-linking strategies applied to skeletal modification with the recently updated literature.

Keywords: chitin, chitosan, marine polysaccharide, drug, modification, pharmaceutics

1. Introduction

Nature is an ancient pharmacy known to human. Life originated over four billion years ago in oceans, besides sea aid ecosystems, and provides source of food for organism and man. Carbohydrates are abundant, essential natural polysaccharides consisting of carbon, hydrogen and oxygen atoms (H:O as 2:1 and empirical formula $C_m[H_2O]_n$ where m is different from n

and usually varies from 200 to2500), having vast health benefits. Polysaccharides contain linear to high heterogeneity linkages of more than 10 monosaccharide/glycosides with slight repeating unit modifications (discrete from their monosaccharide) while oligosaccharides contain 3–10 monosaccharides. Amid carbohydrates, various polysaccharides are found in all marine organisms, which are accountable for an innate bio-active defense mechanism. Plant starch found as both amylose and amylopectin, which resemble cellulose (plant cell wall module), exhibit storage functions while chitin and glycogen in animals (same as starch/ cellulose except NH-substitution) afford structural component enhancement in arthropod exoskeleton and fungus cell wall. Chitins, fucoidan, carrageenan and alginate are few polysaccharides that control cell proliferation and modulate metabolism in marine organisms besides own pharmaceutics utility including antioxidative, antibacterial, antiviral, immune-stimulatory, anticoagulant and anticancer activities poses novel ventures for harnessing potential of oceanic products.

Oceans occupy three-fourths of planet, which covers half of the global biodiversity envelope in certain marine species as the imperative resources for deriving novel bio-chemicals. Marine bacteria, macro/micro algae, sponges and fishes induce defensive actions via such bio-molecules that enable organisms to survive in hostile environment such as different degrees of salinity, pressure, temperature and no/deem light [1], as well as microbial/viral attack. Naturally occurring bio-chemicals viz., terpenoids, polyethers, polyketides, lipo-glycoproteins, and polysaccharides display various functions in nature such as defense against predators, aiding in cell development/differentiation, acting as cell surface receptors and providing innate immunity. Carbohydrates exacted from terrestrial marine organisms/ sediments act as prospective feedstocks for medicine, fertilization, food storage, antioxidant, laxative, smooth/nonirritating hydrated bulk in the digestive tract, tablet ingredient and drug carrier agents, as shown in **Figure 1**.

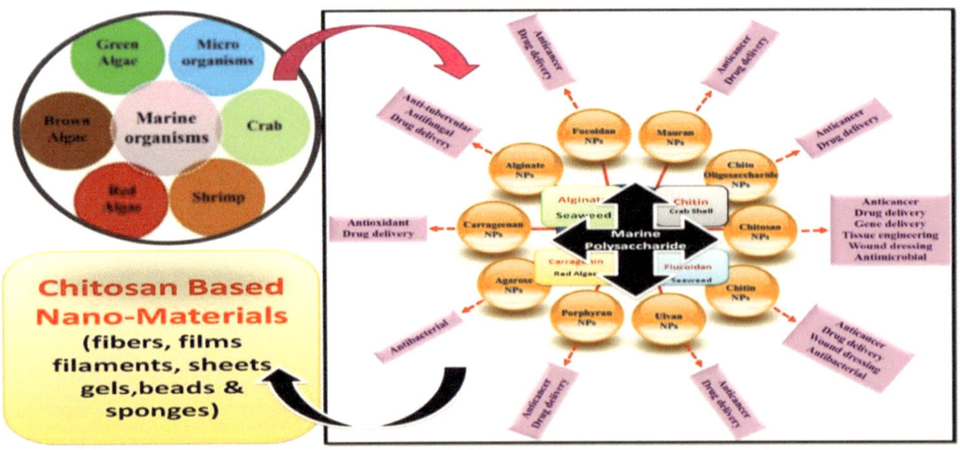

Figure 1. Resources of marine polysaccharides with their wide range of applications [2].

http://dx.doi.org/10.5772/65786

2. Marine biotechnology

Marine biotechnology focuses the following goals:

1. Discovery of bioactive molecules from marine organisms to reveal their functions and actions.

2. Study the environmental parameters, nutritional requirements and genetic factors that control the production of primary and secondary metabolites from marine species.

3. Understand the genetics, biochemistry, physiology and ecology of marine organism/ mariculture.

4. Develop diagnostic tools to improvehuman health.

5. Explore bioremediation for waste processing/disposal, coastal clean-up and oil spilling.

3. Medical potential of marine products

About 20,000 metabolites yielded from marine bacteria, sponge, coral and starfish cater significant untapped promises and act as chemical library database in pharmaceutics due to their intrinsic features. MARPOL-73/78 an international organization is engaged in enzymatic up-gradation/modification of marine polysaccharides being substitute to oil-based polymers in utilizing renewable resource. MARPOL via enzyme technology-modified chitin, alginate, fucoidan, and laminaran from marine with enhanced inherent quality to cater sustainable industrial/pharmacy needs in the below targets (**Figure 2**):

- To derive innovative bioactive chemicals from marine species.
- To increase value creation from biomass by refining/upgrading via enzymology.
- To generate cross-sectorial technology for enzyme evolution and biomaterial design.

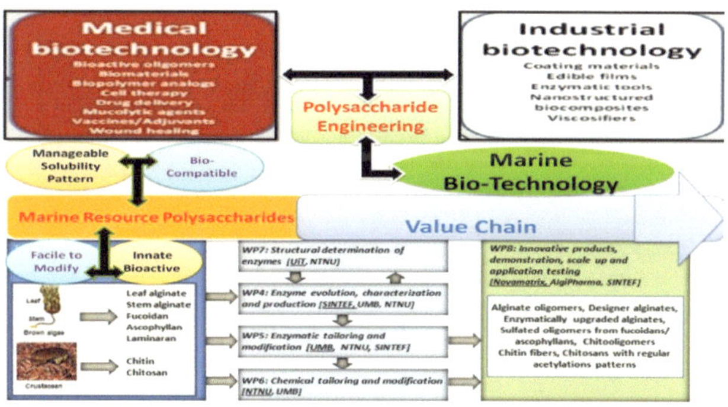

Figure 2. MARPOL via enzyme technology modified marine polysaccharide for wide application.

4. Pharmacological uses of marine polysaccharides

Advance molecular biology has innovated methods for marine organisms to isolate assorted polyunsaturated fattyacids, polysaccharides, minerals, vitamins, enzymes, and peptides. These marine polysaccharides own health benefits besides feedstock for pharmaceutics, nutrition, and pharmaceutics besides cosmetic industries (**Figure 3**) under the strategic activity of Horizon 2020: *Targeted specific activities focus on biodiversity exploration to understand how organism can withstand extremes of temperature-pressure and grows without light can be utilized to develop new industrial enzymes/pharmaceuticals*. Drug obtained from marine polysaccharide caters diverse pharmaceutical challenges by inventing various complex/novel chemicals to be used in cancer, AIDS-HIV treatments besides explicit broad spectrum activity for virus, bacteria, and fungus. Alginate and chitin polysaccharides have an extensive history of use in medicine, particularly, in pharmacy and basic sciences. Majority of carbohydrates in the nature occur as polysaccharides that consist not only glycosidic-link sugars, but also polysaccharides that are covalently linked to amino acids, peptides, proteins, and lipids. Glycan contains d-glucose, d-fructose, d-galactose, l-galactose, d-mannose, l-arabinose, d-xylose along with d-glucosamine, d-galactosamine, N-acetylneuraminic acid, N-acetylmuramic acid, and glucuronic acids. Branch structures are different polysaccharide from protein and nucleic acid polymers. Marine origin tunicin polysaccharide is a cellulose equivalent of invertebrate sea animal utilized for storage and/or structural functions. Exo-polysaccharide found in extracellular of a microbial cell component contains high sulfated and uronic contents to bestow negative charge besides acidity in seawater pH 7.5–8.4, which is used as adhesives, textiles, pharmaceuticals and anticancer agents and food additives [1–3].

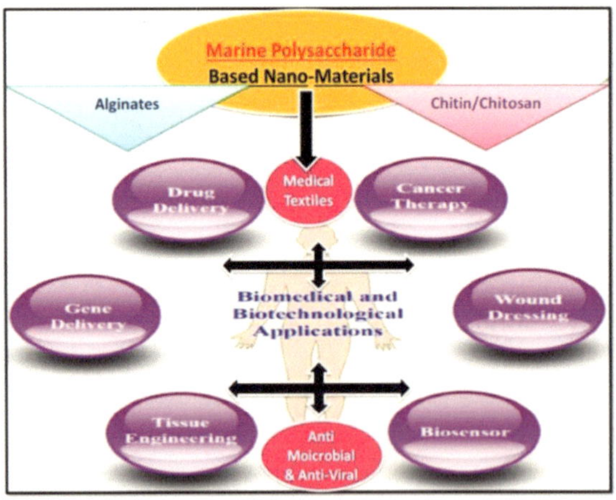

Figure 3. Marine polysaccharide materials for health benefits.

Agar and agarose have less sulfate content, but high uronic contents and corresponding beads are developed that exhibited better optical clarity with improved gel strength exploited in sustainable release of phenobarbital sodium hypnotic drug. Agars are effective for spatial infections such as poliovirus, herpes simplex, dengue viruses and also meats gelling, laxatives and flexible molds in dentistry and criminology. Chitin hydrogel provides cell signaling *in vivo* physiology [2, 3] and hydroxylation enhances its biodegradability; while sulfonation generates heparin-like increased blood compatibility [4] besides amine quaternization imparts high solubility, muco-adhesiveness (via hydrogen bonding) responsible for elevated drug residence-time, inflammation reduction and antimicrobial activity). Nanochitosan-based drug delivery systems easily cross blood-brain barrier results in cell/tissue gap penetration to targeted spleen, spinal cord, liver, lungs, and lymphs.

5. Cosmetic applications

Polysaccharides formulated with vitamins, minerals, botanical extracts, and antioxidants can promote healthy skin, hair, and nails at a cellular level which areused as creams, lotions, ointments, liquids, and pills, as shown in **Figures 2** and **3**. The cosmetic industry and advance biotechnology are mainlyfocused on marine polysaccharides as compared to synthetic chemicals due to perceived inherent prominent effects such as reduce free radical provoked aging, inflammation, and skin degradation. Chitin shows resemblance with living tissues that aid to maintain cutaneous homeostasis, neutralize radical activity, and induces transcutaneous penetration of active drugs. Chitosan nanofibrils induce low TGF-β production results in skin protections by increasing the granular density of epithelial layers and fully compatible with skin cells besides complexing with vitamins, carotenoids, and collagens to facilitate transcutaneous penetration [5]. Chitosan without having amino/hydroxy groups. E.g., E-CE6 typ aid cationic pH-sensitive molds into various shapes such as bead, hydrogel, nanofiber and nanoparticle owning fascination for other biomolecules. Chitosan hydrogel's excellent water absorption property is exploited in making of some moisturizers; it also provides wound healing and exhibits antioxidant and antimicrobial activities against various bacteria, yeast and fungi and metalloproteinase.

6. Nutrition-pharmaceutics applications

Marine-derived products used as food/ingredients are developed in nutrition-pharmaceutics to prevent/treat diseases pertaining anticancer, antiinflammatory, antioxidant, and antimicrobial activities, as shown **Figures 2–4**. Ulva fasciata derive sulfated galactans and fucans possess good anticoagulant, gel stabilizers, preservative, and flavoring agents that reduce LDL-cholesterol, plasma, and triacylglycerols [6]. Align emulsions are used as bio-flocculants in food formulation to obtain certain texture, mouth feel thickening effect, and stabilize suspended dispersed phases. Carrageenans-polyamide hydrocolloid with starch, locust bean gum, and carboxymethyl cellulose are used for milk protein stabilization [7]. Chitins are used in antidiabetic, hypocholesterolemic, adipogenesis inhibiter, food additives, and dietary supplements in nutrition-pharmaceutics which decrease body weight, serum lipids without digestions in GI tract (precipitates fat and reduces absorption via inhibiting pancreatic lipase actions). Cationic chitosan-fatty/bile acid combinations delay cholesterol and steroidal emulsifications via hydrophobic interaction, thus lessening intestinal absorptions.

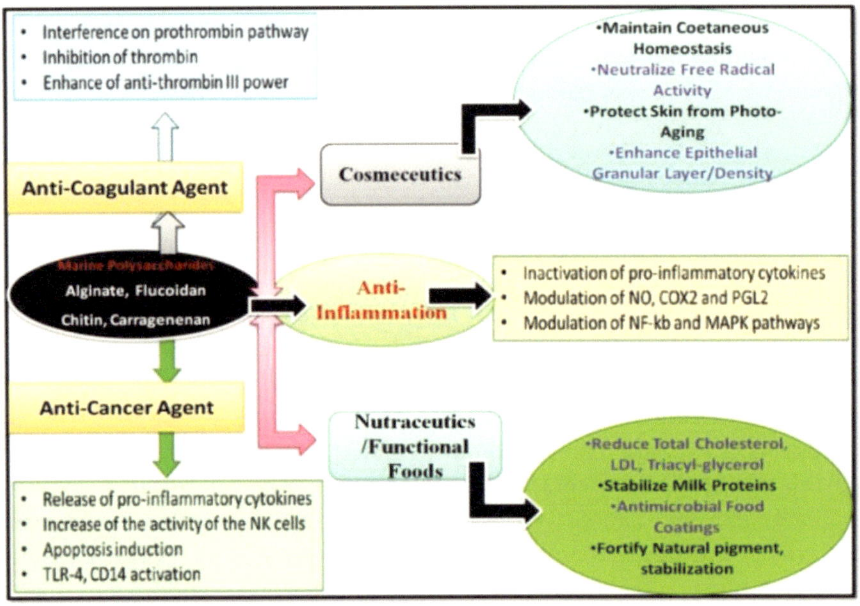

Figure 4. Pharmacology of assorted marine polysaccharides.

7. Chemical structure of chitin/chitosan

7.1. Preparations of chitin/chitosan

About10 gigatons/year of chitins are produced in biosphere which are soft and leathery, encrusted with $CaCO_3$ as the harden mass to obtain translucent, pliable, resilient as a tough exoskeleton of vertebrates, insects, and crustaceans. Chitin exhibited distinct characteristic against fungus pathogens resistance, autolytic, nutritional, and morphogenetic functions in plant, pathogenesis in virus and elicits lysozyme inductions, immunizations and parasitism in bacteria. Chitin is antidiabetic, hypocholesterolemic, adipogenesis inhibiters which decreases body weight/serum lipids without GI digestion prohibiting pancreatic lipase crucial for cholesterols emulsification to precipitate fat and fatty/bile acids [8].

7.2. Chitosan production

Chitin is feedstock for chitosan conversion via enzymatic as well chemical degradation (**Figure 5**) because of its cheap and commercial production, as shown in **Figure 6**.

http://dx.doi.org/10.5772/65786

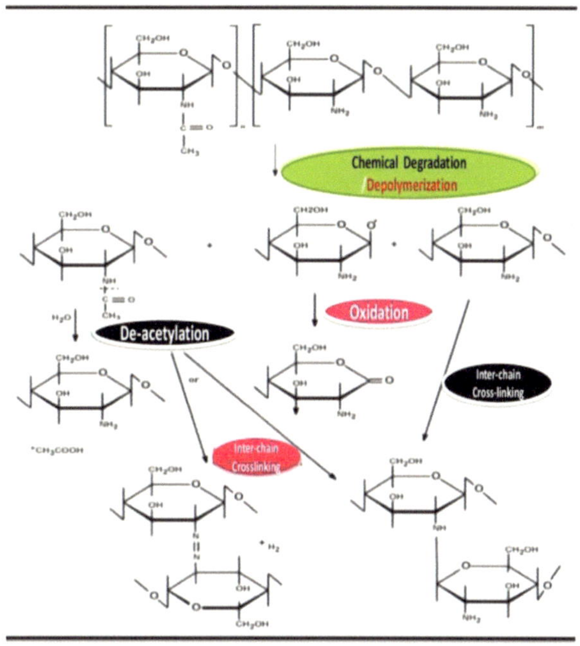

Figure 5. Chitosan chemical degradation.

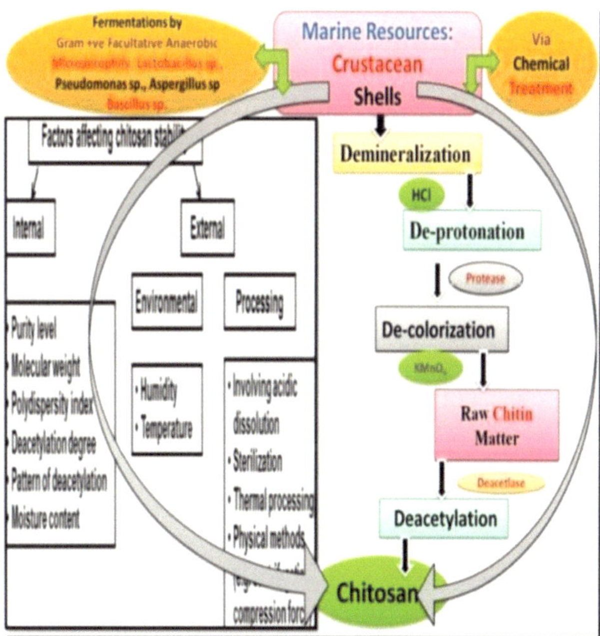

Figure 6. Production of chitosan andfactors affecting stability.

8. Chitosan copolymerization: acetylation, sulfonation, and other cross-link substitutions

Chitosan differs from concordant chitin, cellulose in fraction and distributions of their co-monomers in respective of polymeric length/sequences and found as an alternate, block, and random fashion as shown in **Figure 7**. N-acetyl-glucosamine and N-glucosamine constitutional monomers in chitin and chitosan displayed varied, respectively, solution properties and impart slightly hydrophobic terminal in chitin and ionic character at acidic pH in chitosan. Skeletally different than others amplified in acetyl sequencing that owes vivid outcome in chain conformation and aggregation besides hydrophobic substituents is vital for self-assembling and crumpling as found in liposphere micelles. The modification in chitin/chitosan's skeleton can be made by means of quaternization, glycosylation (via acetyl) and sulphonation (via sulphate groups) techniques that able grafting monomers at -NH2 linkages to yield terpolymer rather intricate matrix as shown in **Figure 8**. Chitosan compositions varied with the degree of acylation as determined by many methods such as titration, circular dichroism, FTIR, UV, NMR, and N-acetyl group hydrolysis [9, 10]. Data from different techniques showed discrepancies in degree of acylation and solubility disparity as no technique points out appropriate clarification for solubility [11]. Differential scanning calorimetry study decomposition of amino/acetyls to provide sovereign composition and molecular weights 0–1 N-acylation degree cheaper than NMR but more accurate than FTIR. NMR easily recognizes O-acetyl groups that aid to obtain degree of acylation by integration/normalization of either anomeric proton or other ring protons.

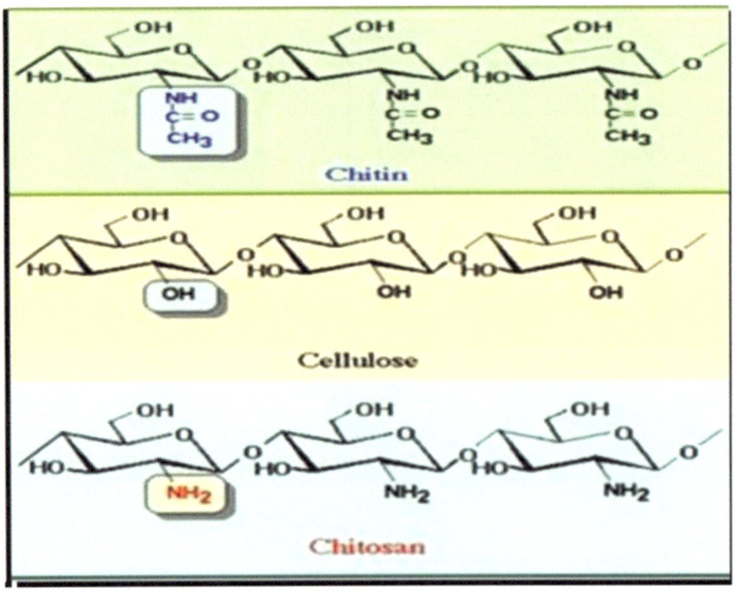

Figure 7. Comparison of chitin, cellulose, and chitosan.

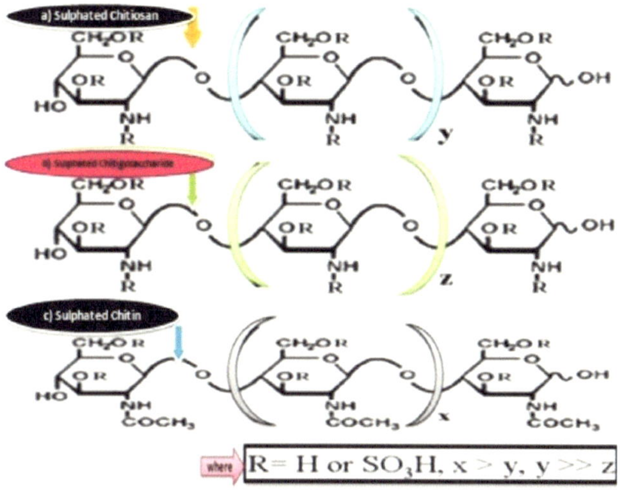

Figure 8. Structures of sulfated chitin, chitosan, and chitigosaccharide.

9. Structure and chemical modifications of chitin/chitosan

9.1. Chemical modification principles

Chitosan chemical modifications improved mechanical properties, biocompatibility, solubility, biodegradability, and shape/size by the following approaches:

- Doping/blending/grafting chemical linkages with synthetic materials;
- Micro/nanosphere surface coating by biocompatible synthetic polymers;
- Cross-linking by means of assorted physical/chemical reagents;
- Hydrophobization via alkylation;
- Modulating guluronic/mannuronic ratio; and
- Varying deacetylation degree.

Dry weight crustacean shells contain protein 30–40%, $CaCO_3$ and $Ca_3\,(PO_4)_2$about30–50%, and 20–30% chitin [11]. Industrially, acid treatment use to dissolve $CaCO_3$ and $Ca_3(PO_4)_2$followed by alkaline extraction to solubilize proteins in chitin processing from crustaceans. Deacetylation removes enough acetyls to yield chitosan with high degree chemical reactive amines that can affect physicochemical properties such as biodegradability and immunological activity [12]. Chitosan's deacetylation degree is determine by the ninhydrin test, potentiometric titration, near-IR, NMR, HBr-titrometry, FTIR, and 1st derivative UV [10] analysis. Chitosan is soluble in dilute acetic acid which has free $–NH_2$ for modifications so supersede chitin.

9.2. Varied chemical modifications

Assorted chitosan matrix modifications at $-NH_2$ in C-2, at –OH in C-3/C-6 carbon are performed via etherification/etherification and amine quaternization [13]:

1. Carboxyalkylation at O-/N: Carboxyalkylation at O-/N of chitosan imparts the amphoteric polyelectrolyte nature needed in biomedical applications such as wound dressings, artificial bone and skin, bacteriostatic agents, and blood anticoagulants [14]. Carboxyl and amino chitosan functionality elicits special biophysical properties for controlled/sustained drug-delivery, e.g., water-soluble O-carboxy methylchitosan microspheres for control pazufloxacin mesilate: antibiotic drug release.

2. Sulfonation: Chitosan sulfonation at amino/hydroxyl groups generates pharmaceutic heterogeneity (analogues to heparin: a natural blood anticoagulant) to pertain desired anticoagulant, antisclerotic, antitumor, and antiviral activities. Free NH_2/OH sulfonation mostly disrupts chitosan crystanality by depleting inherent hydrogen bonding and amphilicity and imparted micelles can act as a drug carrier.

3. Acylation: Chitosan acylation by aliphatic carboxyl, hexanoyl, dodecanoyl acids, and tetradecanoyl chlorides/cyclic esters, e.g., 4-chlorobutyl and decanoylacylation at free NH_2/OH showed higher fungicidal activities via induced hydro-phobicity to prevent particle aggregations by lowering drug irritation in stomach [14]. Such hydrophobic interaction via N-acylation encourages rapid self-expandability in tracheal cartilage, intervertebral discs, menisci, skin, liver, skeletal muscle, neural tissue, and urinary bladder cells. Such acylated chitosan are beneficial such as easy solubility, benign plasma proteins sorption, and drugs selectively with reduce free blood concentration [14].

4. Sugar-modified chitosan: Chitosan reductive N-alkylation by disaccharide or monosaccharide-aldehyde derivatives acts as a liver-specific drug carrier via specific binding at sialoglycoprotein receptors [15]. All these NH_2-alkylated chitosan derivatives are soluble at neutral and basic pH conditions, whereas lactose, maltose, and cellobiose sugars imparted all pH range solubility. Galactosyled chitosan derivatives act as the synthetic extracellular matrix for hepatocyte attachment [14, 15].

5. Graft copolymerization of chitosan: Chitosans are tailored to yield composites so as to improve certain aspects such as inclusion complexation [14], mucoadhesivity retaintion [13, 14], adsorption [15], bacteriostaticity, biocompatibility, and biodegradability [12, 15]. Chitosan grafting with oligol-lactide increases hydrophilicity and control degradation rate as anticipated in wound dressing, drug carrier systems, as micelles hydrophobic core to entrap and control the release of hydrophobic drugs [12]. In last decades, grafting of chitosan with hydroxyethyl-methacrylate, methyl methacrylate, and vinyl monomers copolymers used for control cardiovascular drug release [14], tissue engineering [11, 15], and woundhealing [12].

6. Skeletal cross-linking: Cross-linking at chitosan yields hydrogel with adequate mechanical properties and high-drug-loading capability having potentials in controlled drug delivery systems [11, 15]. Cross-linking by glutaraldehyde, genipin, ethylene glycol,

diglycidyl ether, and diisocyanate at chitosan -NH_2 establishes speciality like nonionic/ionic drug interactions and pH-sensitivity aids swelling in gastric conditions anticipated in site-specific drug release [11]. Chitosan cross-link microspheres are nontoxic, biodegradable oral drug agents [12, 15] without potential deleterious impact. Tripolyphosphate-chitosan ionic gels encapsulate plasmid DNA/dsDNA *in vitro* transfection, cellular uptake, and *in vivo* gene expressions via intratracheal administration imparting high physical stability without DNA release (even after heparin incubation). Confocal studies revealed endocytotic cellular nanoparticle uptake with subsequent cytoplasm fast releases, mediated *in vivo* intratracheal strong β-galactosidase expression. Chitosan-tripolyphosphate nanogel is used in nonviral gene delivery liable to cause steric stabilization and targeting [9, 12]. Calcium sulfate-encapsulated alginate-N-succinylchitosan hydrogel pastes retained structural integrity and found to decrease resorption rate responsible for bone defect healing in bone regenerative techniques [11, 15]. Hydrogels are beneficial due to easy handling, suitable molding, and instant hardening (water release from hydrogel and transform $CaSO_4$ hemihydrate by partial cross-link with alginate to exert cementing via egg-box effect aid in bone regeneration).

10. Physicochemical properties of chitin/chitosan

10.1. Chitosan characterization

Chitosan quality like variable appearance, turbidity, molecular weight, and mechanical stability depends on chitin resource, isolation method [9, 16], and deacetylation degree as manifested in **Figure 6**. Degree of deacetylation >0.5 imparts in aqueous acids soluble to chitosan but not in alkaline/physiological pH as uneven acetyl distribution lowers solubility due to aggregation as determined by FTIR $^{1}H/^{13}C$-NMR (liquid state, solid state) and potentiometric titration. Average molecular weight of chitosan is obtained from stearic exclusion chromatography-viscometer, light scattering detector, matrix-assisted laser desorption, and MS spectrometry as mentioned in **Table 1**.

Physiochemical characteristics	Method of determination
Molecular weight	Viscometry; gel permeation chromatography; light scattering; HPLC; matrix-assisted laser desorption/ionization-mass spectrometer
Deacetylation degree	FTIR; UV; ^{1}H-NMR and ^{13}C-NMR Spectroscopy; conductometry and potentiometry titration; TGA/DTA, differential scanning calorimetry
Crystallinality	X-ray diffraction

Table 1. Methods for physiochemical characteristics of chitosan derivatives.

Chitosan is the only cationic polysaccharide that is nontoxic and biodegradable in body, thus exploited by tissue engineering for wound dressings, drug delivery, and bone graft alternative in orthopaedic beside scaffolds for cartilage, intervertebral disc, and bone tissue. The relevant physicochemical and biological properties of chitosan are presented in **Table 2**.

Physic-chemical parameters	Biological parameters
Cationic polyamine	Biocompatibility
On protonation adheres to negatively charged surface via bio/ mucoadhesionandform hydrogel with polyanions	Secondnaturally abundant polymer after cellulose
Polar salt formations with organic and inorganic acids	Facile biodegradation to normal monomers/oloigomers
It high molecular weight linear,flexible polyelectrolyte	Very safe and nontoxic
Viscosity can be altered (high, moderate to low) depending on degree of deacetylations	Own haemostatic, bacteriostatic, and fungistatic bio-activity
Chelate formations with transition/heavy metal ions	Its spermicidal
Benign to modify, both chemically and bio/enzymatically	Anticancerigen
Own free reactive amino/hydroxyl functionality	Anticholesteremic
High charge density (pH < 6.5)	Versatile, reasonable cost

Table 2. Physic-chemical and biological proprieties of chitosan [16].

10.2. Crystallographic studies of chitosan

X-ray diffraction and crystallography of chitin revealed two polymorphs, namely, hydrated/ tendon and anhydrous/annealed conformers. Crystalline chitosan own ant parallel chains in two-fold helix, zigzag pattern with almost constant pitches of 10.34 A^0 for hydrate and 10.34 A^0 anhydrous forms (similar to cellulose). Chitosan conformations of two-fold extension with moderate flexibility due to salt formation/protonation of free $-NH_2$ with acids imparting tunablehypocholesterolemic and fungicidal activity and metal chelation besides chromatographic uses and gel production [9].

10.3. Critical physico-chemical properties of chitosan

Understanding of relation structure-properties in chitosan become a matter of great interest encountered by regulatory agencies in approving chitosan uses. The critical examination of literature results in correlation between properties of nanoconstructs with component structures as advantageous due to modulating. Thus, chitosan nanoparticles are exploited in modern pharmaceutics/biomedicals by alteration of acylation, molecular weight/its distribution, nature, and fraction of substituents. In general, applications its need to examine molecular characteristics interplay with supramolecular interactions to provide size, shape, and surface-active nanomaterials [12].

http://dx.doi.org/10.5772/65786

11. Biological activities of chitin/chitosan

11.1. Chitosan properties and bio-activity from gel to nanobead in drug delivery

Chitin/chitosan own wide pharmaceutical utility due to excellent biocompatibility, biodegradability, nontoxicity, and adsorption properties which fuels global research interests, e.g., 1150 articles in 1981–1990, about 5700 reviews in 1991–2000, and 23,100 publications in 6 years [9, 15]. Among all natural polysaccharides, chitin/chitosan's outstanding benefited industrial and pharmaceutics significance are depicted in **Figure 1**. Chitin/chitosan is the only high molecular weight cationic polyelectrolyte, while other polysaccharides are either neutral or anionic in nature. Moreover, chitosan owns some undesired features like large variable chemical composition differences in its resultant polymeric chain-size and assorted acetylation degree appropriately classified as "ugly,"frustrating research applicability. Nevertheless, chitosan skeletal optimization yields biomaterials with therapeutic and biological profiles useful in drug delivery formulations and as functional excipients. Chitosan interactions with some cosolutes/biostructures revealed subtle effects in classical solution thermodynamics and in biological utility like interactions with endotoxins.

11.2. Chitosan as antiinflammatory/immunomodulatory agents

Chitoneous administration through vascular system enhances cytokines release by macrophage, and upregulates Th1-immunity and downregulates Th2-immune activity [12]. Chitosan cross-linked resin and chitosan poly-γ-glutamic acid nanoparticles are use as alternative vehicles for oral delivery of NSAID aceclofenac and diclofenac drugs, respectively, which inhibits prostaglandin E2 production to stifle ultimate local inflammatory responses.

11.3. Chitosan as anticoagulants

Sulfation at N-2 and O-3 link of chitin increases thrombin level and activates partial thromboplastin-thrombin time [12] and at C-2, C-3, and C-6 position showed anticoagulant activity at par with therapeutic heparin (mere heparin use infects animal proteins to own antiplatelet activity responsible for abnormal bleeding). Sulfated chitins anticoagulant action inhibits FXa-mediated antithrombin-III activity via N-sulfo/N-acetyls complexation with antithrombin-III, which prolongs thrombin-clotting time. Coagulation interfering factors in fibrin polymerization are assay via antithrombin activity andprothrombin-atroxin time assessment (**Tables 1** and **2**).

11.4. Chitoneousanticancer agents

Chitosan is trusted due to nontoxicity, biodegradability, efficient drug delivery, best cell-permeability, and antiproliferativity against adenocarcinoma HT-29 and caco-2, colon carcinoma HCT-116 in kinase protein of intestinal epithelial cells, which colorectaly inhibits NF-κB transcription and NF-κB-mediated inflammations. Caffeic acid and doxorubicin individually doped chitosan composites revealed strong fluorescence intensity used to vehicle anticancer drugs [17]. Curcumin-coated chitosan nanoparticles possess significant cytotoxicity and reduce human oral cancer cell line.

11.5. Chitosan vehicle drugs delivery systems

Distinctly doped glutamine,methionine, and tryptophan in chitosan matrix exerted protection to C-8166 cells against cytolyticity of HIV-1RF strain and restrain HIV-induced syncytium formation with HIV-1 reverse transcriptase/protease enzymes by suppressing HIV-infected C8166 cells. Similarly, sulfated chitosaccharide-III found to suppress HIV-1 replications,syncytium formation, lytic action, p24 antigen production, and thwarted viral entry/cell fusion by intrusive HIV-1-gp-120 bound CD4-cell receptors (unsulfated chitosan lack such actions). Polyethylated-chitosan vehicle lamivudine curbs HIV-replication more powerfully than lamivudine/no chitosan carrier. Thioglycolic acid-conjugated chitosan carrying tenofovir disoproxil fumarate exhibited high mucoadhesion via intravaginal microspheres stabilization without vagina epithelial cells cytotoxicity and lactobacillus crispatus compared to native nanochitosan. Drug encapsulation and mucoadhesion are profound for chitosan-O-isopropyl-5'-O-d4T-monophosphate vehicle zidovudine since it reduces the diameter to stop HIV transmission. Nanochitosan carrying saquinavir drug imparted excellent antiHIV potential and high cell target efficiency to yield efficient HIV proliferation control [18, 19]. Poly-d,l-lactide-co-glycolide nanochitosan vehicle anti-HIV acyclic nucleoside and phosphonate aids cellular uptakes in target macrophage cells. Broad bioactivity pattern of chitosan vehicle drugs are summarize in **Table 3**.

Targets		**Chitosan composites (antibiotic stock solution minimal inhibitory concentration in ppm)**
Gram-negative bacteria	*Escherichia coli* (facultative anaerobic bacteria)	chitosan-Zinc 0.003%; α/β-chitosan 9 ppm; chitosan-N,N-diethyl-N-methyl 16 ppm; chitosan-N,N-dihexyl-N-methyl 16 ppm; pure chitosan 0.02%
	Escherichia coli (ETEC-K88 type)	chitosan 8 ppm; chitosan nano-particle 0.063 ppm; Copper-chitosan nanoparticle 0.03 ppm
	Escherichia coli (ATCC 25922)	chitosan 8 ppm; chitosan nanoparticle 0.03 ppm; Copper dope chitosan nanoparticle 0.03 ppm
	Escherichia coli (O157 type)	α-chitosan 9 ppm and β-chitosan 9 ppm
	Pseudomonas aeruginosa	chitosan 0.012%; chitosan-Zinc complex 0.0063%; α-chitosan 9 ppm; β-chitosan 9 ppm; N,N-diethyl-N-methylated chitosan 32 ppm
	Proteus mirabilis facultative anaerobic, bacterium	Pure chitosan 0.0251% and chitosan-Zinc complex 0.0063%

Targets		Chitosan composites (antibiotic stock solution minimal inhibitory concentration in ppm)
	Salmonella enterobacteriaceae	chitosan 0.051% chitosan-Zinc solution 0.006%
	Salmonella choleraesuis (ATCC 50020 type)	chitosan 16.0 ppm; chitosan nano-particles 0.063 ppm; Cu-chitosan nano-particle 0.03 ppm
	Salmonella typhimurium	α-chitosan 5.0 ppm and β-chitosan 9 ppm
	Salmonella typhimurium (ATCC 50013 type)	chitosan 16 ppm; chitosan nanoparticles 0.12 ppm; Copper-chitosan nanoparticles 0.063 ppm
	Enterobacter aerogenes (nosocomial and pathogenic bacteria)	chitosan flakes 0.050% and chitosan-Zinc complex 0.0063%
Gram + vebacteria	*Listeria monocytogene*	α-chitosan 9 ppm and β-chitosan 9 ppm
	Staphylococcus aureus (gram-positive coccal bacteria)	chitosan 0.051%; chitosan-Zn complex 0.0063%; α-chitosan 9 ppm; α-chitosan 9 ppm & N-ethyl-N,N-dimethylchitosan 4 ppm
	Staphylococcus aureus (ATCC 25923 type)	chitosan solution 8 ppm; chitosan nanoparticles 0.13 ppm; Cu-chitosan nanoparticles 0.063 ppm
	Corynebacteriaceae (aerobic)	chitosan 0.0251%; chitosan-Zinc 0.031%
	Staphylococcaceae epidermidis	chitosan 0.0251%; chitosan-Zinc complex 0.013%; α-chitosan 5 ppm and β-chitosan 5 ppm
	Enterococcaceae faecalis (commensal bacterium)	chitosan 0.051%; chitosan-Zinc complex 0.013%; N,N-diethyl-N-methyl-chitosan 16 ppm
	Bacillaceae cereus	α-chitosan 9 ppm and β- chitosan 9 ppm
	Bacillaceae megaterium	α-chitosan 9 ppm and β-chitosan 9 ppm
Virus	IC_{50}: Half maximal inhibitory concentration for cyto-pathogenicity by HIV-1 strain	Glutamine-, methionine-, and tryptophan-coatedchitosan composite solution 48 ppm
	IC_{50}:for cyto-pathogenicity by virus HIV-1_{IIIB} strains	Tryptophan, Methionine and Glutamine WMQ-chitosan composite solution 48 ppm
	IC_{50} of luciferase oxidative enzyme to produce bioluminescence for $HIV1_{RF}$	Glutamine-, methionine-,andtryptophan -coatedchitosan solution 68 ppm; Tryptophan-, methionine-, andglutamine-coated chitosan 164 ppm
	IC_{50} of synergistic inhibition for V3 loop of gp41 and target cell CD4 by HIV-1strains	Glutamine-, methionine-, andtryptophan-coatedchitosan solution 39 ppm; Tryptophan-, methionine-, andglutamine-coated chitosan 52 ppm

Targets		Chitosan composites (antibiotic stock solution minimal inhibitory concentration in ppm)
	IC_{50} of HIV-Induce syncytium by HIV-1_{RF} strain	Aminoethyl/sulfated chitosan composite solution 2.2 ppm
	EC_{50} for lysis of HIV-1-infected cells by HIV-1 strains	Carboxylated/sulfated chitosan composite solution 1.5 ppm
	IC_{50} of p24 antigen synthesis by HIV-1_{RF} strains	N-carboxymethylchitosan chitosan composite solution 4.5 ppm
	IC_{50} of antigen p24 synthesis by HIV-$1_{Ba\text{-}L}$ strain	N,O-sulfated chitosan composite solution 7.8 ppm
Fungi	C. albicans/Debaryomycetaceae	Chitosan-Zinc complex 0.11%; chitosan 5.0 ppm
	C. parapsilosis	Chitosan-Zinc 0.051%; chitosan 40 ppm
	C. krusei	Pure chitosan solution 5 ppm
	C. glabrata/Torulopsisglabrata	Pure chitosan solution 20 ppm
	P. digitatum (mesophilic)	Pure chitosan solution 65 ppm
	P. italicum	Pure chitosan solution 58 ppm
	Fusarium moniliforme (mould)	Pure chitosan solution 2.5 ppm
	Penicillium, Talaromyces	Pure chitosan solution 2.5 ppm
	Aspergillus fumigates	Pure chitosan solution 1 ppm
	R. stolonifer (Bread Mold)	Pure chitosan solution 100 ppm
	C. neoformans (yeast)	Pure chitosan solution 5 ppm
	Cryptococus neoformanvargati	Pure chitosan solution 2.5 ppm
	M. phaseolina	Pure chitosan solution 12.5 mg/mL

Table 3. Broad bioactivity pattern of chitosan based drug delivery systems on varied targets.

Pharmaceutically, chitosan delivered tablets, microspheres, micelles, vaccines, nucleic acids, hydrogels, nanoparticles, and conjugates via implantable, injectable oral, nasal, and ocular routes. Chitosan facilitates transmucosal absorption vital in delivery of peptides and in protein vaccination [9, 12, 20] besides an excipient in oral tablet formulation. High molecularweight chitosan is more viscous to impart delayed ingredient release/drug duration activities which improve therapeutic efficiency by lessening side effects of oral tablets [12]. Chitosan-tripoly-phosphate/alginates microspheres can control or protect proteins, drugs, and vaccines absorption in the digestive tract via paracellular route on epithelial layer release in oral and nasal administrations [9]. Hydrophilic chitosan imparts low surface activity that gets improved by glucosidic modifications, hydrophobic substitutions, and by providing hydrophilic shield at hydrophobic centers [12] to protect hydrophobic drugs with improved solubility and bioavailability [20]. Chitosan 3-D hydrogels are structured via diffusion, entrapment, and tethering with hydrophilic polymers to hold up thousands times more fluids than its dry weights is best utilized in drug delivery. Thermosensitive chitosan solution when injected into the body at phisiological conditions forms hydrogels which aids protection of drugs from degradation besides its prolonged-steady release [21]. Biocompatible chitosan drug carriers yield via ionotropic gelation, emulsion, cross-linking, solvent extraction, diffusion/droplet coalescence, reverse microemulsion, and self-assembly techniques. But ionotropic gelation is preferred due to mildness and less time involved in spontaneous aggregation. Chitosan-aided delivery systems protect drugs from chemical/enzymatic degradation in digestive system due to strong mucus binding that enhance drug adsorption in intestinal epithelial cells [20]. Chitosan-glycol nanogel uptakes endocytosis mediated by flotillin-1 with Cdc42 and macropinocytosis attended by actin cytoskeleton participation and internalization mechanisms by folate receptor are useful in drug delivery vectors for targeting different intracellular compartments [9]. Glipizide chitosan-xanthan beads exhibited control-drug release, mucoadhesion, pH-based swell kinetics, good bioadhesiveness, and comparable floating capacity in gastric fluids [22]. Insulin-chitosan-tris-buffer (pH 6.5) nano emulsions showed physical and chemical stability in a reverse micelle system and potent hypoglycemic activity in diabetic rat [20, 22]. Colon prolongs progesterone absorption for much first-pass metabolism with low oral bioavailability, but zinc-pectinate-chitosan vehicle increases drugs oral bioavailability with more residence time in plasma for colonic-specific delivery. Glutaraldehyde-Boswellia-resin doped chitosan composites (phosphate buffer pH 6.8) releases 70% drug load in 7 hours with augmented drug entrapment [9, 12, 20]. Chitosan-catechol carrier imparts higher retention for mussel adhesive proteins found in GI track via irreversible catechol-mucin cross-links aids in mucosal drug delivery [9]. Fish oil-based N-stearoyl O-butylglyceryl-chitosan microcapsules own desirable capacity and carrier encapsulation efficiency for sustained release of fish oil besides higher thermostability [9].

11.6. Chitosan as an antioxidant

Chito-oligosaccharide scavenges hydroxy, carbon-cantered superoxide, alkyl, and 2, 2-diphenyl-1-picrylhydrazyl (DPPH) radicals at free NH_2/OH site of pyranose skeleton and offers stability and *in vitro* antioxidants protection without damaging membrane lipids, protein, and DNA. Chitin-doped carboxyethyl possesses good water solubility above pH 6.5 and potent radical scavenging activity for 2, 2'-azinobis (3-ethylbenzothiazoline-6-sulfonic acid (ABTS) with $EC_{50} < 2$ mg/mL and good *in vivo* bioactivity (**Figure 3**). Methacrylic/sodium-acrylic etherified chitosans owing to their high surface area with mico-porosity and tensile strength can be molded into different shape, size/forms films, fibers, sponges, beads, powder, gel, and solutions for therapeutic antioxidant activity.

11.7. Chitosan in tissue engineering

Flexible mechanical and structural features of chitin is explored in tissue engineering to procure materials which impart improved bio-functions to be used to repair tissues like bone, cartilage, blood vessels, bladder, skin, and muscles. Tissue engineering induces varied dimensional/shape/size chitin forms like fiber, filament, film, sponge, and gel to provide instant mechanical support and compatibility to bio-fluids/tissues via cell, scaffold, and cell scaffold interaction as shown in **Figure 9**. The 3-D chitosan scaffold act as an artificial extracellular matrix as reabsorbed by body with time till new tissue forms to aid to integrate new tissues [20]. Chitosan interacts with cellular glycosaminoglycans to enhance cell attachment and proliferation results for cell growth via mechanical enhancement which resembles replaceable hard/soft tissue like bones, cartilage, muscles, and blood vessels.

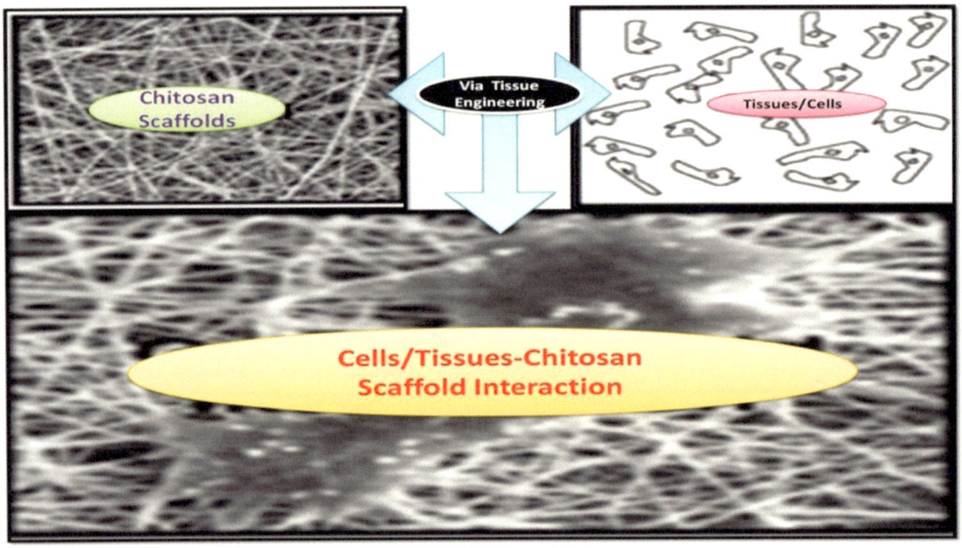

Figure 9. Schematic depiction of highly compatible cell-chitosan scaffold interactions [20].

Medical textiles or healthcare textiles is arapidly expanding technical textile market to prepare materials for medical and healthcare products like simple gauzes, bandages, tissue culturing scaffolds, and prostheses for permanent body implants [23]. Chitosan acquire basic requirements of textile material for medical applications like biocompatible, resistance to alkali, acids, and microorganisms, high-dimensional stability, elasticity, free from contaminates/impurities, absorption/repellence, and air permeability [24].

11.8. Chitosan in wound healing

Low immunogenicity characteristics of chitosan aids to provide 3-D tissue growth matrix with profound activities such as enhancing macrophage activity, stimulating cell proliferation to heal wound and facilitating polymorphonuclear leukocytes, macrophages besides fibroblasts induced granulation responsible for tissue repairs [25]. *N*-acetyl-β-D-chitosan degradation imparts peculiarity like fibroblast proliferation, collagen deposition, and hyaluronic acid stimulation at wound to accelerate healing without scar formation [26]. Nanochitosan adhesive mats own high porosity, tensile strength, surface area, and ideal H_2O vapour-O_2 transmission rate compatible with adipose-derived stem cells accountable for wound dressing (**Figure 10(A–D)**). Nanochitosan surgical dressing provides fast wound healing through adhesiveness with strong sealing strength without sutures/staples and prohibits blood vessel bleeding and air leakage in lung surgery [26].

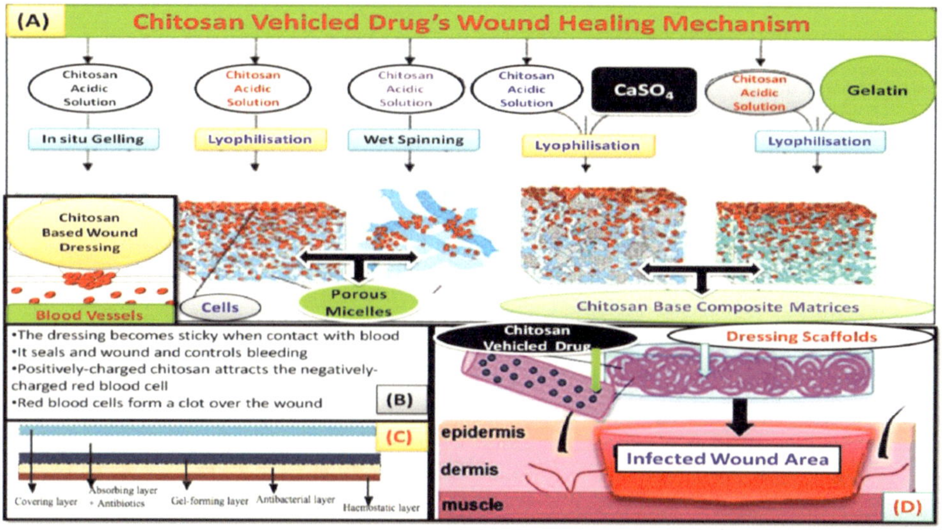

Figure 10. (A) Schematic pathway for chitosan vehicle drug's wound healings, (B) wound healing steps, (C) multilayered nanochitosan-fiber-based wound dressings, (D) chitosan compliance for fabrication: at pH < 6 amines get protonated to polycations and at pH > 6.5 amines get deprontonated to undergo interpolymer alliance yielding fiber/globule.

Clinically, chitosan-based nanofibers, composites, films, and sponges used for wound healing in plastic surgery [9, 25], skin grafting [9, 27], and endoscopic sinus surgery [26, 27]. HemCon® hemostatic latex-free bandages derived from chitosan-coatings acts as extreme adherent on blood contact to seal wound and controls bleeding (via affinity to red blood cells) and effectively reduces hemostasis span. ChiGel, Chitopack-C®, Trauma-Stat™, Tegasorb™ and Tegaderm™ Guarda-Care®, Chito-Flex®, and Chito-Gauze® are chitosan-based products used for dressings in surgical protective thickness, dermal, limb trauma, ulcers, injury, abrasions, and burns where chitosan swells with exudates to gel inaid healings. Celox™ granules/flakes are hemostatic gauze that controls emergency bleeding via swelling to gel-clot on contact with blood and induce hemostasis in penetrating limb trauma in contrast to conventional pressure bandages [9]. Topical chitosan-coated pads promote vascular hemostasis percutaneous catheters/tubes interventionally put on puncture site found to aggregate red blood cells and platelets, thus shortening the clot formation-5. Carboxypolyvinylalcohol-chitosan hydrogel film improved swelling ratio to maintain moisture over wound besides sustaining drug release and effective suppression of bacterial proliferations [27]. Curcuminbioglass-encapsulated chitosan found many uses like wound healing dressing, quenchingactivity of DPPH and superoxide, inhibit *Staphylococcus aureus* bacteria, and reduce tumor necrosis [9, 27]. Varied ionic cross-linkers aid in chitosan drug delivery as shown in **Table 4**.

Ionic cross-linker	Types of agents		
Metal cations	Fe(III)		
	Pt(II)		
	Mo(VI)		
Smaller anions/molecules	citric acid		
	Butanedioic acid		
	Glauber's salt		
	Ionic phosphate compound	tripolyphosphate (TPP)	
		sodium beta glycerophosphate	
		Sodium salt ofcori ester/glucose 1-phosphate	
		Sodium salt Robison ester/glucose-6-phosphate	
Anionic polymers	Natural	Kappa/Iota-Carrageenin (linear sulfated polysaccharide	
		Collagens (partial hydrolysis protein)	
		Hyaluronan (nonsulfated glycosaminoglycan)	
		Acidic *gum* (high uronic acid % natural exudates)	
		D-galacturonic acid/Heteropolysaccharide	
		Gamma PGA (amino acid glutamic acid polymer)	
		Alginate/algin from seaweed	
		Dextran sodium sulfate (DSS)	
		E 415 gum from Xanthomonascampestris	
	Synthetic	Acrylic acid polymer	polyacrylic acid-divinyl glycol cross-link polymer
		Methacrylate polymer	Eudragit
		PNIPA/NIPApolymer	Synperonic/Pluronic/Koliphor

*Relationship of polyolphosphate with chitosan is unclear and not been elucidated.

Table 4. Ionic cross-linkers used for chitosan-based drug delivery in biomedical devices [28].

11.9. Chitosan in water treatment

Chitosan acts as a natural adsorbent due to free amino and hydroxyl groups responsible for adsorptive interactions with water pollutants like dyes [29], metals [30–33], and organic compounds, etc. Functionality of chitosan is facile for modifications, viz., cross-linking and grafting so as to enhance its inherent absorption efficiency and specificity. Cross-linking of chitosan's functionality improved its sorption efficiency at low pH while grafting with sulfur/

nitrogen improves specificity and capacity for heavy metals [30–33]. Dye adsorption by unmodified chitosan is good; but its low stability prompted researchers to modify/graft at amino, carboxyl, sulfur, and alkyl groups. Chitosan can be cross-linked with epichlorohydrin, ethylene glycol diglycidyl ether, glutaraldehyde, and tripolyphosphate to improve its sorption efficiency besides mechanical and physical properties. Chitosan alteration is best for adsorption of dyes, phenols, polycyclic aromatic, pesticide, herbicides, and metal ions. Metal cations are chelated specifically at protonated amines in acidic conditions whereas anions by electrostatic interactions. Chitoneous adsorptive interaction includes partition, diffusion, chelation, trapping, scavenging, cation exchange, hydrogen bonding, Van der waals force, dipole–dipole, and electrostatic interactions [29–33]. Quaternary tetra-alkylammoniumchitosans own permanent +ve charge at -OH/NH_2 to boost antimicrobial activity (at wide pH) in orthopedic, wound dressing in surgery, anion exchange cartridge, dental implants, colorimetric analysis, and perchlorate removal of water [9, 20]. Fe-chelated-chitosan granules found to promote selective fluoride sorption over chloride from water in defluoridation technique. Protonated polyamidoamine-grafted chitosan-Zr (IV) beads selectively removed fluoride than other ions in spontaneous and endothermic way [34]. Magnetic hydroxypropyl-chitosan multiwalled-carbon nanotubes adsorbs lead (II) from water. Chitosan-glutaraldehydenanofibers exhibited double adsorption capacity (than chitosan) forCr (VI) removal from water. Chitosan-polyphenol-oxidase quinine nanobeads rapidly eliminate bisphenols from water [33]. Chitosan-pyruvic acid composites are found to adsorb Cd (II) from wastewater. Chitosan-modified soil cyanobacterial breakdowns harmful algae blooms and microcystins via flocculation and inhibiting algal cells and sequester liberate toxins to promote biodegradation [9]. Chitosan-doped sodium tripolyphosphate nanorods mitigate toxic Cr^{+6} from water via multilayer adsorption and consequent oxidation to ambient Cr^{+3} ions [34]. These chitosan-based commercialized viable adsorbents own peculiar characters like high specific surface area, low internal diffusion resistance, biodegradability, quantum size effect, ecofriendliness, versatility, low cost, and high adsorption capability and selectivity.

11.10. Chitosan in regenerative medicines

N-methacryloyl chitosan possessesdesirable features like hydro-solubility, UV-cross-linkability, and injectability facilitating cell-loaded microhydrogels and quick transdermal curing *in vivo* needed in localizing/sustaining protein delivery [20]. Chitosan-α-tricalcium phosphate exhibited histocompatibility for Beagle mesenchymal stem cells without effecting cellular growth and proliferation, further manifesting efficacy by enhancing osteogenesis and vascularization to repair bone defects in conjunction with mesenchymal stem cells [12, 20]. Silk reinforced chitosan promotes redifferentiation of caprine chondrocytes and retained more glycosaminoglycan to improve aggregate modulus construction, whichresembles with native tissues. Bone morphogenic protein-chitosan scaffolds burst sustainable drug release and biocompatibility which is necessary for cartilage tissue [9, 20]. Chitosan hollow tubes regenerates repair sciatic and damage phrenic nerves via improved diaphragm movement besides slow phrenic nerve transfer by granulation in beagle-dogs [12]. Cell encapsulated chitosan gels/porous chitosan fibrous matrix with biocompatible materials like $CaPO_4$, gelatin modifying biomechanical stiffness, and cell-matrixinteraction properties. These chitosan adaptations optimize cell/tissue differentiation and tailor transplantation to different clinical cell delivery by improving adherent ability for seeding cells to allow encapsulations.

11.11. Chitosan in obesity treatment

Chitosan dietary supplement/nutraceutical lowers serum cholesterol besides controlling obesity imparted to no digestion in our gastrointestinal tract. Chitosan gets swelled up to feel satiety by physically filling the stomach [9, 20] and inhibition of pancreatic lipase enzyme chitosan reduces dietary fat absorption in intestines. Chitosan precipitates fat in intestines via anionic binding with carboxyls of fatty/bile acids and hinder neutral cholesterol/sterol emulsification through hydrophobic interaction, thus it remains reduced/unabsorbed in GI tract for obesity and hypercholesterolemia treatments [35].

11.12. Chitosan in cardiovascular treatment

Chitooligosaccharides entry in gastric gavages of mice in the treatment of apo-lipoprotein-E deficiency along with high-fat diet feeding showed peculiar changes like lowered triglycerides and cholesterols, undermined atherosclerosis, increased atherosclerotic plaque stability, unregulated hepatic expression of low density lipoprotein receptor, macrophage scavenger receptor BI, and ATP binding cassette transporter-A1. But in wild mice with low density lipoprotein receptors deficiency and high cholesterol absorption found no change in plasma lipid levels of LDL-R. Chitooligosaccharides aid in hypercholesterolemia, i.e., remove low-density lipoprotein (LDL) oxidized products: cholesterol which causes coronary atherosclerosis as toxic for endothelial cells. Chitosan increase binding of LDL to endothelium and smooth muscle by mediating inflammation such as TNF-α, IL-1, and macrophage colony-stimulating factors. Hypocholesterolemic effect lowers lipid and media-milled chitosan treatments found to decrease serum triacylglycerol, total, and LDL cholesterol which is highest than pure chitosan. The elevated serum cholesterol causing cardiovascular diseases, since 10% blood cholesterol reduction using chitosan consumptions reduces the risk of coronary heart disease by 30%.

11.13. Chitosan in aging treatments

Degenerative aging diseases like cardio-/cerebrovascular, diabetes, osteoporosis, and cancer are common in old people that are diet-affected [9]. Old people showdeficiency of Zn, Fe, Se, Cu-metals, Vit-A, B, C, and E, which impacts immune responses or impair immunity. But chitosan-ascorbate can compensate such deficit and thrust neutrophils, NK/NKT/dendritic cell, monocyte/macrophage, and mediate initial pathogen interactions link to compromise signal transduction T cells pathway and gut microbiota homeostatic regulation to reduce low-grade inflammation in age-related diseases via triggering intestinal activity. Chitosanoligosaccharides used as functional food and aging disease therapy/treatment, pathophysiology via affecting oxidative stress, low-density lipoprotein oxidation, enhance tissue stiffness, govern protein conformational changes, and chronic inflammations [9, 20, 35].

11.14. Chitosan in mucosal immunity enhancer

Chitosan -C48/80 nanoparticle carried Bacillus anthracis protective antigen in mice have produced elevated serum titers of antibodies against protective antigen and a more balanced Th1/Th2 pattern then mere C48/80 solution and nanochitosan/alginate – C48/80 composite. C48/80 within chitosan found to promote a stronger mucosal immunity than other adjuvant groups indicating action in concert with a mast cell activator to affect nasal immunity [9, 20].

11.15. Chitosan in dry mouth syndrome therapy

Chitosan-thioglycolic-mercaptonicotinamide conjugates are nontoxic andare useful against Caco-2 cells that remarkably improved swelling and cohesive characteristics that are promising for dry mouth syndrome therapy where lubrication and mucoadhesiveness of mucosa is needed [9, 12, 20] than that of unmodified chitosan. About 10% of older people have dry mouth syndrome/xerostomia, i.e., not enough saliva/spit in the mouth. Treatment of dry mouth syndrome includes chitin-based products that moisten themouth, e.g., electrospun chitosan fibers decrease microorganisms, molds, yeast, to yield lighter appearance, and less muscle denaturation in comparison with traditional dryageing. Test disks were compressed out of unmodified chitosan-TGA (thiomers) and/or TGA-MNA conjugates to investigate cohesive properties, cytotoxicity assays, and mucoadhesion studies. Immobilized-MNA achieved higher swelling and cohesion for chitosan-TGA-MNA conjugates compared to unmodified chitosan. Preactivated chitosan thiomer exhibited higher stability among all conjugates and nontoxic against Caco-2 cells.

11.16. Chitosan in gene silencing in disease vector mosquito larvae

The vector mosquitoes inflict diseases in humans, such as malaria, dengue, and yellow fever, which cause death of more than one million people per year compared to any other living organism. RNA interference mediated gene silencing/targeting the interested/responsible gene for disease using chitosan nanoparticles combined with food and ingested by larvae. Thus, a technically straightforward, high-throughput, and cheap methodology is compatible for mosquitoes, insects, agricultural pests, and nonmodel organisms deals with long double-stranded/small interfering RNA. Chitosan nanofibers can potentially inhibit gene functions in disease vector mosquitoes ingested by larvae when mixed with food [12, 20]. Chitosan/siRNA nanoparticles used to target semaphorin-1-a during olfactory system development in dengue and yellow fevers arise by vector Aedesaegypti mosquitos. Chitosan/AgCHS-dsRNA-basenanocrystals own repression of AgCHS-1 and AgCHS-2 chitin synthase genes via feeding larval in Anopheles gambiae.

12. Conclusion

Chitinorchitosan hasdistinctive biological/physicochemical properties and are researchedin the fields of biotechnology, medicine, cosmetics, food technology, and textiles. Marine crabs, lobsters, shrimps, and mollusks are a resource of chitin as well as arthropod, crustacean, fungi, yeasts, algae, and squid pen exoskeletons. Comparatively, chitosan is a widely used in various industries,viz., biochemical, food, drugs, as dietary fiber, wastewater treatments, surgical threads, wound healing, immune response to allergy, plant immune inducer/defense against pathogens, and is also used in antimicrobial, anticholesterol, and antitumor activities. Chitosan alteration/designing via tissue engineering is also used in various areas such as bone scaffold, drug delivery, wound healing, and metal/dye absorbents. The marine crustacean shells are rich chitooligosaccharides with calcium carbonate (20–50%), proteins (20–40%), and chitin (15–40%), contents that can be separated by using an integrated biorefinery with mechano/chemical processes for their distinct uses. This chapter presents a variety of chitooligosaccharides with specific characteristics, such as skeletal modifications, biocompatibility, and antimicrobial and antiinflammatory activities, in relation to possible solution properties.

Author details

Rajendra S. Dongre

Address all correspondence to: rsdongre@hotmail.com

Department of Chemistry, RTM Nagpur University, Nagpur, India

References

[1] Aneiros, A.; Garateix, A. Bioactive peptides from marine sources: Pharmacological properties and isolation. J. Chromat. B Anal. Technol. Biomed. Life Sci. 2004, 15, 41–53.

[2] Rasmussen, R.S.; Morrissey, M.T. Marine biotechnology for production of food ingredients. Adv. Food Nutr. Res. 2007, 52, 237–292.

[3] Montaser, R.; Luesch, H. Marine natural products: A new wave of drugs? Future Med. Chem. 2011, 3, 1475–1489.

[4] Ahmed, A.B.; Adel, M.; Karimi, P.; Peidayesh, M. Pharmaceutical, cosmetic, and traditional application of marine carbohydrates. Adv. Food Nutr. Res. 2014, 73, 197–220.

[5] Kim, S.K.; Ravichandran, Y.D.; Khan, S.B.; Kim, Y.T. Prospective of cosmetic derived from marine organisms. Biotechnol. Bioprocess Eng. 2008, 13, 511–523.

[6] Wijesinghe, W.; Jeon, Y.J. Biological activities and potential industrial applications of fucose rich sulfated polysaccharides and fucoidans isolated from brown seaweeds: A review. Carbohydr. Polym. 2012, 88, 13–20.

[7] Marin, J.; Zee, J.; Kouaouci, R. Effect of hydrocolloids on potential availability of calcium. Le Lait, INRA Editions. 1990, 70 (56), 467–473.

[8] Wang, H.; Fu, Z.-M.; Han, C.C. The potential applications of marine bioactives against diabetes and obesity. Am. J. Mar. Sci. 2014, 2, 1–8.

[9] Martins, A.; Vieira, H.; Gaspar, H.; Santos, S. Marketed marine natural products in the pharmaceutical and cosmeceutics: Tips for success. Mar. Drugs. 2014, 12, 1066–1101.

[10] d'Ayala, G.G.; Malinconico, M.; Laurienzo, P. Marine derived polysaccharide for biomedical application: A review chemical modification approaches. Molecules. 2008, 13(9), 2069–2106. Doi: 10.3390/molecules13092069

[11] Bayat, A.; Sadeghi, A.M.M.; Avadi, M.R.; Amini, M.; Rafiee-Tehrani, M.; Shafiee, A.; Junginger, H.E. Synthesis of N, N-dimethyl N-ethyl chitosan as a carrier for oral delivery of peptide drugs. Bioact. Compat. Polym. 2006, 21, 433–444.

[12] Ruocco, N.; Costantini, S.; Guariniello, S.; Costantini, M. Polysaccharides from the marine environment with pharmacological, cosmeceutical and nutraceutical potential. Molecules. 2016, 21(5), 551. Doi:10.3390/molecules21050551.

[13] Liu, Y.; Huang, K.; Peng, D.M.; Ding, P.; Li, G. Preparation and characterization of glutaraldehyde cross-linked O-carboxymethylchitosan microspheres for controlled delivery of pazufloxacinmesilate. Int. J. Biolog. Macromol. 2007, 41 (1), 87–93.

[14] d'Ayala, G.G.; Malinconico, M.; Laurienzo, P. Marine derived polysaccharides for biomedical application: chemical modification approach. Molecules. 2008, 13(9), 2069–2106.

[15] Bellich, B.; D'Agostino, I.; Semeraro, S.; Gamini, A. Cesàro, A. Review the Good, Bad and Ugly of Chitosans. Mar. Drugs. 2016, 14 (5), 99.

[16] Tanodekaew, S.; Prasitsilp, M.; Pateepasen, R. Preparation of acrylic grafted chitin for wound dressing. Biomaterials. 2004, 25 (7–8), 1453–1460.

[17] Wimardhani, Y.S.; Ikeda, M.A. Chitosan exerts anticancer activity through induction of apoptosis and cell cycle arrest in oral cancer cells. J. Oral. Sci. 2014, 56 (2), 119–126.

[18] Cheung, R.C.F.; Wong, J.H.; Chan, W.Y. Review chitosan: Update on potential biomedical & pharmaceutical applications. Mar. Drugs. 2015, 13 (8), 5156–5186.

[19] Lee, D. W.; Lim, H.; Chong, H. N.; Shim, W. S. Advances in Chitosan Material and its Hybrid Derivatives: A Review *Open Biomater. J.* 2009, 1, 10–20.

[20] Vishnu A. Dorugade International journal on Textile Engineering and Processes., 2015 Vol.1 (3), 89–95.

[21] Liu, B.; Sun, Y.Y. Antidiabetic effects of chitooligosaccharideson pancreatic islet cells in streptozotocin-induced diabetic rats. World J. Gastroenterol. 2007, 13, 725–731.

[22] Maeda, M.; Hiraoka, A. Heparinoid-active sulphat polysaccharide from Monostroma-nitidum and their distribution in the Chlorophyta. Phytochemistry 1991, 30, 3611–3614.

[23] Roadmap to Sustainable Textiles and Clothing, Subramanian S. Muthu, Hong Kong, SAR. Book. 2014. pp, 104. DOI 10.1007/978-981-287-065-0. ISBN 978-981-287-064-3.

[24] Kim, K.N.; Jeon, Y.J. Effect of chitosaccharide on cholesterol level & antioxidant activities in hypercholesterolemic rat. J. Korean Soc. Food Sci. Nutr. 2005, 34, 36–41.

[25] Matsubara, K.; Miyazawa, K. Anticoagulant properties of sulfate galactan preparation from a marine green alga, Codium cylindricum. Int. J. Biol. Macromol. 2001, 28, 395–399.

[26] Ilium, L. Chitosan use as pharmaceutic excipient. Pharm. Res. 1998, 15, 1326–1331.

[27] Szymańska, E.; Winnicka, K. Review stability of chitosan – a challenge for pharmaceutical and biomedical applications. Mar. Drugs. 2015, 13 (4), 1819–1846.

[28] Vakili, M., Amouzgar P. Application of chitosan as adsorbents for dye removal from water-wastewater. Rev.Carbohyd. Poly. 2014, 113, 115–130.

[29] Gedam, A.H.; Dongre, R.S.; Bansiwal, A.K. Synthesis and characterization of graphite doped chitosan for adsorption of lead (II) from aqueous solution. Adv. Mat. Let. 2014, 6 (1), 59–67.

[30] Gedam, A.H.; Dongre, R.S. Adsorption characterization of Pb (II) onto iodate doped chitosan composite: equilibrium and kinetic studies. RSC Adv. 2015, 5 (67), 54188–54201.

[31] Gedam, A.H.; Dongre, R.S. Activated carbon from Luffa cylindrica doped chitosan for mitigation of lead (ii) from an aqueous solution. RSC Adv. 2016, 6 (27), 22639–22652.

[32] Dongre, R.S. Adsorption of hexavalent chromium by graphite–chitosan binary composite, Bulletin of Mater. Sci. 2016, 39 (3), 865–887.

[33] Dongre, R.S.; Ghugal, D.N.; Meshram, J.S.; Ramteke, D.S. Fluoride removal from water by zirconium (IV) dope chitosan biocomposite. Afric. J. Env. Sci. Tech. 2012, 6 (2), 130–141.

[34] Jianglian, D.; Shaoying, Z. Application of chitosan based coating in fruit & vegetable preservation: Review. J. Food Process. Tech. 2013, 4 (5), 1–4. Doi: 10.4172/2157-7110.1000227.

[35] Kim, S.-K..; Mendis, E. Bioactive compounds from marine processing byproducts – A review. Food Res. Intern. 2006, 39 (4), 383–393.

Section 3

Marine Polysaccharides and Industries

Chapter 10

Applications of Chitosan in Wastewater Treatment

Petronela Nechita

Additional information is available at the end of the chapter

http://dx.doi.org/10.5772/65289

Abstract

In the last time, the use of natural additives that are biocompatible, are biodegradable, have low toxicity and are from renewable resources attracted attention of many researchers due to their high ability to retain different pollutants from wastewaters. In this context, there are many research studies that highlight the biosorbent ability of chitosan and their composites for the pollutants from wastewaters such as heavy metal ions, organochloride pesticides, suspended solids, turbidity, organic oxidised substances, fatty and oil impurities or textile wastewater dyes. Furthermore, the increase of adsorption ability of chitosan by chemical modifications leading to the formation of chitosan derivatives, grafting chitosan and chitosan composites gained much attention, being extensively studied and widely reported in the literature. In this chapter the research studies regarding the chitosan application in wastewater treatments as well as the preliminary results on its chemical modification to obtain and utilisation of zeolite-chitosan composites in adsorption of organic pollutants from industrial wastewaters are presented.

Keywords: Chitosan, Chitosan composites, Wastewaters, Chitosan-magnetite, Chitosan-zeolites, Adsorption isotherm, Adsorption kinetics, Pollutants, Total suspended solids, Chemical oxygen demand, Heavy metal ions, Removal efficiency

1. Introduction

In the last time, different wastewater decontamination methods that include chemical precipitation, nanofiltration, solvent extraction, ion exchange, reverse osmosis and adsorption have been extensively studied. Out of these methods, adsorption is particularly attracting scientific focus mainly because of its high efficiency, low cost and easy handling and high availability of different adsorbents [1].

Chitosan is a versatile polysaccharide widely distributed in nature (second most abundant biopolymers after cellulose) produced by alkaline N-deacetylation of chitin. Many application fields are described in scientific publications regarding the use of chitin, chitosan and their derivatives. Wastewater treatment using chitin or chitosan is an important application. According to this, there are many research studies that highlight the biosorbent ability of chitosan and their composites to remove the pollutants from wastewater. They could be used as coagulating/flocculating agents for polluted wastewaters [2, 3], in heavy metal or metalloid adsorption (Cu(II), Cd(II), Pb(II), Fe(III), Zn(II), Cr(III), etc.) [4, 7, 8] for the removal of dyes from industrial wastewater (i.e. textile wastewaters) [6], as well as for the removal of other organic pollutants such as organochloride pesticides, organic oxidised or fatty and oil impurities.

Due to the high performances, chitosan derivatives are used as adsorption additives [5] in many research investigations. Some examples are derivatives that contain heteroatoms based on nitrogen, phosphorous and sulphur or complex combinations of chitosan with ethylenediaminetetraacetic acid (EDTA) and diethylenetriaminepentaacetic acid (DTPA).

In the last time, the chitosan composites have been tested in wastewater treatments for adsorption of dyes [6] and heavy metals [7– 9]. To form composites with chitosan, different substances have been used, such as montmorillonite, polyurethane, activated clay, bentonite, zeolites, oil palm ash, calcium alginate, polyvinyl alcohol, cellulose, magnetite, sand, cotton fibres, perlite and ceramic alumina [10–14].

This chapter highlights the application of chitosan and their composites (zeolite-chitosan composites) as adsorbents and flocculants in wastewater treatments including the method of preparation, mechanisms/kinetics and factors that can affect their efficiency in the pollutant adsorption capacity (pH, biosorbent dosage, contact time). Some experimental results obtained on static adsorption methods applied on industrial and municipal wastewaters are presented.

2. Chitosan in wastewater treatments

2.1. Chitosan structure and properties

Chitosan is a partially deacetylated polymer obtained by the alkaline deacetylation of chitin, a biopolymer extracted from shellfish sources. It is a linear hydrophilic amino polysaccharide with a rigid structure containing both glucosamine and acetylglucosamine (poly-β-(1 → 4)-2-amino-2-deoxy-D-glucose) units (**Figure 1a**). Chitin (poly-β-(1 → 4)-*N*-acetyl-D-glucosamine) can be characterised as one of the most abundant natural biopolymers (**Figure 1b**) [15].

Native chitosan is insoluble in water or organic solvents, but at acidic pH (below pH 5), when the free amino groups are protonated, chitosan becomes a soluble cationic polymer with high charge density [16–18].

The infrared (IR) absorption spectra of chitosan and chitin are presented in **Figure 2**. It can be observed that three types of absorption bands exist: the amide I bands of chitosan characterised

http://dx.doi.org/10.5772/65289

by absorption at approximately 1655 and 1630 cm^{-1}, the amide II bands of chitin at approximately 1560 cm^{-1} and the absorption bands for –OH groups at 3450 cm^{-1} [19].

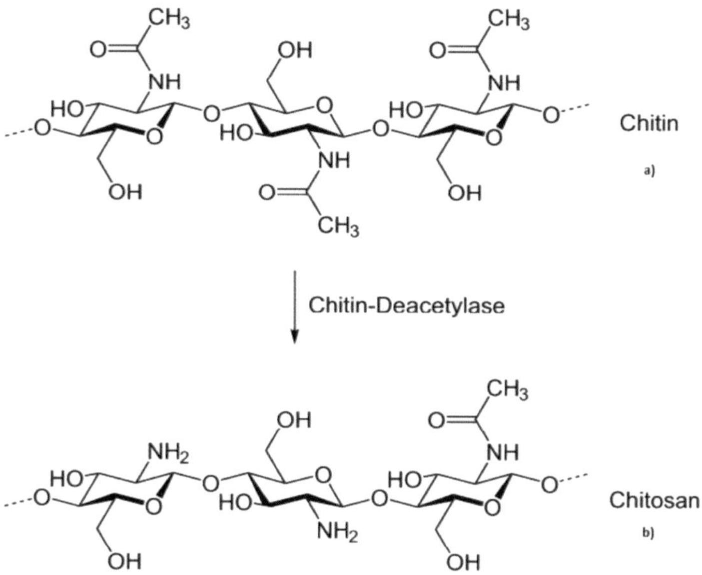

Figure 1. Chemical structural representation of chitin (a) and chitosan (b).

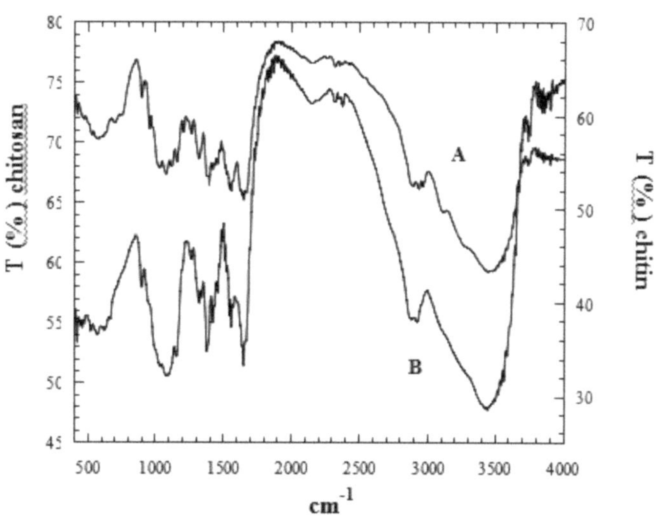

Figure 2. Infrared spectra of chitin (A) and chitosan (B) [19].

Chitosan has many attractive properties such as hydrophobicity, biocompatibility, biodegradability, non-toxicity and the presence of very reactive amino (–NH_2) and hydroxyl (–OH) groups in its backbone, which makes chitosan to be used as an effective adsorbent material for the removal of wastewater pollutants.

The main advantage of chitosan over other polysaccharides (cellulose or starch) is their chemical structure that allows specific modifications to design polymers for selected applications. On the one hand, their *reactive groups* are able to develop composites with different compounds that have proven to have better capacity to adsorb the wastewater pollutants and to resist in acidic environment. Some examples include bentonite, kaolinite, oil palm ash, montmorillonite, polyurethane, zeolites, magnetite, etc. On the other hand, their *cationic charge* (chitosan is single cationic biopolymer) is able to neutralise and successfully flocculate the anionic suspended colloidal particles and reduce the levels of chemical oxygen demand, chlorides and turbidity in wastewaters [2, 3].

2.2. Chitosan as flocculant/coagulant of pollution colloid impurities

Flocculation is an essential phenomenon in industrial wastewater treatment. Organic polymeric flocculants are widely used nowadays due to its remarkable ability to flocculate efficiently with low dosage compared with inorganic coagulants (salts of multivalent metals) that are being commonly used (due to its low cost and ease of use) but have low flocculating efficiency and present the residual concentration of metal in the treated water. In this context, the coagulation and flocculation properties of chitosan (given by their cationic charge) can be exploited to remove the negative-charged colloidal organic or inorganic impurities from wastewaters [16].

Since most pollution colloids are negatively charged, cationic polymers or polyelectrolytes are of particular interest as potential coagulants/flocculants. Due to its cationic unique feature, chitosan is one of the most promising biopolymers for extensive application in wastewater treatment, and its coagulative action is very effective compared with the mineral coagulants such as aluminium sulphate, polyethyleneimine and polyacrylamide in removing different pollutants from aqueous solution [20, 21].

The protonated amine groups along the chain obtained by dissolving of chitosan in acids facilitate electrostatic interactions between polymer chains and the negatively charged contaminants (metal anions, dyes, organic compounds, etc.) [16]. Due to the presence of primary amino groups, the biopolymer has a high cationic charge density [22] and long chains with high molecular weight, being an effective coagulant and/or flocculant for the removal of contaminants in the suspended and dissolved state [16, 23,24]. The active amino groups (NH_2) in the chitosan molecule can be protonated with H+ in water into a cationic polyelectrolyte [25] such that the molecule has characteristics of static attraction and adsorption. Chitosan coagulation produces better quality floaters, namely, larger floaters with faster settling velocity. The effectiveness of chitosan for coagulating mineral suspensions can be improved due to the presence of inorganic solutes or due to the addition of materials extracted from soils at high pH [26].

Based on the high affinity of chitosan for different contaminants, there are many studies where these properties of chitosan for removing of dyes from solution [22] or textile wastewater [27, 28], organic matter (e.g. lignin and chlorinated compounds) in pulp and paper mill wastewater [29], heavy metals and phenolic compounds in cardboard-mill wastewater [30, 31] and inorganic suspensions in kaolinite suspension are demonstrated [32].

In this context, Abu Hassan et al. in their research, have emphasised that a chitosan dosage between 12 mg/l and 30 mg/l lead to an important decreasing of chemical oxygen demand and turbidity (**Figures 3** and **4**) as result of high charge density of chitosan that has effect to quickly destabilisation of colloidal equilibrium [33].

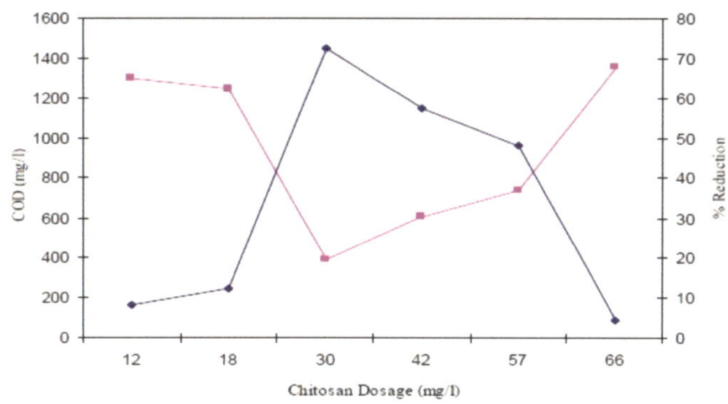

Figure 3. Effects of chitosan dosage on COD level and the percentage of COD level reduction: -■-, COD level and -♦-, % COD level reduction [33].

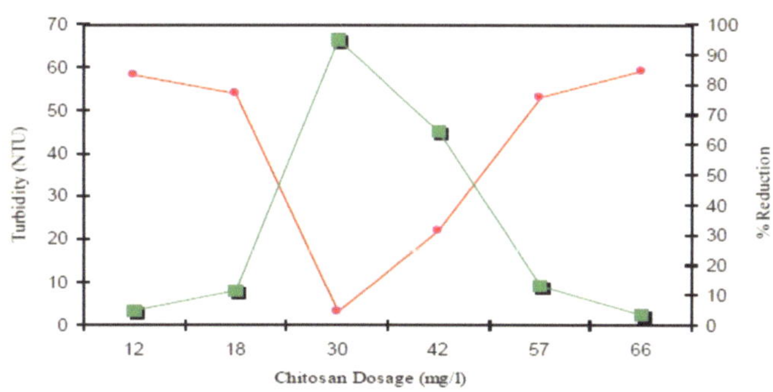

Figure 4. Effects of chitosan dosage on turbidity level and the percentage of turbidity level reduction by using chitosan, ●-, turbidity level and -■-, % turbidity level reduction [33].

Zeenat et al. [3] have extracted the chitosan from Indian prawn and used it in treatment of wastewaters from ghee industry. They examined the flocculation process regarding the influence parameters such as chitosan dosage, optimum pH and mixing times. The chitosan showed significant difference by successfully flocculating the negatively charged suspended particles, thereby reducing chemical oxygen demand with 80.1 %, turbidity with 91.8 %, total dissolved solids with 72.5 % and conductivity of analysed wastewaters with 73.7 % (**Table 1**).

Parameters	Wastewater from ghee industry	Chitosan-treated wastewater	Reduction, %
COD, mg/l	1934	383	80.1
TDS, mg/l	3108	852	72.5
Conductivity, μS/cm	1987	521	73.7
Turbidity, NTU	69	5.6	91.8

Table 1. Effect of chitosan treatment on the ghee wastewater quality [3].

In our studies [34, 35] we have been using the chitosan (with high molecular mass and 20.8 % acetylation degree) to remove the pollutants from textile wastewaters. The obtained results highlighted a high efficiency for suspended solid removal (99.65 %), for chemical oxygen demand (70.5 %) and about 79.57 % for grease impurity removal (**Table 2**).

Pollution indicator	Before the treatment with chitosan	After the treatment with chitosan	The wastewater pollutant removal efficiency, %
Total suspended solids, mg/l	5798	20.0	99.65
COD, mg/l	51.5	15.17	70.5
Sulphides, mgH_2S/l	5.64	4.2	25.53
Greases and oil content (extractives in organic solvents), mg/l	1351	276	79.57

Table 2. The wastewater pollutant removal efficiency after chitosan treatment.

For a dosage of about 1 g chitosan/100 ml wastewater and mixing duration of about 60 min, a high efficiency for dye removal from textile wastewaters was registered (**Figure 5**).

a) b) c)

Figure 5. Variation of textile wastewater colour after treatment with (a) initial wastewater, (b) synthetic resin Purolite C100 and (c) chitosan.

2.3. Chitosan as adsorbent of metal ions

Adsorption has been proven to be a reliable and economical alternative to remove the pollutants from wastewaters, and the use of chitosan as biosorbent for heavy metal ions is reported in a large quantity of literature studies.

The bond between the metal ion and chitosan functional groups in the biosorption process involves different phenomena as complexation, electrostatic attraction, micro-precipitation and ion exchange. The mechanism of complex formation between chitosan and metal ions during adsorption process can be developed in two ways:

Bridge model: metal ions are bonded with various amino groups from the same chain or from different chains through complex inter- or intramolecular reactions.

Pendant model: metal ions are bonded with amino groups in a hanging manner.

In our experiments, we tested the adsorption ability of chitosan for heavy metal ions from textile wastewaters (Cu(II) and Pb(II)) and aqueous solutions (Zn(II) and Fe(III)).

The removal efficiency was 91.67 % for Pb ions and 54.15 % for Cu ions, being influenced by pH value which is a very useful parameter that affects the surface charge of wastewater solutions in terms of stability of suspensions. In our studies the maximum adsorption of Cu(II) and Pb(II) ions from textile wastewaters was obtained at pH value of 8 (**Figure 6 (a)** and **(b)**).

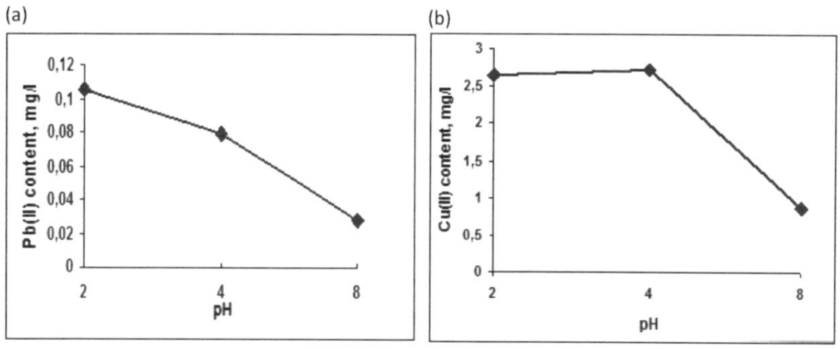

Figure 6. The influence of pH value on the adsorption of metal ions from wastewater: (a) Pb(II) content and (b) Cu(II) content.

The main functional groups of chitosan which are potential points for adsorption of metal ions are –OH and $–NH_2$. In pH acid conditions, these groups are protonated ($–OH_2^+$, $–NH_3^+$) and make the adsorption of metal ions to be reduced. As pH value increases, the degree of protonation of functional groups decreases. This influences the process forming of complex coordination bonds between the metal ions and functional groups.

In the experiments regarding the adsorption capacity of Zn^{2+} ions in aqueous solution on the chitosan, we used three different samples of chitosan, a pure chitosan (having a 55.94 % acetylation degree) and two commercial products [36]. The influence of temperature (20, 40, 60 °C) on the dissociation process was verified by pH and conductivity measurements. The effects of chitosan dosage (0.1 g, 0.2 g), heavy metal concentration (10–20 g L^{-1} Zn^{2+}), contact time (15–60 min) and mixing rate (450 rpm) on adsorption efficiency were studied. During the adsorption process, Zn^{2+} ions were chelated by chitosan through the amine groups in the fibres. The duration of action of the metal ions on chitosan is the main factor affecting the final product. The complex formed between chitosan and Zn was analysed by spectrophotometric method at λ of 264 nm (**Figure 7(a)** and **(b)**).

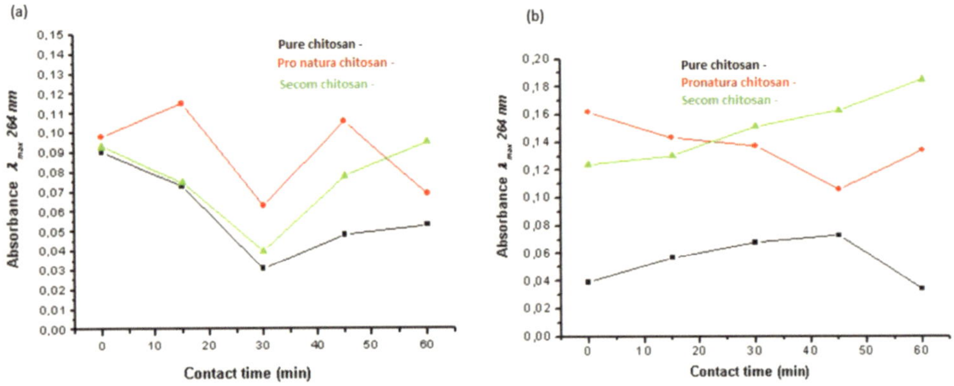

Figure 7. Chitosan-Zn^{2+} ion interaction effects: (a) 0.1 g chitosan dosage and (b) 0.2 chitosan dosage.

High quantity of chitosan in complex formed between chitosan and Zn involves longer contact time to reach the adsorption equilibrium. The obtained results showed that chitosan can be used to retain Zn^{2+} ions in certain systems (e.g. farmlands and industrial wastewater). Chitosan as adsorbent could be regenerated and reused, being an effective adsorbent for zinc ions and other metal ions from wastewater [36].

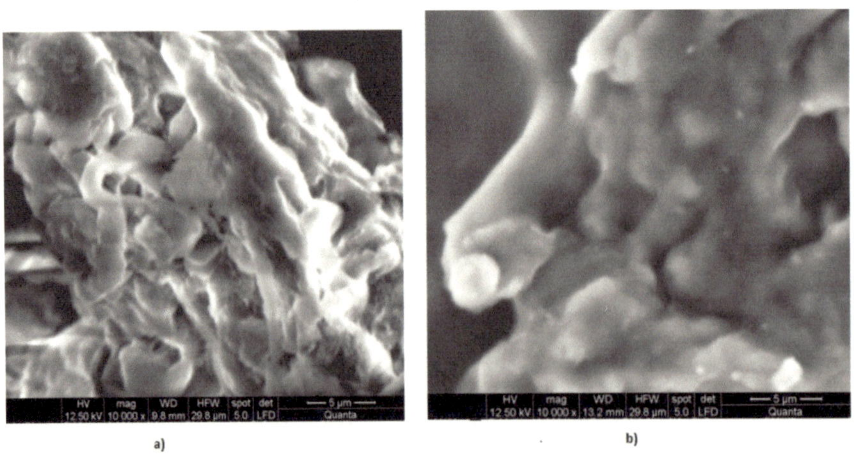

Figure 8. SEM structural analysis of chitosan (a) and chitosan-Fe(III) chelated complex (b).

In other research studies, the removal of Fe(III) ions from aqueous solution using chitosan was studied [37]. Retention of Fe(III) ions on chitosan is influenced by factors as contact time, the concentration ratio of the phases, pH and mixing rate. Retention of Fe(III) ions on chitosan is strongly dependent on the pH changes. For pH lower than 3, chitosan is dissolved, and for pH higher than 4.5, a colloid solution is obtained. The equilibrium of the chitosan adsorption is established relatively quickly, after 60 min, and the adsorption capacity for retaining the Fe(III) ions is more than 80 % (mg/g). Scanning electron microscope (SEM) images showed the formed complexes and the chemical modification of chitosan depends on the ion concentration. Structural analysis by SEM provides an indication that the mechanism of adsorption of Fe(III) ions on chitosan is a complex phenomenon involving the formation of nodosities on the chitosan structure. The mechanism of retention of Fe(III) ions on chitosan is a complex phenomenon and involves the formation of *lumps* on the structure of chitosan through the surface adsorption of metal ions and strong coordination with functional groups (**Figure 8**).

3. Chitosan composites in wastewater treatment

Chitosan is a very promising adsorbent, which can be modified in many ways (grafting, cross linking, functionalisation for forming composites, etc.). Because chitosan is very sensitive to pH, forming either gel or dissolve depending on pH values, some cross linking reagents such as glyoxal, formaldehyde, glutaraldehyde, epichlorohydrin, ethylene glycon diglycidyl ether and isocyanates have been used to improve its performance as adsorbent [38]. This process of cross linking stabilises chitosan in acid solutions becoming insoluble and enhances its mechanical properties [39]. A large volume of works has been published during the last three years, presenting results of chitosan-modified adsorbents for the removal of various pollutants (dyes, metals/ions, others).

Recently, chitosan-based metal particle composites have been studied increasingly as an alternative adsorbent in water treatment, such as using metals [40], metal oxides [41], magnetite [42] and bimetals [43], to adsorb heavy metals and dyes from wastewater. For example, chitosan-coated magnetite nanoparticles (CMNP) were prepared and used as bactericidal agent to remove organic contaminants and bacteria from water [14].

Moradi Dehaghi et al. [44] in their research have developed the chitosan-ZnO nanoparticle composites by sonication process. The dissolution and swelling studies were performed on these composites, and crystallinity and surface morphology characterisation using X-ray diffraction, Fourier transform infrared spectroscopy (FT-IR) and scanning electron microscope of nanocomposite samples were studied. The testing of sorbent performance of these materials was carried out on pesticide adsorption from aqueous solution of permethrin 25 % using a column of chitosan-ZnO nanocomposite beads. It was demonstrated that the CS/ZnONP beads had an excellent adsorption performance. In comparison with chitosan beads, the removal efficiency of CS/ZnONP beads was increased from 49 % to 99 %. The adsorption and regeneration studies of permethrin demonstrated that the CS/ZnONP beads could be reused effectively with 56 % regeneration after three cycles in on-line column. Based on the high sorbent capacity, CS-ZnONP beads could explore a new biocompatible and eco-friendly strategy for pesticide removal and could be used in water treatment process.

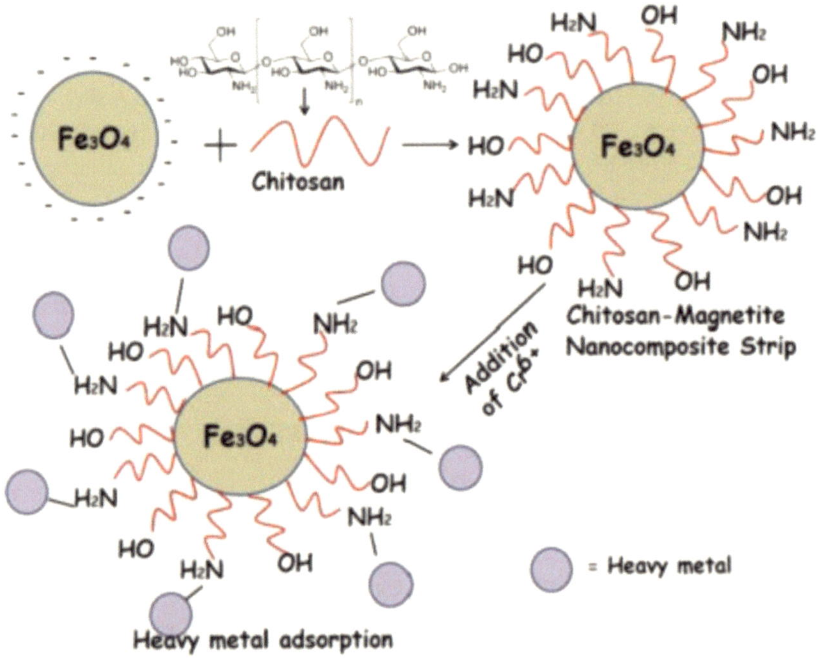

Figure 9. Schematic representation of removal mechanism of chromium ions by chitosan-magnetite nanocomposite strip [42].

In their studies, Sureshkumar et al. [42] have synthesised magnetite nanoparticles (Fe_3O_4) using co-precipitation method. After UV-VIS, X-ray diffraction and atomic force microscopy characterisation, these nanoparticles were mixed with chitosan solution to form hybrid nanocomposites. The affinity of hybrid nanocomposite for chromium was studied using $K_2Cr_2O_7$ (potassium dichromate) solution as the heavy metal solution containing Cr(VI) ions. Adsorption tests were carried out using hybrid nanocomposite strips at different time intervals compared with chitosan-only strip (**Figure 9**).

The chromium removal efficiency of chitosan strip is 29.39 % and that of the chitosan-magnetite nanocomposite strip is 92.33 %. Based on these results, the chitosan-magnetite nanocomposite strips are highly efficient for chromium removal from tannery wastewaters.

Abd-Elhakeem et al. [14] have prepared chitosan-coated magnetite nanoparticles and used it as adsorbent, chelating agent or bactericidal agent to remove organic contaminants, heavy metals and bacteria from water. In their research they find that the adsorption capacities of the different contaminants considerably increased with chitosan-magnetite nanoparticle concentration. The highest affinity was found for petroleum impurities, where 1 g of CMNP removed about 98 % from it, while the lowest capacity was recorded by heavy metals (**Figure 10**).

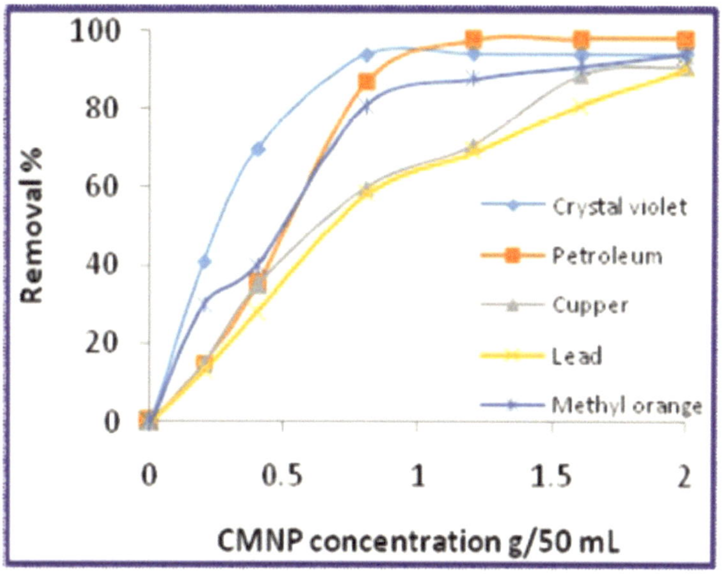

Figure 10. The influence of CMNP concentration on the removal % of different contaminants [14].

In the same studies, the influence on the bacterial growth was partially inhibited at concentration 0.1 g/ml of CMNP. The complete growth inhibition has occurred at concentration of 0.5 g/10 ml.

Hritcu et al. [45] in their research have evaluated the performances of composite chitosan-magnetite microparticles (Fe-Cc) as a new adsorbent for cobalt and nickel ion separation from aqueous solutions. Their sorption batch experiments were conducted for optimising the pH, initial target ion concentration and adsorbent amount. The experimental data have emphasised that Langmuir isotherm model is the best fit; the material has a maximum adsorption capacity of 588.24 mg/g for cobalt ions and, respectively, 833.34 mg/g for nickel ions. Regeneration study demonstrated that Fe-Cc particles might be reused up to three times without significant loss in adsorption capacity.

Saifuddin and Dimara [46] have investigated the potential and effectiveness of applying chitosan-magnetite nanocomposite particles as a primary coagulant and flocculants compared with chitosan for pretreatment of palm oil mill effluent (POME). The experiments were carried out under different conditions of dosage and pH, and the performance was assessed in terms of turbidity, total suspended solids (TSS) and chemical oxygen demand (COD) reductions. At the optimum conditions of pH and chitosan-magnetite, dosage was obtained about 98.8% reduction of turbidity, 97.6% of TSS and 62.5% for COD level. The synergistic effect of cationic character of both the chitosan amino group and the magnetite ion in the pretreatment process for POME brings about enhanced performance for effective agglomeration, adsorption and coagulation. It is important to mention that the chitosan-magnetite nanocomposite has a high efficiency (about 99%) to remove the oil residue from POME, better than chitosan only. The results showed that coagulation with chitosan-magnetite or chitosan was an effective and environmentally friendly pretreatment technique for palm oil mill effluent wastewater compared to alum and alum polychloride-PAC which creates hazardous residual waste.

3.1. Chitosan-zeolite composites in wastewater treatment

Due to their thermal and chemical stability and great potential for the separation of ions by cation exchange, zeolites are especially appealing among all kinds of inorganic fillers. The cation exchange capacity of the zeolites therefore depends on the framework Si/Al ratio and decreases with an increase of the Si/Al ratio.

Chitosan-zeolite composites have shown good adsorption properties for different pollutants such as dyes, phosphates, nitrates, ammonium and humic acids [47–49] as well as for the removal of heavy metal cations [50, 51].

Nesic et al. [52] in their research presented data about synthesis of chitosan-zeolite composite films and its application on adsorption of anionic dye (Bezactive orange 16 (BO16)) from aqueous solutions. The equilibrium data were fitted to Langmuir and Freundlich models, presenting $Qm = 305.8$ mg/g. This value was taken at $pH = 6$, which based on preliminarily results was found to be optimum.

Wan Ngah et al. [53] in their studies have prepared chitosan-zeolite (CZ) composite adsorbent to remove Cu(II) ions from aqueous solutions. The kinetic, adsorption isotherm and desorption studies have been completed. The optimum pH value was 3 and the best isotherm was fitted by the Redlich-Peterson and Langmuir models. The percentage of Cu(II) desorption was only 47.97%, which indicated that the Cu(II) ions were strongly bonded to the CZ surface.

Our studies were focused on obtaining of chitosan-zeolite (CZ) composites using commercial chitosan and zeolites from local volcanic tuff deposits with 71–83.3% of clinoptilolite contents. These composites were applied on organic impurities adsorption from poultry farm wastewaters. Chitosan-zeolite composites have been prepared by the encapsulation method according to the procedure described by Wan Ngah et al. [53] as follows: chitosan and zeolite were mixed in acetic acid solution (5%) and stirred for 2 h. At this mixture acetic acid solution was added (5%) and stirring continued about 1 h. Aiming to form the composite beads, the obtained suspension was added dropwise into the precipitation bath containing NaOH, and the mixture was stirred for 3 h. The formed beads were filtered and washed with distilled water to remove excess of NaOH and finally air-dried. After drying the beads were grinded to obtain the desired size (<200 μm). After this, the beads of chitosan-zeolite composite were structurally characterised by SEM image analyses and EDX spectral analyses and used as adsorbent for the organic impurities from wastewater (COD and greases and oil impurities).

From the SEM micrograph presented in **Figure 11**, chitosan-zeolite composite has rough and flaky surface. Zeolite is present as loose aggregates of micrometric octahedral crystals included in cavities of a continuous polysaccharide matrix, in the case of evaporative drying; the shrinkage of the polysaccharide gel has led to a physical separation between polymer and embedded zeolites.

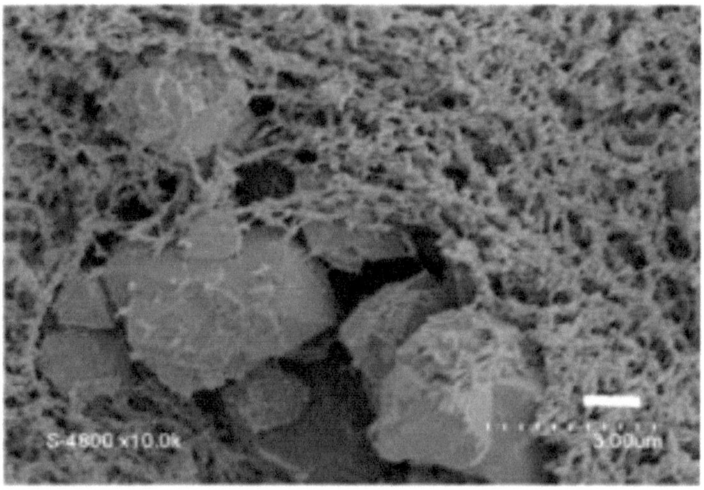

Figure 11. Chitosan-zeolite composite SEM images.

The EDX spectra (**Figure 12**) show the presence of sodium, which is originated from zeolite where the sodium ions counterbalance the negative charge of zeolite. Carbon, nitrogen, oxygen, aluminium and silicon were found in chitosan-zeolite composites since they are the major components of chitosan and zeolite.

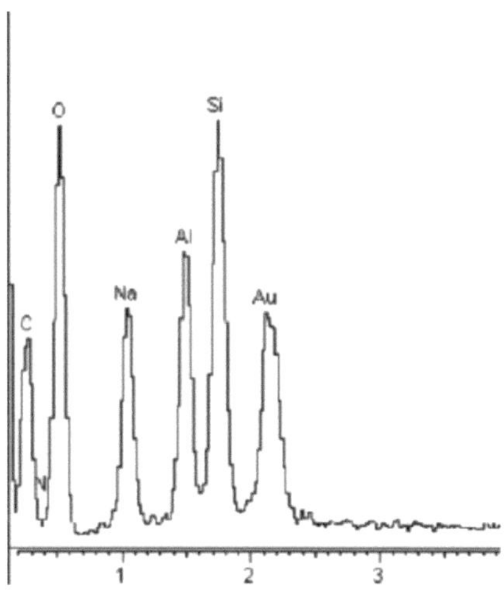

Figure 12. EDX spectra of chitosan-zeolite composites.

Experiments were carried out at 25 °C where different amounts of chitosan-zeolite composite ranging from 30 to 150 mg were mixed with 50 ml wastewater and stirred at 250 rpm for 60 min. After adsorption, the mixture was filtered, and the removal percentage of chemical oxygen demand (COD) and fatty impurities was calculated using Eq. (1) formula.

$$\varepsilon = \frac{Ci - Cf}{Ci} \times 100\% \tag{1}$$

where Ci is the pollutant content before treatment [mg/l] and Cf is the pollutant content after treatment [mg/l].

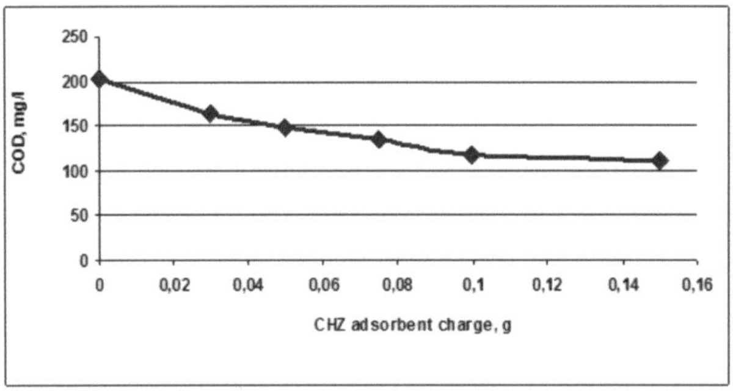

Figure 13. The effect of chitosan-zeolite composite dosage on the COD reduction.

The effect of adsorbent dosage on the removal of COD is shown in **Figure 13**. The quantity of COD removed increases as the chitosan-zeolite dosage increased. This was due to the increase in the number of active sites on chitosan-zeolite composites. The dosage of 0.10 g can be selected as the optimum dosage for further experiments. It can be observed that over this dosage, no further increase exists in the percentage removal of COD.

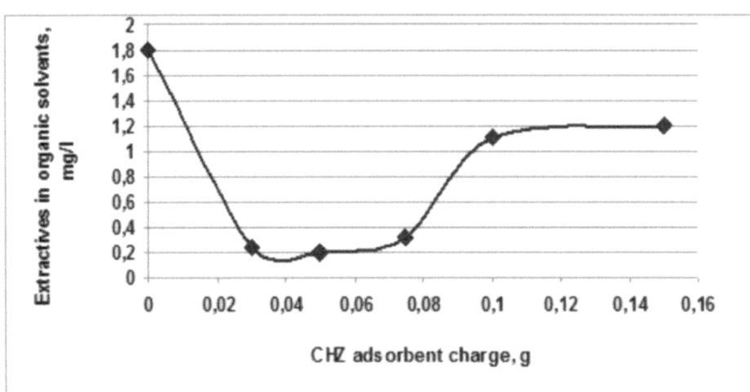

Figure 14. The effect of chitosan-zeolite composite dosage on the fatty and oil reduction.

The reduction of fatty and oil impurities increases with the chitosan-zeolite composite dosage, as it can be observed in **Figure 14**. It considers that the optimum dosage is at 0.05 g when the highest quantity of fatty impurities from wastewaters is eliminated.

Compared with the use of chitosan and zeolites only as adsorbents, it can be observed from **Figure 15** that the chitosan-zeolite composite performance is better. At the 60 min of adsorption time and the 0.3 g adsorbent dosage, the COD reduction is 45% for chitosan-zeolite composites than 20–21% for chitosan and zeolites only [54].

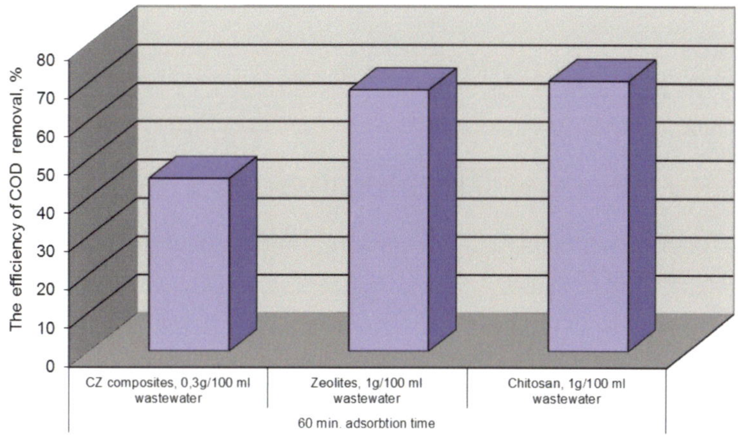

Figure 15. The pollutant reduction performances for different natural adsorbents.

4. Conclusions

Based on their proved properties, chitosan can be a very promising adsorption additive for wastewater pollutants. Aiming to improve their adsorption performances, chitosan can be modified by grafting, cross linking, functionalisation for forming composites, etc. Based on its origin product (chitin which can be found abundant in marine media, i.e. in the exoskeleton of crustaceans or cartilages of mollusks), the potential of chitosan to be used as (bio) adsorbent for wastewater pollutants is strong. However, the main drawback that limits the use of chitosan at industrial level is their low solubility in aqueous media. In this respect, the performances of chitosan can be improved by cross linking with different reagents, allowing chitosan composites to be used in acidic condition. To form composites with chitosan, different kinds of substances have been used.

The cationic nature of chitosan influences the adsorption mechanism of chitosan composites. In acid pH conditions, the amino groups of chitosan form protonated amines able to retain the metal ions or dye molecules from solutions or wastewaters. The interaction mechanisms can be through electrostatic attractions or/and complexation.

Although there is a wide range of chitosan derivatives with adsorption properties (raw chitosan, chitosan derivatives, chitosan composites, etc.), choosing the most suitable type of adsorbent in pollutant adsorption processes is still developed.

This field of research has a great area for improvement, and based on a large quantity of promising results, it is the hope that chitosan and their composites can be applied commercially instead of only at laboratory scale.

Author details

Petronela Nechita

Address all correspondence to: pnechita@yahoo.com; petronela.nechita@ugal.ro

Department of Environmental, Applied Engineering and Agriculture, Lower Danube University of Galați, Romania

References

[1] Mudasir A, Shakeel A, Babu Lal S and Saiqa I: Adsorption of heavy metal ions: Role of chitosan and cellulose for water treatment. Int. J. Pharmacogn. 2015; 2:6

[2] Zha F, Li S, Chang Y: Preparation and adsorption property of chitosan beads bearing β-cyclodextrin cross-linked by 1,6-hexamethylene diisocyanate. Carbohydr. Polym. 2008; 72(3): 369-570

[3] Zeenat MA, Abdul Jabbar L, Abdul Khalique A, Mohammad Y K: Extraction and characterization of chitosan from Indian prawn (*Fenneropenaeus indicus*) and its applications on waste water treatment of local ghee industry. IOSR J. Eng. 2013; 3:10

[4] Wang L, Xing R, Liu S, Cai S, Yu H, Feng J, Li R, Li P: Synthesis and evaluation of a thiourea-modified chitosan derivative applied for adsorption of Hg(II) from synthetic wastewater. Int. J. Biol. Macromol. 2010; 46524–46528

[5] Zuo X: Preparation and evaluation of novel thiourea/chitosan composite beads for Copper(II) removal in aqueous solutions. Ind. Eng. Chem. Res. 2014; 53: 1249–1255

[6] Mouzdahir Y, Elmchaouri A, Mahboub R, Gil A, Korili S A: Equilibrium modeling for the adsorption of methylene blue from aqueous solution on activated clay minerals. Desalination 2010; 250: 335–338

[7] Tran H V, Tran L D, Nguyen T N: Preparation of chitosan/magnetite composite beads and their application for removal of Pb(II) and Ni(II) from aqueous solution. Mater. Sci. Eng. 2010; 30: 304–310

[8] Wan M W, Kan C C, Buenda D R, Maria L P D: Adsorption of copper(II) and lead(II) ions from aqueous solution on chitosan-coated sand. Carbohydr. Polym. 2010; 80: 891–899

[9] Dinu M V, Dragan ES: Evaluation of Cu^{2+}, Co^{2+}, and Ni^{2+} ions removal from aqueous solution using a novel chitosan/clinoptilolite composites: Kinetics and isotherms. Chem. Eng. J. 2010; 160: 157–163

[10] Wan Ngaha W S, Teonga LC, Hanafiaha MAKM: Adsorption of dyes and heavy metal ions by chitosan composites: A review. Carbohydr. Polym. 2011; 83: 1446–1456

[11] Shahram MD, Bahar R, Ali Mashinchian M, Parviz A: Removal of permethrin pesticide from water by chitosan–zinc oxide nanoparticles composite as an adsorbent. J. Saudi Chem. Soc. 2014; 18: 348–355

[12] Kyzas G Z, Bikiaris D N: Recent modifications of chitosan for adsorption applications: A critical and systematic review. Mar. Drugs 2015; 13: 312-337

[13] Vaishnavi S: Fabrication of chitosan–magnetite nanocomposite strip for chromium removal. Appl. Nanosci. 2016; 6:277–285

[14] Abd-Elhakeem M A, Alkhulaqi T A: Simple, rapid and eefficient water purification by chitosan coated magnetite nanoparticles. J. Environ. Nanotechnol. 2014; 3(4): 17-20

[15] Renault F, Sancey B, Badot PM, Crini G: Chitosan for coagulation/flocculation processes —an eco-friendly approach. Eur. Polym. J. 2009; 45: 1337–1348

[16] Szyguła A, Guibal E, Palacín MA, Ruiz M, Sastre A M: Removal of an anionic dye (Acid Blue 92) by coagulation–flocculation using chitosan. J. Environ. Manage. 2009; 90: 2979–2986

[17] Rinaudo M: Chitin and chitosan: Properties and applications. Prog. Polym. Sci. 2006; 31(7): 603-632

[18] Andres Y, Giraud L, Gerente C, Le Cloirec P: Antibacterial effects of chitosan flakes: Approach of mechanism and applications to water treatments. Environ. Technol. 2007; 28(12):1357-63. doi: 10.1080/09593332808618893

[19] Ganjidoust H, Tatsumi K, Wada S, Kawase M: Role of peroxidase and chitosan in removing chlorophenols from aqueous solution. Water Sci. Technol. 1996; 34(10): 151–159

[20] Siah Lee C, Robinson J, Chonga M F: A review on application of flocculants in waste-water treatment. Process Safety Environ. Protect. 2014; 92: 489–508

[21] Guibal E, Roussy J: Coagulation and flocculation of dye-containing solutions using a biopolymer (chitosan). React. Funct. Polym. 2007; 67: 33–42

[22] Guibal E, Van Vooren M, Dempsey B A, Roussy J: A review of the use of chitosan for the removal of particulate and dissolved contaminants. Sep. Sci. Technol. 2006; 41: 2487–2514

[23] No H, Meyers S: Application of chitosan for treatment of wastewaters. Ware G. (Ed.), Reviews of Environmental Contamination and Toxicology 2000; Springer, New York, 1–27

[24] Jaafari K, Ruiz T, Elmaleh S, Coma J, Benkhouja K: Simulation of a fixed bed adsorber packed with protonated cross-linked chitosan gel beads to remove nitrate from contaminated water. Chem. Eng. J. 2004; 99:15

[25] Pan J R, Huang C P, Chen S C, Chung Y C: Evaluation of a modified chitosan biopolymer for coagulation of colloidal particles. Colloids Surf. A. 1999; 147: 359-3643–160

[26] Szyguła A, Guibal E, Palacín MA, Ruiz M, Sastre AM: Removal of an anionic dye (Acid Blue 92) by coagulation–flocculation using chitosan. J. Environ. Manage. 2009; 90: 2979–2986

[27] Sye W F, Lu L C, Tai J W, Wang C I: Applications of chitosan beads and porous crab shell powder combined with solid-phase microextraction for detection and the removal of colour from textile wastewater. Carbohydr. Polym. 2008; 72: 550–556

[28] Rodrigues A C, Boroski M, Shimada N S, Garcia J C, Nozaki J, Hioka N: Treatment of paper pulp and paper mill wastewater by coagulation–flocculation followed by heterogeneous photocatalysis. J. Photochem. Photobiol. A: Chem.. 2008; 194: 1–10

[29] Renault F, Sancey B, Charles J, Morin-Crini N, Badot P M, Winterton P, Crini G: Chitosan flocculation of cardboard-mill secondary biological wastewater. Chem. Eng. J. 2009; 155:775–783

[30] Li J, Bai R: Mechanisms of lead adsorption on chitosan/PVA hydrogel beads. Langmuir 2002; 18: 9765–9770

[31] Li J, Jiao S, Zhong L, Pan J, Ma Q: Optimizing coagulation and flocculation process for kaolinite suspension with chitosan. Colloids Surf. A: Physicochem. Eng. Aspects 2013; 428: 100–110

[32] Mohd A, Abu H, Tan Pei L, Zainura Z N: Coagulation and flocculation treatment of wastewater in textile industry using chitosan. J. Chem. Nat. Resour. Eng. 2009; 4(1): 43-53

[33] Nechita P, Negreanu V, Axinti AM: Utilisation of (bio) additives in the adsorption processes of industrial wastewater pollutants. In Proceedings of The 8th International Symposium on Advanced Technologies For The Pulp, Paper And Corrugated Board Industry; 15–18 September 2015, Braila, Romania, p. 18.

[34] Nechita P: Natural additives used in adsorption of pollutants from textile wastewaters. Am. J. Environ. Eng. 2015; 5(2): 39-42. DOI: 10.5923/j.ajee.20150502.01

[35] Patriche S, Parfene G, Nechita P, Dinică RM, Cârâc G: Adsorption capacity of the chitosan for Zn(II) ions in aqueous solution and antibacterial activity. In Proceedings

of The 8th International Symposium on Advanced Technologies For The Pulp, Paper And Corrugated Board Industry; 15–18 September 2015, Braila, Romania, p. 21.

[36] Ghinea I O, Cârâc C, Cantaragiu A, Nechita P, Dinică R M, Cârâc G: Adsorption behaviour of the Fe(III) ions from aqueous solution on chitosan. In Proceedings of The 8th International Symposium on Advanced Technologies For The Pulp, Paper And Corrugated Board Industry; 15–18 September 2015, Braila, Romania, p. 26.

[37] Crini G, Badot P M: Application of chitosan, a natural aminopolysaccharide, for dye removal from aqueous solution by adsorption process using batch studies: A review of recent literature. Prog. Polym. Sci. 2008; 33:399-447

[38] Chiou M S, Ho P Y, Li HY: Adsorption of anionic dyes in acid solutions using chemically cross-linked chitosan beads. Dyes Pigment 2004; 60: 69–84

[39] Gupta A, Chauhan V S, Sankararamakrishnan N: Preparation and evaluation of iron–chitosan composites for removal of As(III) and As(V) from arsenic contaminated real life groundwater. Water Res. 2009; 43: 3862–3870

[40] Zainal Z, Hui L K, Hussein M Z, Abdullah A H, Hamadneh I R: Characterization of TiO_2–chitosan/glass photocatalyst for the removal of a monoazo dye via photodegradation–adsorption process. J. Hazard. Mater. 2009; 164:138–145

[41] Vaishnavi Sureshkumar SCG, Kiruba D, Ruckmani K: Fabrication of chitosan–magnetite nanocomposite strip for chromium removal. Appl. Nanosci. 2016; 6:277–285

[42] Thakre D, Jagtap S, Sakhare N, Labhsetwar N, Meshram S, Rayalu S: Chitosan based mesoporous Ti–Al binary metal oxide supported beads for defluoridation of water. Chem. Eng. J. 2010; 158: 315–324

[43] Moradi Dehaghi S, Rahmanifar B, Mashinchian Moradi A, Aberoomand Azar P: Removal of permethrin pesticide from water by chitosan–zinc oxide nanoparticles composite as an adsorbent. J. Saudi Chem. Soc. 2014; 18: 348–355

[44] Hritcu D, Dodi G, Popa M I: Heavy metal ions adsorption on chitosan-magnetite microspheres. Int. Rev. Chem. Eng. 2012; 4: 3

[45] Saifuddin N, Dinara S: Pretreatment of palm oil mill effluent (POME) using magnetic chitosan. E-J. Chem. 2011; 8: S67-S78

[46] Lin J, Zhan Y: Adsorption of humic acid from aqueous solution onto unmodified and surfactant-modified chitosan/zeolite composites. Chem. Eng. J. 2012; 200–202, 202–213

[47] Xie J, Li C, Chi L, Wu D: Chitosan modified zeolite as a versatile adsorbent for the removal of different pollutants from water. Fuel 2013; 103: 480–485

[48] Arora M, Eddy N K, Mumford K A, Baba Y, Perera J M, Stevens G W: Surface modification of natural zeolite by chitosan and its use for nitrate removal in cold regions. Cold Reg. Sci. Technol. 2010; 62: 92–97

[49] Dragan E S, Dinu M V, Timpu D: Preparation and characterization of novel composites based on chitosan and clinoptilolite with enhanced adsorption properties for Cu^{2+}. Bioresour. Technol. 2010; 101: 812–817

[50] Zhang Y, Yan W, Sun Z, Pan C, Mi X, Zhao G, Gao J: Fabrication of porous zeolite/chitosan monoliths and their applications for drug release and metal ions adsorption. Carbohydr. Polym. 2015; 117: 657–665

[51] Nešić AR, Veličković S J, Antonović D G: Modification of chitosan by zeolite a and adsorption of bezactive orange 16 from aqueous solution. Compos. B: Eng. 2013; 53: 145–151

[52] Wan Ngah W S, Teong L C, Toh R H, Hanafiah M A K M: Utilization of chitosan–zeolite composite in the removal of Cu(II) from aqueous solution: Adsorption, desorption and fixed bed column studies. Chem. Eng. J. 2012; 209: 46–53

[53] Nechita P: Studies regarding the use of natural zeolites in the industrial wastewaters treatment, Metall. Mater. Sci. 2015; Special issue: 64

[54] Nechita P: Studies regarding the use of natural zeolites in the industrial wastewaters treatment, The Annals of "Dunarea de Jos" University. Fascicle IX. Metallurgy and materials science 2015; Special issue: 64

Chapter 11

Flow Properties of Lambda Carrageenan in Aqueous Systems

Andrea Rivera del Rio, Mariana Ramírez-Gilly and
Alberto Tecante

Additional information is available at the end of the chapter

http://dx.doi.org/10.5772/65785

Abstract

Small amplitude oscillatory and steady shear measurements at 25°C were used to investigate the rheological behavior of λ-carrageenan solutions at pH 7.0 ± 1.0 without and with added sodium counterion. The dynamic moduli, $G'(\omega)$ and $G''(\omega)$, show the typical behavior of macromolecular solutions in which the viscous character predominates. The steady shear flow exhibits a Newtonian zero-shear viscosity (η_0) region followed by a shear-thinning zone. Viscosity data can be well described by the Carreau-Yasuda model. Without added Na^+, the intrinsic viscosity, $[\eta]$, and the critical overlap concentration, C^*, are 204 dL/g and 0.21%, respectively. With 20 mmol/dm^3 Na^+, $[\eta]$ = 14.7 dL/g and C^* = 0.38%. For concentrations below C^*, the viscous character is more sensitive to the presence of added Na^+, and the opposite occurs when the concentration exceeds C^*. The dynamic moduli and viscosity increase with the increase of polysaccharide concentration, but they decrease with added Na^+, confirming the polyelectrolyte nature of λ-carrageenan. Empirical shift factors were used to obtain master curves for the dynamic moduli and apparent viscosity for different polysaccharide and added Na^+ concentrations.

Keywords: carrageenan, polysaccharides, rheology, viscoelasticity, viscosity

1. Introduction

Recently, the nondigestible polysaccharides of natural origin have received great interest from the perspective of human health although they have been widely used for a long time [1, 2]. Carrageenans are among these polysaccharides. They are nondigestible linear sulfated polysaccharides extracted from red algae (*Rhodophyta*) of the genera *Chondrus*, *Gigartina*, *Iridaea*, *Eucheuma*, and *Hypnea* [3]. They are useful in the food industry as glazing, gelling, emulsifying, thickening, stabilizing, wetting, and bulking agents in a wide range of processed foods such as dairy products, meat, drinks, condiments, infant formula, and animal feed [3, 4]. Carrageenans are in the list of authorized additives at the level of Good Manufacturing Practices [4] and should be used in sufficient quantity to act as emulsifiers, stabilizers, or thickeners in food [5]. The kappa and iota types of carrageenans form gels under particular conditions, while lambda carrageenan will not form a gel but can only be used as a thickener and stabilizer. λ-Carrageenan is used to clarify beverages and improve the texture of cosmetic products as creams. It is used as a thickener in dairy products and as a stabilizer in water-based systems. Also, it can act as an asymmetric selector in the separation by capillary electrophoresis of the enantiomers of tryptophan derivatives and racemic mixtures of beta blockers, used as medications for heart disease [6]. In particular, λ-carrageenan oligosaccharides with a low degree of polymerization, e.g., λ-carraheptose, have shown high activity against blood vessel growth, i.e., angiogenesis, tumor proliferation, and endothelial cell invasion [1]. However, there is a fierce controversy because oligosaccharides with molecular weights in the range of 10–40 kDa obtained from hydrolysis of λ-carrageenan have been reported to be potentially carcinogenic [2].

The functionality of carrageenans resides mainly in their structure and polyelectrolyte properties that allow them to interact with other components; they are anionic polysaccharides highly unstable in their free acid form and are commonly commercialized as a mixture of sodium, potassium, and calcium salts. Carrageenans are composed of alternating units of β-D-galactopyranoside linked to position 3 (G units) and α-D-galactopyranoside attached to position 4 (D units) or 3,6-anhydrogalactose attached to position 4 (DA units). They form repetitive disaccharide "ideal" units, commonly called carrabioses, giving rise to each type of carrageenan. According to IUPAC rules, the name of λ-carrageenan is carrageenase 2,6,2′-trisulfate (G2S-D2S, 6S) (**Figure 1**). The λ-carrageenan lacks DA units and has a $^4C_1$ conformation that causes the creation of twisted zones or segments in the polysaccharide chain. As a consequence, the formation of helices is not possible and gels are not formed but only viscous solutions. As it occurs with the other types, the solubility of λ-carrageenan depends on temperature, pH of the solvent, type, and concentration of counterions, and other solutes. λ-Carrageenan is soluble in water at room temperature because

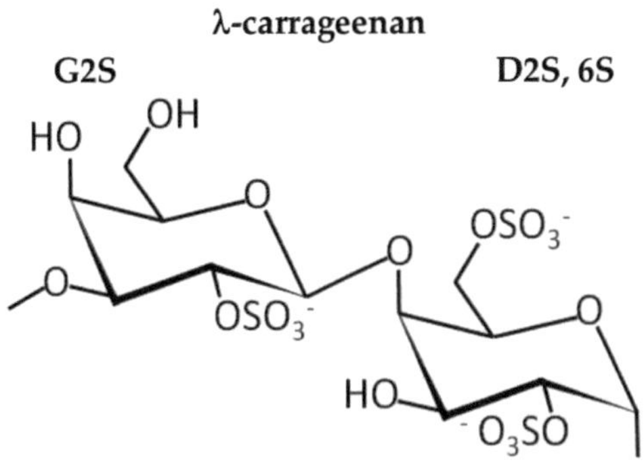

Figure 1. Ideal structure of the repeating unit of λ-carrageenans in its anionic form. G2S = 3-linked β-D-galactopyranose 2-sulfate; D2S, 6S = 4-linked α-D-galactopyranose 2,6 disulfate.

http://dx.doi.org/10.5772/65785

the polysaccharide-polysaccharide interactions are weak [7]. Most of the early investigations were focused on the characterization of the structure and size of λ-carrageenan by infrared spectrometry, nuclear magnetic resonance, and light scattering [8–10]. The synergistic effects on the rheological properties of λ-carrageenan combined with other polysaccharides, such as locust bean gum [11], whey protein concentrate [12], and inulin [13], have been studied in complex food systems. However, to the best of our knowledge, the different aspects investigated in this work for λ-carrageenan have not been previously reported.

Rheological techniques are widely used for the characterization of products and food additives. The variation of shear stress with shear rate or apparent viscosity with shear rate indicates if the flow behavior is Newtonian, shear-thinning, or both. Small-amplitude oscillatory tests within the zone of linear viscoelasticity (ZLV) are used to determine the variation of the storage modulus, $G'(\omega)$, and the loss modulus, $G''(\omega)$, with angular frequency and allow the viscoelasticity of the material to be characterized.

The objective of this work is to discuss the flow behavior and viscoelastic properties of λ-carrageenan in aqueous solution, without or with the addition of sodium to characterize the polyelectrolyte behavior of the polysaccharide, and evaluate the effect of its concentration and that of the counterion on these behaviors. The overlap concentrations separating the dilute from the semidilute regimes were also determined. The flow behavior was described with empirical models. Shift factors for generating master curves of the mechanical spectra and flow curves are proposed for different λ-carrageenan and added Na^+ concentrations. These results are useful to understand better the thickening properties of this important polysaccharide.

2. Experimental

A food-grade commercial preparation of λ-carrageenan as provided by a local supplier (FMC Biopolymer, Mexico) was used without further treatment. Other materials included sodium chloride ACS reagent grade (Mallinckrodt Baker, Mexico) and deionized water. The content (ppm) of sodium, potassium, calcium, and magnesium ions in the commercial preparation of the polysaccharide, determined by atomic absorption, was Na^+ – 30,863; K^+ – 15,595; Ca^{2+} – 972, and Mg^{2+} = 1882.

2.1. Preparation of λ-carrageenan solutions

λ-Carrageenan solutions with concentrations of 0.002, 0.006, 0.01, 0.08, 0.5, 0.8, 1.0, 1.5, and 2.0% by weight were prepared without and with added Na^+ considering the moisture content of the polysaccharide. The necessary amount of λ-carrageenan to make 50 g of a solution was dispersed as fine rain with a vibrating spatula in the appropriate solvent at 70°C under magnetic stirring at 1000 rpm (Barnstead International, model Super-Nuova SP131825, USA) until complete dissolution. The water that evaporated was compensated with the corresponding solvent. The solutions were stored in refrigeration at 3°C until analysis. Solutions with 0.002–2.0% λ-carrageenan were prepared in 0, 20, 30, 50, 70, 80, 100, 120, and 140 mmol/dm^3 NaCl. Besides, 1.5 and 2.0% λ-carrageenan solutions were prepared in 160, 180, 200, 220, 240, and 260 mmol/dm^3 Na^+. Only for 2.0% λ-carrageenan, solutions with 300, 350, and 400 mmol/dm^3 Na^+ were prepared too. The pH was 7.0 ± 1.0.

2.2. Ionic strength of λ-carrageenan solutions without and with added sodium counterion

Given the anionic nature of λ-carrageenan, the ionic strength of the aqueous environment is crucial. **Table 1** shows the range of total ionic strength of all solutions without and with added Na^+, the range of contribution of added Na^+ to the total ion strength, the amount of sodium ion required to neutralize the charges in the different polysaccharide solutions, and the concentration of Na^+ present in the commercial preparation.

The contribution of the internal counterions to the total ionic strength becomes significant only for high levels of the polysaccharide. Below 0.5% λ-carrageenan, the total ionic strength is practically given by the added Na^+, regardless of the concentration of added Na^+. Above 0.5% λ-carrageenan, the contribution of added Na^+ depended on the concentrations of the polysaccharide and added Na^+. The contribution of the internal counterions is always greater than that of added Na^+ for concentrations of added $Na^+ \leq 30$ mmol/dm^3. On the other hand, the Na^+/λ-carrageenan stoichiometric ratio is 4.76 meq/g. This value can be obtained considering a molecular weight of 630.41 for the sodium salt of the repeating unit of λ-carrageenan and indicates the amount of sodium ion needed to neutralize the charges in 1 g of polysaccharide.

2.3. Rheometry

The rheological behavior of λ-carrageenan solutions without and with added Na^+ was determined in a rheometer (ARES-RFS III, TA Instruments, Delaware, USA) using the Couette double-wall concentric cylinders fixture with a diameter ratio of 0.95, and 1.0 mm gap between

$C_{\lambda C}$ (%)	I_{min}–I_{max} [a]	Contribution (%)[b]	Na^+_{nc} (meq)	Na^+_{cp} (meq/mL)
0.002	0.024–70.0	99.8–99.9	0.00952	2.98×10^{-5}
0.006	0.072–70.1	99.3–99.9	0.0286	$8.95 \cdot \times 10^{-5}$
0.01	0.119–70.1	98.8–99.8	0.0476	1.49×10^{-4}
0.08	0.95–71.0	91.3–98.7	0.381	1.19×10^{-3}
0.50	5.97–76.0	62.6–92.1	2.38	7.46×10^{-3}
0.80	9.50–79.5	51.3–88.1	3.81	1.19×10^{-2}
1.0	11.9–81.9	45.7–85.5	4.76	1.49×10^{-2}
1.5	17.9–87.9	35.8–79.6	7.14	2.24×10^{-2}
2.0	23.9–93.9	29.5–74.5	9.52	$2.98 \cdot \times 10^{-2}$

[a]For 0–140 mmol/dm^3 added Na^+.
[b]For 20–140 mmol/dm^3 added Na^+.

Table 1. Range of total ionic strength (mmol/dm^3) of λ-carrageenan solutions, range of contribution (%) of added Na^+ to the total ionic strength, the amount (meq) of Na^+ necessary to neutralize the charges in λ-carrageenan solutions (Na^+_{nc}), and concentration (meq/mL) of Na^+ present in the commercial preparation (Na^+_{cp}).

http://dx.doi.org/10.5772/65785

cylinders. All determinations were performed in duplicate at 25 ± 0.5°C. Rheological data are presented as means of at least two repetitions with a standard deviation not greater than 5.0%.

The viscoelastic properties were determined by small-amplitude oscillatory shear tests. Strain sweeps were run from 0.1 to 100% strain (γ) at a constant angular frequency (ω) of 6.28 rad/s to determine the zone of linear viscoelasticity. Frequency sweeps were done in the ZLV from 0.1 to 100 rad/s to determine the storage modulus, $G'(\omega)$, and the loss modulus, $G''(\omega)$.

The flow properties were determined from steady angular shear tests in the range of 0.03–300 s^{-1}. Flow curves (η vs. $\dot{\gamma}$) for each concentration of λ-carrageenan and added Na^+ were obtained. Experimental data were fitted to the Cross model [14]

$$\frac{\eta-\eta_\infty}{\eta_0-\eta_\infty}=\frac{1}{1+(K\dot{\gamma})^{1-n}} \tag{1}$$

and the Carreau-Yasuda equation [15]

$$\frac{\eta-\eta_\infty}{\eta_0-\eta_\infty}=(1+\left(\lambda\dot{\gamma}\right)^a)^{\frac{n-1}{a}} \tag{2}$$

using the nonlinear regressions routines of SigmaPlot© programming software. Only regressions with $r^2 > 0.9960$ were considered.

2.3.1. *Intrinsic viscosity and critical concentration*

The intrinsic viscosity was determined from the graphical representation of the Huggins and Kraemer equations, given by Eqs. (3) and (4), respectively,

$$\eta_{\text{red}}=\left[\eta\right]+k'[\eta]^2C \tag{3}$$

$$\eta_{inh}=\left[\eta\right]+k''[\eta]^2C \tag{4}$$

In these equations $\eta_{\text{red}} = \eta_{\text{sp}}/C$ is the reduced viscosity and $\eta_{\text{inh}} = (\ln \eta_{\text{rel}})/C$ is the inherent viscosity. The specific viscosity is defined as $\eta_{\text{sp}} = \eta_{\text{rel}}-1$ and the relative viscosity as $\eta_{\text{rel}} = \eta/\eta_s$, i.e., the ratio of solution to solvent viscosities. The reduced and the inherent viscosities were calculated from the zero-shear viscosity, η_0. The intrinsic viscosity is determined by extrapolation to zero concentration and the corresponding constants, k' and k'', are determined from the respective slopes. The critical concentration, C^*, was determined from the graph η_{sp} vs. $C[\eta]_0$ as the point where there is a change in slope of the linear relationship between the plotted variables.

2.3.2. *Master curves*

Concentration-dependent shift factors on the ordinate and the abscissa of the flow curves (η vs. $\dot{\gamma}$) and frequency sweeps, $G'(\omega)$ and $G''(\omega)$ vs. ω, were determined empirically to obtain the corresponding master curve for each added Na^+ concentration and different λ-carrageenan concentrations. The 0.5% polysaccharide solution was used as a reference.

3. Results and discussion

3.1. Viscoelastic behavior of λ-carrageenan solutions without added sodium counterion

The variation with frequency of $G'(\omega)$ and $G''(\omega)$ for 0.006, 0.5, and 2.0% λ-carrageenan solutions is shown in **Figure 2**. The λ-carrageenan concentrations, 0.01, 0.08, 0.8, 1.0, and 1.5%, lie between 0.5 and 2.0%. The behavior displayed is characteristic of macromolecular solutions. The loss modulus is greater than the storage modulus, both depend on the angular frequency with a continuous rise, in the terminal zone $G'(\omega) \propto \omega^{1.6}$ and $G''(\omega) \propto \omega^{0.8}$, and the viscous character dominates. In solutions with 0.006, 0.01, and 0.08% λ-carrageenan moduli overcrossed, i.e., $G' = G''$, at frequencies of 10.0, 15.9, and 39.8 rad/s, respectively. These overcrossing points separate the behavior into a region predominantly viscous in which the loss modulus is superior, and an elastic domain in which the behavior is governed by the storage modulus. The reciprocal of the overcrossing frequency, ω_c, is related to the relaxation time of the macromolecular chains. This last time indicates the magnitude of the elastic character of the material. When ω_c becomes larger, the relaxation time becomes shorter, and the contribution of the elastic character diminishes.

For a given constant frequency both moduli increase with the increase of polysaccharide concentration, but this increase is not proportional. For example, at 10 rad/s $G'(\omega)$ increased 50 times when the polysaccharide concentration increased from 0.5 to 2.0%, while $G''(\omega)$ increased 20 times. The inverse occurs when λ-carrageenan concentration goes from 0.006 to 0.5%; $G'(\omega)$ increases about 5 times and $G''(\omega)$ does it 30 times. Therefore, at high polysaccharide concentrations, the contribution of the elastic character becomes more significant.

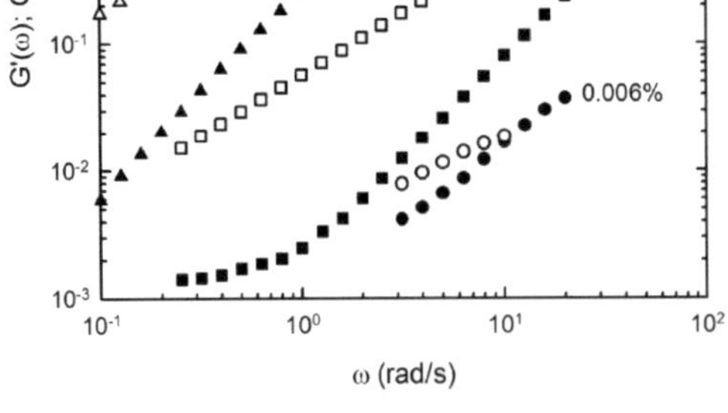

Figure 2. Variation of the storage modulus, $G'(\omega)$ (full symbols) and the loss modulus, $G''(\omega)$ (empty symbols) with angular frequency for solutions with 0.006, 0.5 and 2.0% λ-carrageenan without added Na^+. Strains are 50, 40, and 20% for 0.006, 0.5, and 2.0%, respectively.

3.2. Steady flow behavior of λ-carrageenan solutions without added sodium counterion

The variation of apparent viscosity with a shear rate and λ-carrageenan concentration is shown in **Figure 3**. The dependence between these two quantities for a given shear rate is approximately $\eta \alpha C^2$. The 0.002% λ-carrageenan solution exhibits Newtonian behavior. All other polysaccharide concentrations exhibit Newtonian behavior at low shear rates, a transition zone at intermediate shear rates and shear-thinning behavior at high shear rates. The end of the zero-shear Newtonian zone is displaced to lower shear rates when λ-carrageenan concentration increases. This behavior is typical of many polysaccharides used in the food industry and is closely related to the polymer-polymer and polymer-solvent interactions. The intrinsic viscosity and the overlap concentration are useful parameters to explain this behavior.

3.3. Intrinsic viscosity and critical concentration of λ-carrageenan solutions without added sodium counterion

The intrinsic viscosity, $[\eta]$, of a polymer is a measure of its contribution to the viscosity of a solution when the polymer concentration, C, tends to zero and depends on the conformation and molecular weight of the polymer in solution. The determination of this quantity from Huggins and Kraemer equations requires dilute solutions and a linear dependence of η_{sp}/C and $(\ln \eta_{rel})/C$ with concentration. The intrinsic viscosity for λ-carrageenan in aqueous solution without added Na^+ is around 204 dL/g. The repeating unit of λ-carrageenan has three sulfate groups. In the absence of external counterion, these groups repel each other and hence the polysaccharide chains are primarily unfolded. Therefore, this can explain the high intrinsic viscosity observed. The intrinsic viscosity is related to the hydrodynamic volume of a polymer in solution. A dimensionless quantity, called the coil-overlap parameter, is the product of intrinsic viscosity and polymer concentration ($C[\eta]_0$). This parameter is a measure of the total volume occupied by the polymer in solution [16].

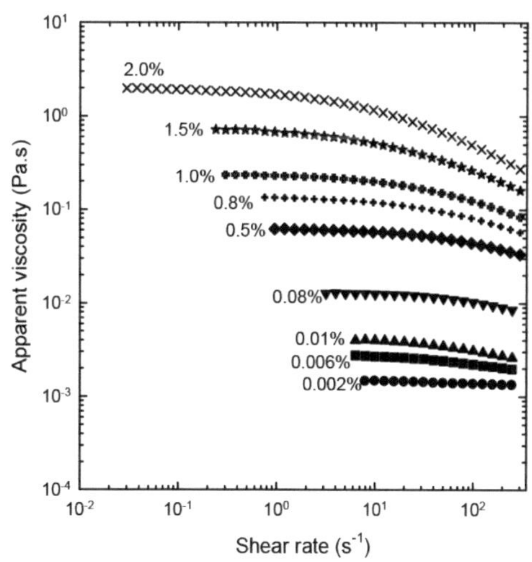

Figure 3. Variation of apparent viscosity with shear rate and λ-carrageenan concentration in the range of 0.002–2.0% without added Na^+.

On the other hand, the critical concentration, C^*, indicates the transition from dilute to semidilute regimes and is closely related to the formation of entanglements between macromolecules. The critical concentration of λ-carrageenan in aqueous solution without added Na^+ is about 0.21%. Solutions with 0.002, 0.006, 0.01, and 0.08% λ-carrageenan can be considered dilute. These solutions exhibit a slight shear-thinning behavior (**Figure 3**); the zero-shear viscosity decreases about 30–35% along the range of shear rate. At low shear rate, polysaccharide strands are in a highly extended conformation due to electrostatic repulsion. When shear rate increases, the strands are aligned or oriented in the direction of flow and occupy a smaller volume, this explains the decrease in viscosity.

Solutions with 0.5, 0.8, 1.0, 1.5, and 2.0% λ-carrageenan are semidilute. The shear-thinning behavior becomes more evident than for dilute solutions (**Figure 3**). In these solutions, polysaccharide-polysaccharide interactions, in the form of overlaps and interlocks, become increasingly significant while polysaccharide-water interactions decrease. When the chains are deformed, the entanglements are disturbed. At low shear rate, there is enough time for new entanglements to be formed and their number to remain constant over time. Therefore, viscosity remains practically constant. This phenomenon gives rise to the zero-shear Newtonian region. The transition to the shear-thinning zone occurs when the shear rate is greater than the rate of formation of new entanglements [16]. As a consequence, there are fewer polymer-polymer interactions which facilitate the flow of the solution and cause the viscosity to decrease.

The rate of formation of new interactions is related to the concentration of polymer in solution. The space occupied by the polymer and particularly the available space are related to $C[\eta]_0$. The unoccupied space allows the chains to move freely, but when $C[\eta]_0$ increases motion becomes more restricted. That is the reason why upon increasing the concentration of polymer, the beginning of the transition to the shear-thinning region is displaced to lower shear rates [16].

3.4. Flow models of λ-carrageenan solutions without added sodium counterion

The Cross and the Carreau-Yasuda are two of the more traditional flow models. The regression parameters of the two models for each λ-carrageenan solution without added Na^+ are shown in **Table 2**. In both models, the increase in the zero-shear viscosity (η_0) is attributed to the increase in polysaccharide concentration. The parameters K and λ of the Cross and Carreau-Yasuda models, respectively, have the same meaning. They represent the time for which the shear-thinning region appears. They are different for each model but the Carreau-Yasuda model describes better the start of the shear-thinning region. The parameter n indicates the proximity to Newtonian behavior. The indirect relationship between polysaccharide concentration and n for the Carreau-Yasuda model means that for a higher polysaccharide concentration, the flow behavior deviates from Newtonian. This behavior is not observed with the Cross model, and the values of n do not explain the observed shear-thinning behavior.

The parameter a in the Carreau-Yasuda equation indicates how long and smooth the transition is from the Newtonian to the shear-thinning zone. This parameter takes bigger values for more dilute solutions, except for the 0.002% solution, in which the flow transition is not observed. In all cases, this change is long and smooth as observed experimentally. The

Cross					Carreau-Yasuda				
C_{AC} (%)	η_0 (Pa·s)	K (s)	n	r²	η_0 (Pa·s)	λ (s)	a	n	r^2
0.002	0.0030	0.2789	0.944	0.9879	0.0017	0.0193	0.2047	0.939	0.9883
0.006	0.0030	0.0011	0.526	0.9967	0.0028	0.0466	1.3320	0.860	0.9984
0.01	0.0044	0.0022	0.396	0.9962	0.0041	0.0415	1.7380	0.813	0.9995
0.08	0.0131	0.0019	0.189	0.9976	0.0129	0.0160	1.2420	0.734	0.9990
0.5	0.0631	0.0029	0.343	0.9997	0.0627	0.0069	0.7295	0.531	0.9998
0.8	0.1392	0.0061	0.384	0.9997	0.1377	0.0107	0.6746	0.489	0.9999
1.0	0.2414	0.0095	0.363	0.9997	0.2387	0.0225	0.7329	0.525	0.9999
1.5	0.7602	0.0305	0.408	0.9997	0.7593	0.0323	0.6028	0.413	0.9998
2.0	2.0080	0.0633	0.377	0.9998	2.0110	0.0581	0.6147	0.361	0.9999

Table 2. Regression parameters of the Cross and Carreau-Yasuda models for different λ-carrageenan concentrations without added Na^+.

Carreau-Yasuda model describes better the flow behavior of the λ-carrageenan solutions without added Na^+. In addition, it provides more information because it contains the parameter a. However, both models have some limitations. If the experimental data do not allow the zero-shear and shear-thinning regions to be observed, the values of the parameters K, λ, and a will be somehow unrealistic because the models are unable to predict the beginning of the shear-thinning region and its shape. It is because of this the values of the parameters for the 0.002% λ-carrageenan solution are less in agreement, as shown by the corresponding regression coefficients, with experimental observations.

The power-law model is used to describe only the shear-thinning region. The description of the flow curves with this model requires the experimental data to be adjusted to Eq. [5].

$$\eta = k\dot{\gamma}^{n-1} \tag{5}$$

Table 3 shows the parameters obtained with such fitting and the good correlations obtained. This proficiency of the model is not surprising, but the limitations in comparison with the Cross

C_{AC} (%)	k (Pa·s^n)	n	r^2
0.5	0.1331	0.76	0.9971
0.8	0.2881	0.72	0.9955
1.0	0.5350	0.68	0.9960
1.5	1.708	0.59	0.9982
2.0	4.556	0.51	0.9987

Table 3. Regression parameters of the power-law model for different λ-carrageenan concentrations without added Na^+.

and Carreau-Yasuda equations are evident. The consistency index, *k*, is directly related to viscosity. As the polysaccharide concentration is increased, the values of *k* increase, and if the behavior becomes Newtonian, the consistency index is the viscosity. The parameter *n*, called flow behavior index, has the same meaning that in the Cross and Carreau-Yasuda models. Values close to 1 indicate a close-to-Newtonian behavior. Lower polysaccharide concentrations result in a behavior closer to Newtonian, as expected and predicted by the Cross and Carreau-Yasuda models.

3.5. Viscoelastic behavior of λ-carrageenan solutions with added sodium counterion

Figure 4 shows the effect of added Na^+ on the dynamic moduli. The ratios $G'(\omega)/G'_0(\omega)$ and $G''(\omega)/G''_0(\omega)$ are plotted against the total ionic strength for different polysaccharide concentrations at a given angular frequency. In these ratios, $G'(\omega)$ and $G''(\omega)$ are the storage and loss modulus, respectively, of the solutions with added Na^+, while $G'_0(\omega)$ and $G''_0(\omega)$ are the corresponding moduli without added Na^+. Drastic reductions in the dynamic moduli occur when the total ionic strength increases. Reductions are so drastic for 20, 30, 50, 70, 80, 100, 120, and 140 mmol/dm^3 added Na^+ that the dynamic moduli for solutions with polysaccharide concentrations in the dilute regime, i.e., 0.002, 0.006, and 0.01%, cannot be detected by using the rheometer. Only for 0.08% λ-carrageenan dynamic moduli can be determined, and the characteristic behavior of macromolecular solutions in which the loss modulus is greater than the storage modulus is observed. The viscous character dominates and moduli depend on frequency. The increase in ionic strength by addition of Na^+ reduced $G'(\omega)$ 39% and $G''(\omega)$ 81%. The previous value represents an almost three-fold reduction in $G'(\omega)$ for the same polysaccharide concentration but without added Na^+. This behavior illustrates the polyelectrolyte character of λ-carrageenan which is responsible for the sensitivity of viscoelasticity to the addition of the counterion. In addition, the viscous character is the most sensitive.

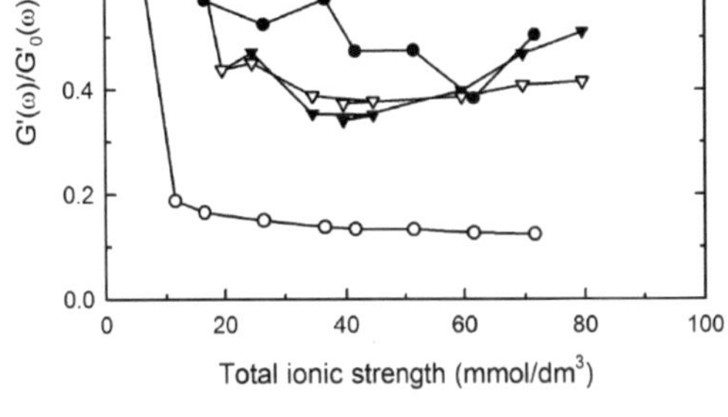

Figure 4. Variation of the ratio of storage moduli with added Na^+ to storage moduli without added Na^+ (full symbols), and the ratio of loss moduli with added Na^+ to loss moduli without added Na^+ (empty symbols) with the total ionic strength for 0.08 and 0.5% λ-carrageenan solutions (ω = 6.31 rad/s). Lines are included as a visual guide.

http://dx.doi.org/10.5772/65785

After an initially noticeable decrease, the dynamic moduli remain practically constant with the increase in the total ionic strength, that is, for concentrations greater than 20 mmol/dm^3 Na^+. On the other hand, this concentration is superior to the stoichiometric counterion/polysaccharide ratio and is sufficient to screen all the charges in the macromolecule.

The effect of the total ionic strength on 0.5% λ-carrageenan solutions can also be seen in **Figure 4**. In this case, the rate of diminution due to the increase in the total ionic strength is 60% for both moduli in comparison with a solution with the same polysaccharide concentration but without added Na^+. Moduli decrease and then become practically independent of Na^+ concentration. However, above 45 mmol/dm^3 approximately, the moduli increase around 10%. At this point, it is possible that ions start to compete for water associated with the polysaccharide because ions are not hydrated enough as they are added to the solution which for 0.5% λ-carrageenan is in the semidilute regime. The addition of Na^+ can lead to the formation of a new more elastic structure than the one observed when there is enough water to hydrate all molecules and ions in solution.

The effect of the increase in ionic strength on 0.8% λ-carrageenan solutions is observed in **Figure 5**. The dynamic moduli decrease about 25–30% for 20 mmol/dm^3 Na^+ (I_{total} = 19.5 mmol/dm^3) as compared with systems without added Na^+ (I_{total} = 9.50 mmol/dm^3), and a minimum is observed when the ionic strength due to the counterions is 44.5 mmol/dm^3. At this point $G'(\omega)$ decreases 35% and $G''(\omega)$ decreases 45%. From a total ionic strength of 44.5 mmol/dm^3, the storage and loss moduli increased up to 86 and 65% of their values without added Na^+, respectively. These changes mean that $G'(\omega)$ increases 21% from its low to its maximum value and $G''(\omega)$ increases 11%. Also, $G'(\omega)$ decreases less and recovers more than $G''(\omega)$.

For 1.0% λ-carrageenan solutions, the diminution in the presence of added Na^+ is almost 30% as compared with solutions without added Na^+. The minimum $G'(\omega)$ occurs for a total ionic

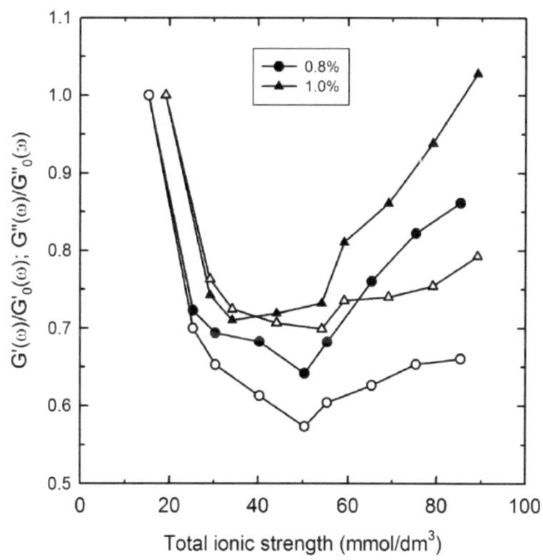

Figure 5. Variation of the ratio of storage moduli with added Na^+ to storage moduli without added Na^+ (full symbols), and the ratio of loss moduli with added Na^+ to loss moduli without added Na^+ (empty symbols) with the total ionic strength for 0.8 and 1.0% λ-carrageenan solutions (ω = 6.31 rad/s). Lines are included as a visual guide.

strength of 26.9 mmol/dm^3 and decreases 27% as compared with 11.9 mmol/dm^3. The minimum $G''(\omega)$ takes place for 46.9 mmol/dm^3 and decreases 30% compared with 11.9 mmol/dm^3. $G'(\omega)$ and $G''(\omega)$ increase up to 103 and 79%, respectively, from these minimum values as compared with those without added Na^+. This means that $G'(\omega)$ increases 30% and $G''(\omega)$ increases 7% from their corresponding minimum to maximum values. The increase in storage modulus is more significant than the loss modulus, so the presence of added Na^+ affects the elastic character more than the viscous one in the semidilute regime, unlike the dilute regime in which the viscous character is more sensitive than the elastic one. This behavior may be due to the above-stated reason, i.e., the competition between the counterions and the macromolecules for water to become hydrated that occurs when the solution begins to saturate. This competition increases the association of the polysaccharide chains that form structures with a certain degree of rigidity that favors the elastic nature of the behavior.

In the case of 1.5 and 2.0% λ-carrageenan solutions with added Na^+, a respective decrease of 10% for I_{total} = 32.9 mmol/dm^3 and 7% for I_{total} = 38.9 mmol/dm^3 in the dynamic moduli is also observed in comparison with 0 mmol/dm^3 Na^+ (I_{total} = 17.9 mmol/dm^3 for 1.5% λ-carrageenan and I_{total} = 23.9 mmol/dm^3 for 2.0% λ-carrageenan) (**Figure 6**). The λ-carrageenan is a highly flexible polyelectrolyte which means that by screening its charges, the hydrodynamic volume of the polymer is significantly reduced causing a corresponding decrease in its dynamic moduli. Although this occurs, the polysaccharide concentration is so high that the reduction of the moduli is smaller than the one observed for lower polysaccharide concentrations.

The decrease in dynamic moduli for 1.5% λ-carrageenan solutions continues until the total ionic strength is 32.9 mmol/dm^3. From this point $G'(\omega)$ and $G''(\omega)$ increases up to 189 and 113%, respectively, regarding a solution with the same concentration of polysaccharide but without added Na^+. Also, in the 2.0% λ-carrageenan solutions, the minimum dynamic moduli are achieved for a total ionic strength of 33.9 mmol/dm^3, and $G'(\omega)$ and $G''(\omega)$

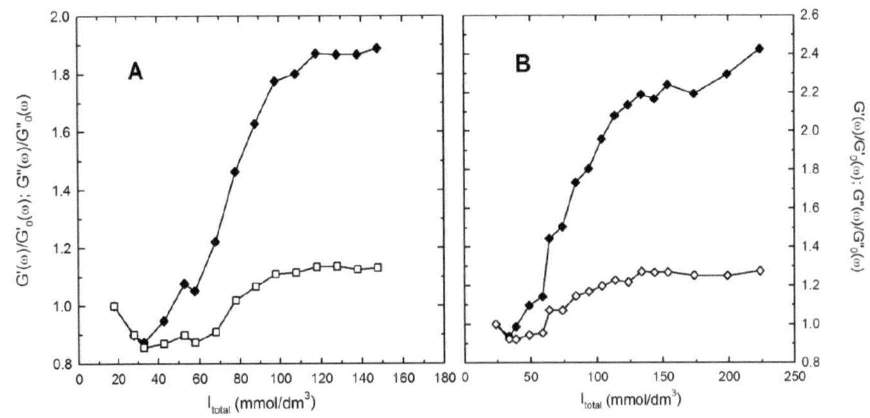

Figure 6. Variation of the ratio of storage moduli with added Na^+ to storage moduli without added Na^+ (full symbols), and the ratio of loss moduli with added Na^+ to loss moduli without added Na^+ (empty symbols) with the total ionic strength for 1.5% (A) and 2.0% (B) λ-carrageenan solutions (ω = 6.31 rad/s). Lines are included as a visual guide.

http://dx.doi.org/10.5772/65785

increase to 242 and 127%, respectively, regarding the corresponding solutions without added Na^+.

3.6. Steady flow behavior of λ-carrageenan solutions with added sodium counterion

Solutions were also examined under steady angular shear to observe the effect of added Na^+ on their thickening properties. Newtonian behavior with a viscosity close to 1 mPa s was found in the range of 6–300 s^{-1} for 0.002, 0.006, and 0.01% λ-carrageenan solutions with 20 mmol/dm^3 Na^+ ($I_{total} \approx 10$ mmol/dm^3). This behavior indicates that the ionic strength is sufficient for screening the charges of all chains and their hydrodynamic volume decreases in such a way that all the flow resistance of the pure solvent is not altered. The λ-carrageenan concentration is low enough for available water to interact with the polysaccharide and added Na^+. The viscosity remains constant with the increase in ionic strength to a level close to 70 mmol/dm^3.

In the case of the 0.08% λ-carrageenan solution with 20 mmol/dm^3 Na^+ ($I_{total} = 10.95$ mmol/dm^3), the presence of the external counterion causes a decrease of 80% in the viscosity regarding the same polysaccharide concentration without added Na^+ ($I_{total} = 0.95$ mmol/dm^3). However, this polysaccharide concentration is sufficiently high for the viscosity not to drop close to that of the solvent. This decrease is maintained constant regardless of the amount of added Na^+ as shown in **Figure 7**. In this figure, the viscosity ratio is plotted against the total ionic strength. This ratio is the quotient of the zero-shear viscosity of the solution with added Na^+ and the zero-shear viscosity of the solution without added Na^+ for the same polysaccharide concentration and a given shear rate.

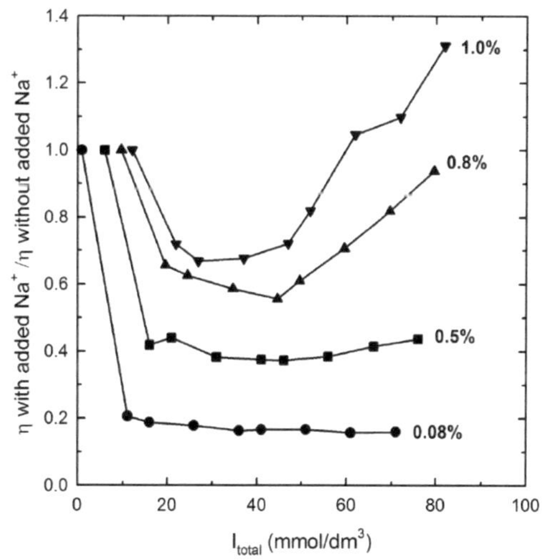

Figure 7. Variation of the ratio of apparent viscosities with added Na^+ to apparent viscosities without added Na^+ with the total ionic strength for 0.08 ($\dot{\gamma} = 10\ s^{-1}$), 0.5, 0.8, and 1.0% $\dot{\gamma} = 3.8\ s^{-1}$) λ-carrageenan solutions. Lines are included as a visual guide.

In solutions with 0.5% λ-carrageenan, a similar behavior is observed. The viscosity decreases 60% when the concentration of added Na^+ is 20 mmol/dm^3 (I_{total} = 15.97 mmol/dm^3), compared with a solution of equal carrageenan concentration without added Na^+ (I_{total} = 5.97 mmol/dm^3). This decrease remains constant until the ionic strength is 55.97 mmol/dm^3. Beyond this value, a slight increase in viscosity of about 5% is observed.

For 0.8 and 1.0% λ-carrageenan solutions, viscosity decreases of 34 to 28%, respectively, are observed with the addition of 20 mmol/dm^3 Na^+; I_{total} = 19.5 and 21.90 mmol/dm^3, respectively. The viscosity reduction continues until a minimum is reached when the ionic strength is 44.5 and 26.9 mmol/dm^3, for 0.8 and 1.0% λ-carrageenan, respectively. From these minimum viscosity ratios, the viscosity of the 0.8% λ-carrageenan solutions increases linearly (r^2 = 0.9983) with the increase in the concentration of added Na^+, and recovers 96% of the viscosity without added Na^+.

For 1.0% λ-carrageenan solutions, viscosity increases from the minimum ratio up to 132% of its value without added Na^+. However, this increase in viscosity is not linear (r^2 = 0.9697). The decrease in viscosity with added Na^+ highlights the polyelectrolyte property of λ-carrageenan and is due to charge screening of the polysaccharide chains. In the absence of external Na^+, the chains are highly unfolded due to electrostatic repulsion between the sulfate groups. In the presence of external Na^+, the negative charges of the polysaccharide chain are screened, and a more compact conformation can be adopted. The hydrodynamic volume decreases, and so does the viscosity of the solution. The maximum increase in viscosity observed primarily for 0.8 and 1.0% λ-carrageenan solutions with the addition of more Na^+, which in turn implies that an increase in viscosity in the shear-thinning region is due to the presence of excess external Na^+, i.e., additional to that necessary to screen all the charges of the sulfate groups. The excess Na^+ is hydrated and competes with the polysaccharide for water molecules, which results in a decrease in polysaccharide-water interactions and increases the polysaccharide-polysaccharide interactions forming semiordered and more elastic structures that increased the viscosity. The amount of polysaccharide in 1.0% λ-carrageenan solutions becomes important. A minimum viscosity appears when the ionic strength is 26.9 mmol/dm^3. Competition for the solvent begins probably at this point. Therefore, viscosity increases from this ionic strength to the extreme concentration of added Na^+ for the same reason explained previously.

For solutions with concentrations higher than 1.0% polysaccharide, a different behavior is observed (**Figure 8**). The viscosity of 1.5% λ-carrageenan solutions without added Na^+ decreases 12% when the ionic strength is 32.9 mmol/dm^3 in comparison with I_{total} = 17.9 mmol/dm^3 without external Na^+. That is the minimum viscosity observed.

For the 2.0% λ-carrageenan solutions, a minimum is observed for an ionic strength of 33.9 mmol/dm^3 which corresponds to a 12% decrease (**Figure 8B**) regarding 23.9 mmol/dm^3 without added Na^+. From these minimum values, viscosity suddenly increases with increasing the addition of Na^+. Hydration of the Na^+ counterions and the formation of semiordered and more elastic structures might explain the growth in viscosity to 200 and 270% for 1.5 and 2.0% λ-carrageenan, respectively. It is possible that for higher external Na^+ concentrations (I_{total} > 87–94 mmol/dm^3) more ordered structures are formed which have a greater resistance to flow and remain up to very high added Na^+ concentrations. It is presumed that if the addition of

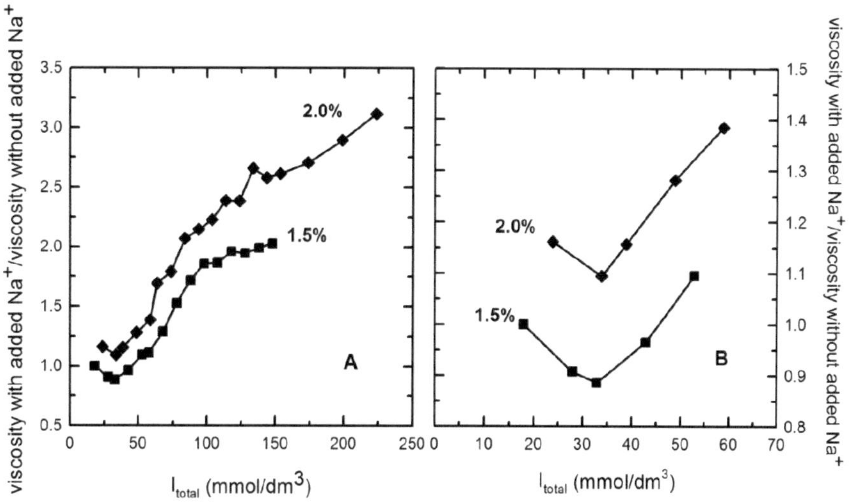

Figure 8. (A) Variation of the ratio of apparent viscosities with added Na^+ to apparent viscosities without added Na^+ with the total ionic strength for 1.5 and 2.0% λ-carrageenan solutions ($\dot{\gamma}$ = 9.5 s^{-1}). (B) Initial section of (A) corresponding to I_{total} = 18–59 mmol/dm^3 (0–70 mmol/dm^3 added Na^+). Lines are included as a visual guide.

Na^+ continued, a state would be reached in which precipitation of the polysaccharide chains would occur. It is worth noting that only the shear-thinning region was observed, and the infinite-shear Newtonian region was not observed. It is for this reason that the viscosities in **Figure 8** correspond to a shear rate in the former zone.

3.7. Intrinsic viscosity and critical concentration of λ-carrageenan solutions with added sodium counterion

The intrinsic viscosity of λ-carrageenan in a 20 mmol/dm^3 Na^+ solution is 14.7 dL/g. Values reported in the literature are 9.5 dL/g in 100 mmol/dm^3 NaCl for M_w = 614 kDa [9], and 10.8 (M_w = 870 kDa), 9.60 (M_w = 730 kDa), 3.40 dL/g (M_w = 340 kDa) in 100 mmol/dm^3 NaCl, pH 7.0 [17]. The intrinsic viscosity of λ-carrageenan without added Na^+ decreased almost 93% with added Na^+ (20 mmol/dm^3). This reduction in [η] suggests high flexibility of λ-carrageenan chains, since otherwise, the charge screening would not have been sufficient to decrease the hydrodynamic volume. The intrinsic viscosities were not determined for concentrations higher than 20 mmol/dm^3 Na^+ because the viscosity of solutions with lower polysaccharide concentrations remained practically constant, regardless of the concentration of external Na^+. It is possible to assume that [η] of λ-carrageenan with added Na^+ is constant regardless of the amount of sodium ion present because in a 20 mmol/dm^3 solution, the stoichiometric ratio of sulfate groups/Na^+ is satisfied and all charges in the polysaccharide chains are screened.

The critical concentration, C^*, of λ-carrageenan with 20 mmol/dm^3 added Na^+ was 0.38%. This value is higher than that for the solutions without added Na^+. This behavior means that more polysaccharide chains are required for overlapping to occur. This phenomenon is essentially due to the decrease in hydrodynamic volume of the strands and compaction from a highly expanded form to a coiled one. The physical state of the systems can be depicted as follows. In the dilute and semidilute regimes, without added Na^+ carrageenan chains are extended and stiff to some extent because of electrostatic repulsion between unscreened charges. Under an applied shear rate, the chains align themselves in the flow direction and overlap to a degree dependent on the solution regime. From 0.5% λ-carrageenan, semidilute regime, the chains overlap and under the action of shear rate, they align in such a way that the shear-thinning zone becomes important. In this flow region, the rate of formation of new entanglements would be lower than the shear rate. Therefore, the chains are oriented and isolated, which results in a decrease in viscosity. When Na^+ is added to solutions with low polysaccharide concentrations, e.g., from 0.002 to 0.01%, charge screening results in a reduction in hydrodynamic volume and the viscosity drops practically to that of the solvent. For solutions with 0.08% λ-carrageenan, the polysaccharide concentration is sufficient for the viscosity not to fall to that of the solvent. In the semidilute regime, 0.5, 0.8, and 1.0% λ-carrageenan, the chains fold and viscosity decreases. Further addition of external Na^+ makes cations accumulate in the medium without significant interaction with the polysaccharide. The effect of added Na^+ on the solutions with 1.5 and 2.0% λ-carrageenan is different from the rest. From very low concentrations of added Na^+, viscosity, and dynamic moduli increase suggesting the formation of a semiordered, rigid, and elastic structure, due to the increase in the polysaccharide concentration and counterion and the intermolecular interactions of the polysaccharide with water and with added Na^+.

3.8. Flow models of λ-carrageenan solutions with added sodium counterion

The Carreau-Yasuda model describes better the flow curves of λ-carrageenan without added Na^+. The same happens with added Na^+. However, it is not always possible to obtain proper fittings because, in most cases, the zero-shear region is not observed in the semidilute regime or the shear-thinning zone in the dilute regime. **Table 4** shows the experimental values of the parameters η_0 and a. However, λ and n values do not always follow the expected trend. For instance, in solutions with 0.8% λ-carrageenan one might expect that as the concentration of added Na^+ increases the non-Newtonian behavior becomes more accentuated, and n would

$C_{\lambda C}$ (%)	Na^+ (mmol/dm³)	η_0 (Pa·s)	λ (s)	a	n	r^2
0.5	70	0.022	0.018	1.426	0.82	0.9945
	80	0.023	0.014	1.139	0.79	0.9994
0.8	20	0.087	0.014	0.801	0.61	0.9996
	30	0.080	0.023	0.990	0.69	0.9995
	50	0.077	0.021	0.931	0.67	0.9999
	70	0.073	0.020	0.888	0.65	0.9998
1.0	20	0.166	0.030	0.847	0.61	0.9990
	30	0.159	0.009	0.643	0.42	0.9986
1.5	20	0.658	0.046	0.691	0.47	0.9997

Table 4. Parameters of the Carreau-Yasuda model for different λ-carrageenan concentrations and added Na^+.

decrease progressively. On the other hand, the power-law model cannot be applied to the flow curves of the solutions with added Na^+ because the concavity exhibited by most of them in the shear-thinning region does not allow only this model to describe the entire flow region.

3.9. Master curves for mechanical spectra and steady flow

The mechanical spectra and flow curves can be expressed with their corresponding master curves for a constant concentration of added Na^+ and λ-carrageenan concentrations of 0.002–2.0%. Two empirical shift factors were determined to produce each master curve. The factor

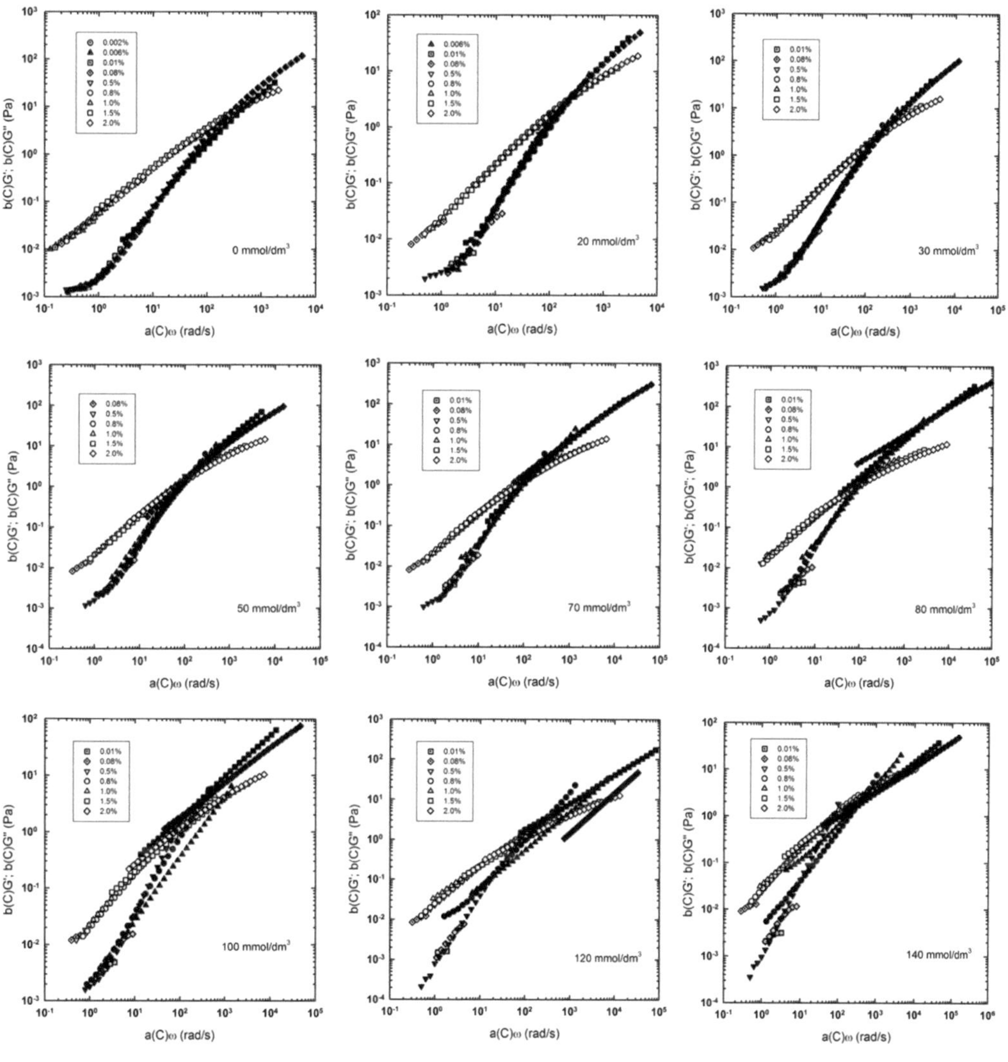

Figure 9. Master curves for the modified dynamic moduli as a function of the modified angular frequency for all λ-carrageenan solutions and different concentrations (mmol/dm³) of added Na^+. The shift factors were determined as explained in the text. The complete range of total ionic strength is 0.024–94 mmol/dm³.

$a(C)$ modifies the x-axis, ω, or $\dot{\gamma}$, and the factor $b(C)$ the ordinate axis, $G'(\omega)$, $G''(\omega)$, or $\eta\ (\dot{\gamma})$. Here, (C) indicates that factors depend on polysaccharide concentration. These factors were determined by choosing 0.5% λ-carrageenan as a reference. The factors applied to the curves are the following:

$$a(C_1) = \frac{1}{n}\sum_{\omega_{max}}^{\omega_{min}} \frac{G'(C)}{G'(0.5\%)};\ b(C_1) = \frac{G'(0.5\%)\forall\omega}{G'(C)\forall\omega} \tag{6}$$

$$a(C_2) = \frac{1}{n}\sum_{\omega_{max}}^{\omega_{min}} \frac{G''(C)}{G''(0.5\%)};\ b(C_2) = \frac{G''(0.5\%)\forall\omega}{G''(C)\forall\omega} \tag{7}$$

The modulus G' is superposed by applying $a(C_1)$ to each frequency and $b(C_1)$ to G'. The same procedure is followed for the loss modulus using $a(C_2)$ and $b(C_2)$. In **Figure 9** the master curves of the dynamic moduli are presented. A proper overlapping of the loss moduli for concentrations of added Na^+ in the range 0–70 mmol/dm^3 is obtained along six decades of modified frequency. However, for higher added Na^+ concentrations (80–140 mmol/dm^3), the highest concentrations of polysaccharide show some deviation from the reference curve. This same behavior is observed for the storage modulus for 70 mmol/dm^3 added Na^+. Superposition was difficult for high concentrations of added Na^+ (100–140 mmol/dm^3). The more the values of $\tan\delta = G''(\omega)/(G'(\omega)$ approach one, the more difficult is to produce a master curve for $G'(\omega)$. This behavior is only observed when the polysaccharide concentration is high (1.5 and 2.0%), but not when it is 0.5%.

The following shift factors are used to superimpose the flow curves:

$$a(C) = \frac{1}{n}\sum_{\dot{\gamma}_{max}}^{\dot{\gamma}_{min}} \frac{\eta(C)}{\eta(0.5mM)};\ b(C) = \frac{\eta_0(0.5mM)}{\eta_0(C)} \tag{8}$$

As it happens with the mechanical spectra, the factor $a(C)$ modifies the shear rate and $b(C)$ the viscosity. The reference concentration was 0.5% because it was possible to observe the zero-shear and transition regions together with the shear-thinning zone. As shown in **Figure 10**, good superpositions are obtained for the flow curves from concentrations in the range of 0–70 mmol/dm^3 added Na^+ along seven decades of modified shear rate. As it happens with the master curves for $G''(\omega)$, it is hard to superimpose the flow curves for the highest polysaccharide concentrations (1.5–2.0%) with appropriately calculated shift factors along thirteen decades of modified shear rate. It was easier to superimpose the dynamic moduli and flow curves when systems were in the dilute regime; without added Na^+, $C^* = 0.21\%$, and 20 mmol/dm^3 added Na^+, $C^* = 0.38\%$, and low concentrations of added Na^+. It is possible to assume that for high polysaccharide concentrations (1.5–2.0%) with added Na^+, the dynamic moduli and viscosity increase as compared with systems without added Na^+ due to the entanglement of λ-carrageenan chains. This behavior is not observed for the reference concentration (0.5%) until concentrations higher than 100 mmol/dm^3 added Na^+. This situation may explain the difficulty to superimpose the curves of such systems.

http://dx.doi.org/10.5772/65785

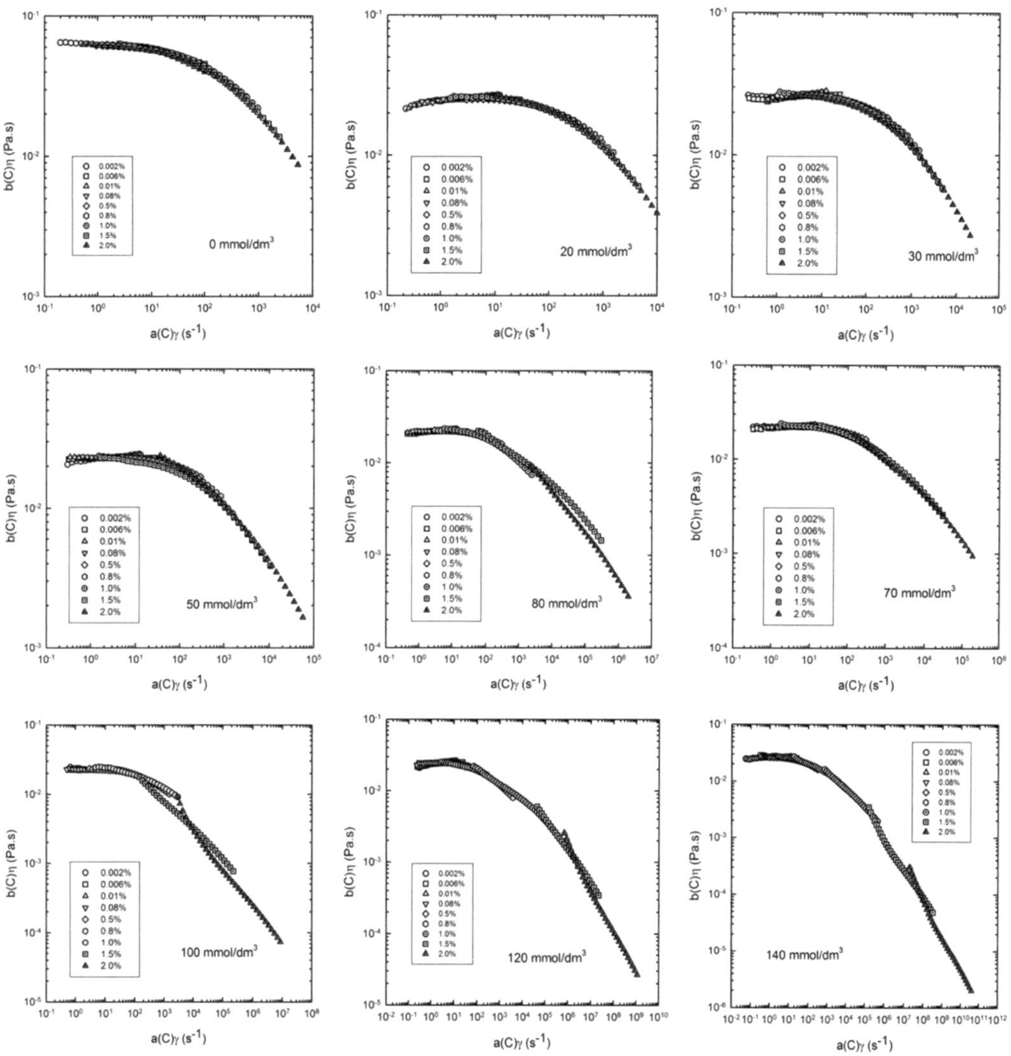

Figure 10. Master flow curves for all λ-carrageenan solutions and different concentrations (mmol/dm^3) of added Na^+. The shift factors were determined as explained in the text. The complete range of total ionic strength is 0.024–94 mmol/dm^3.

4. Conclusion

The results of this investigation make evident the high sensitivity of λ-carrageenan to the ionic strength of the aqueous environment, mainly given by the added Na^+. This sensitivity is primarily attributed to the polyelectrolyte character of the polysaccharide, but hydration effects and competition for the solvent between the polyanion and sodium counterions can also play a role. The solutions of the commercial preparation of λ-carrageenan without and with added Na^+ are viscoelastic fluids with a dominant viscous behavior. However, with added Na^+ the elastic character becomes more important when the polysaccharide concentration increases. In the dilute regime, the viscous character is considerably more sensitive to the addition of Na^+. In the semidilute regime, the opposite occurs as the addition of Na^+ affects more the elastic nature. Also, with added Na^+, it is possible clearly to distinguish three types of behaviors of the apparent viscosity and the dynamic moduli. In the dilute regime and the low-concentration region of the semidilute regime, moduli decrease drastically when the total ionic strength increases slightly over that without added Na^+ and then remain substantially constant regardless of the increase in the total ionic strength. In the semidilute regime for moderate carrageenan concentrations, the moduli reach a minimum for intermediate total ionic strengths and then recover when the total ionic strength increases without necessarily achieve their values without added Na^+. Finally, in the same semidilute regime but for large concentrations of polysaccharide, the moduli decrease as the total ionic strength increases slightly over that without added Na^+ and show a high recovery, which in the case of the storage moduli and the apparent viscosity significantly exceed their level without added Na^+.

Considering all this evidence, the existence of different levels of structural organization of the polysaccharide chains together with their close interactions with them and with the solvent and the added Na^+ mainly can be postulated. Therefore, the results presented here allow the polyelectrolyte behavior of λ-carrageenan to be better understood for a significant range of polysaccharide concentrations under broad conditions of total ionic strength. These characteristics take relevance in the case of commercial preparations that under normal circumstances are used without further treatments, e.g., separation of accompanying counterions to produce a particular salt form of the polysaccharide. Such purified forms are used for more fundamental studies and would be the next step in the investigation of the flow properties of the polysaccharide.

The authors would like to acknowledge the financial support of Programa de Apoyo a la Investigación y el Posgrado (PAIP) of Facultad de Química-UNAM (Grant 5000-9098).

Author details

Andrea Rivera del Rio, Mariana Ramírez-Gilly and Alberto Tecante*

*Address all correspondence to: tecante@unam.mx

Department of Food and Biotechnology, Faculty of Chemistry, National Autonomous University of Mexico, Cd. Universitaria, CdMx, México

References

[1] Niu TT, Dong-Sheng Z, Hai-Min C, Xiao-Jun Y. Modulation of the binding of basic fibroblast growth factor and heparanase activity by purified λ-carrageenan oligosaccharides. Carbohydrate Polymers. 2015;125:76–84. DOI: 10.1016/j.carbpol.2015.02.069

[2] Chen H, Wang F, Mao H, Yan X. Degraded λ-carrageenan activates NF-κB and AP-1 pathways in macrophages and enhances LPS-induced TNF-α secretion through

AP-1. Biochimica et Biophysica Acta. 2014;1840(7):2162–2170. DOI: 10.1016/j.bbagen.2014.03.011

[3] McHugh DJ. Carrageenan: A Guide to the Seaweed Industry. FAO Fisheries Technical Paper 441. [Internet]. 2003. Available from: http://www.fao.org/docrep/006/y4765e/y4765e0a.html [Accessed: 2016-05-30]

[4] Commission of Codex Alimentarius. GSFA Online. Carrageenan (407). [Internet]. 2012. Available from: http://www.codexalimentarius.net/gsfaonline/additives/details.html?id=49&lang=es [Accessed: 2016-05-30]

[5] Code of Federal Regulations. Title 21: Food and Drugs, Part 172 Food Additives Permitted for Direct Addition to Food for Human Consumption, Subpart G—Gums, Chewing Gum Bases and Related Substances. §172.620 Carrageenan [Internet]. 2013. Available from: www.wcfr.gov [Accessed: 2016-05-30].

[6] van de Velde F, De Ruiter GA. Carrageenan. In: Steinbúchel A, Rhee SK, editors. Polysaccharides and Polyamides in the Food Industry. Weinheim: Wiley–Blackwell; 2005. pp. 85–111.

[7] Eliasson A. Carbohydrates in Food. 2nd ed. Boca Ratón: CRC Press; 2006. pp. 244–252.

[8] De Lestang Bremond G, Quillet M, Bremond M. λ-Carrageenan in the gametophytes of *Chondrus crispus*. Phytochemistry. 1987;26(6):1705–1707.

[9] Slootmaekers D, van Dijk JAPP, Varkevisser FA, Bloys van Treslong CJ, Reynaers H. Molecular characterization of κ- and λ-carrageenan by gel permeation chromatography, light scattering, sedimentation analysis and osmometry. Biophysical Chemistry. 1991;41:51–59.

[10] Noseda MD, Cerezo AS. Room temperature, low-field ^{13}C-N.M.R. spectra of degraded carrageenans: Part III. Autohydrolysis of a lambda carrageenan and of its alkali-treated derivative. International Journal of Biological Macromolecules. 1993;15:177–181.

[11] Camacho MM, Martínez-Navarrete N, Chiralt A. Influence of locust bean gum/λ-carrageenan mixtures on whipping and mechanical properties and stability of dairy creams. Food Research International. 1999;31:653–658.

[12] Lizarraga MS, De Piante D, González R, Rubiolo A, Santiago LG. Rheological behaviour of whey protein concentrate and λ-carrageenan aqueous mixtures. Food Hydrocolloids. 2006;20:740–748.

[13] Bayarri S, Chuliá I, Costell E. Comparing λ-carrageenan and an inulin blend as fat replacers in carboxymethyl cellulose dairy desserts. Rheological and sensory aspects. Food Hydrocolloids. 2010;24:578–587.

[14] Cross MM. Rheology of non-Newtonian fluids: a new flow equation for pseudoplastic systems. Journal of Colloid Science. 1965;20:417–437.

[15] Bird RB, Armstrong CR, Hassager O. Dynamics of Polymeric Liquids. Vol. 1 Fluid Mechanics. New York: John Wiley & Sons Inc.; 1987. 670 p.

[16] Graessley W. The entanglement concept in polymer rheology. Advances in Polymer Science. 1974;16:164–179.

[17] Almutairi FM, Adams GG, Kök MS, Lawson CJ, Gahler R, Wood S, Foster TJ, Rowe AJ. An analytical ultracentrifugation based system study on the conformation of lambda carrageenan in aqueous solution. Carbohydrate Polymers. 2013;97:203–209.

Chapter 12

Chitosan-Based Sustainable Textile Technology: Process, Mechanism, Innovation, and Safety

Jagadish Roy, Fabien Salaün, Stéphane Giraud,
Ada Ferri and Jinping Guan

Additional information is available at the end of the chapter

http://dx.doi.org/10.5772/65259

Abstract

This chapter reviews relevant findings regarding the activities and contributions of chitosan in different textile processing following the varieties of process, mechanism, and applications. Chitosan is a better candidate in both aspects of biodegradability and efficiency instead of synthetic polymers. The technical and scientific discussions behind the role of chitosan in all the processes and treatments have been explored in the chapter. Over the last few years, enormous efforts and challenges are being practiced in research and industry to design and development of eco-friendly and sustainable technologies. Therefore, the chapter emphasizes on chitosan-based formulations of fibers, fabrics, coatings, and functional textiles.

Keywords: chitin, chitosan, textile, fiber, fabric, antibacterial, nanofibers, electrospun

1. Introduction

The first discovery of chitin was traced by Braconnot, a French professor, in 1811 in research on the mushroom. Later on, Rouget discovered the solubility test through chemical and thermal treatment of chitin fiber in 1859 [1]. Ledderhose identified the chitin molecular structure, which consists of glucosamine and acetic acid, in 1878 and Hoppe-Seyler announced the name 'chitosan' of the modified chitin in 1894.

Chitin is a very common bio product produced naturally in the exoskeleton of crustaceans and insects (spiders, shrimps, crabs, and lobster), the radulae of mollusks and the cell wall of mushrooms, algae, and fungi. The amount of chitin produced naturally each year is 10^{10} ton of which 70% comes from oceans. Therefore, the by-product of the fishing industry is enough to support the huge commercial demand of chitin and chitosan with a long-term potential supply. The main advantages of using chitosan are biodegradable [2, 3], biocompatible [4, 5], and nontoxic because it degrades as sugar after being metabolized.

Chitosan is a biomaterial which possesses the superior mechanical property and very diverse biological activity. The chitosan-derived products have a slow-growing market for poor reproducibility in product development. The first- and second-generation chitosans have been used as a biomaterial and biological functionalities for wastewater treatment and agricultural applications, but the scenario alters now. The third phase of chitosan exhibits strong relationship in molecular structure-function. Up to date, studies and investigations are approaching to understand the cellular mode of action to reveal every biological and biomedical activity of chitosan which will open the door for new material applications. Chitin and chitosan have very diversified use in applications (**Table 1**) and in the form of nanofibers, membranes, micro-/nanoparticles, scaffolds, beads, hydrogels, and sponges. Due to the growing concern about health, environment, and economics, enzymatic synthesize of chitosan is continuously studied as an alternative to the hazardous process and nonspecificity for chitosan fictionalization. Besides, the sustainability and safety in textile technology are being practiced and equipped by innovating new process, machinery, and designs from raw materials to the end of life cycle.

Application	System
Antibacterial cotton fibers	Fiber-reactive chitosan derivative
Cell culture	Chitosan-based nanofiber
Biodegradable and biocompatible nonwoven wound dressing	Chitin and chitosan nanofibers [spinning solvent: 1,1,1,3,3,3-hexafluoro-2-propanol (HFIP)]
Bioactive and biodegradable Wound dressing	Dibutyrylchitin fibers

Table 1. Chitosan fibers and their applications.

2. Chemistry of chitin and chitosan

According to the chemical structure of chitin (**Figure 1**) and chitosan (**Figure 2**), they are the modified form of cellulose (**Figure 3**) in which an acetamide and an amine group takes the place of C-2 hydroxyl group, respectively. Chitosan is derived from natural chitin which consists of *N*-acetyl glucosamine and glucosamine units distributed randomly in a linear polysaccharide chain. In nature, it is very rear to exist 100% acetylated amino groups and 100% deacetylated free amine groups. Therefore, less than 10% degree of deacetylation (DD) may exist in chitin structure, while 40–98% DD is found for chitosan. The average molecular weight of chitin and chitosan mainly depends on the degree of deacetylation and degree of polymerization. Besides, the achieved properties from different applications and products vary according to the two parameters. The polymorphic structures of chitin and chitosan change due to different packing and orientation of polysaccharide chain. Every individual chain undergoes a full rotation along 10.1–10.5 Å of the chain axis due to the presence of chiral glycosidic units. These units link between C-1 oxygen atoms of one unit and C-4 of the adjacent unit in parallel and anti-parallel directions. The characterization of those structures using X-ray diffraction and NMR spectroscopy reveals three types of allomorphs: α, β, and γ form where a higher percentage of α form exists [6].

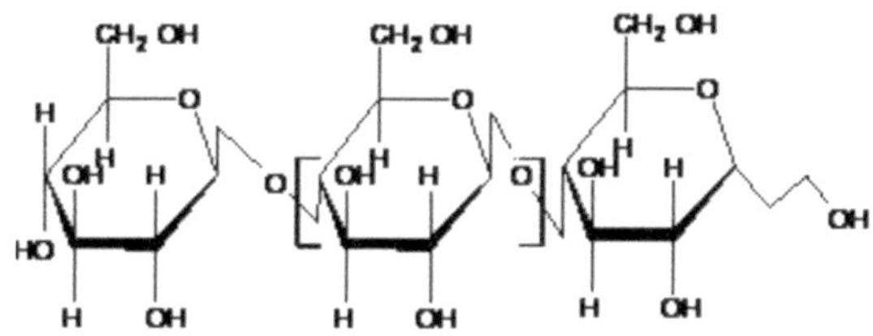

Figure 1. Molecular structure of chitin.

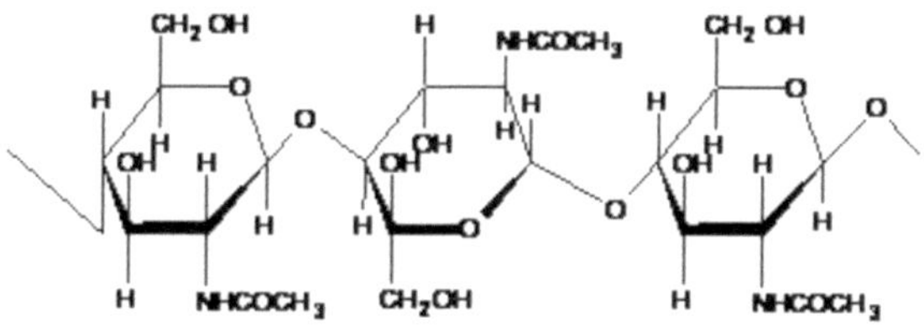

Figure 2. Molecular structure of chitosan.

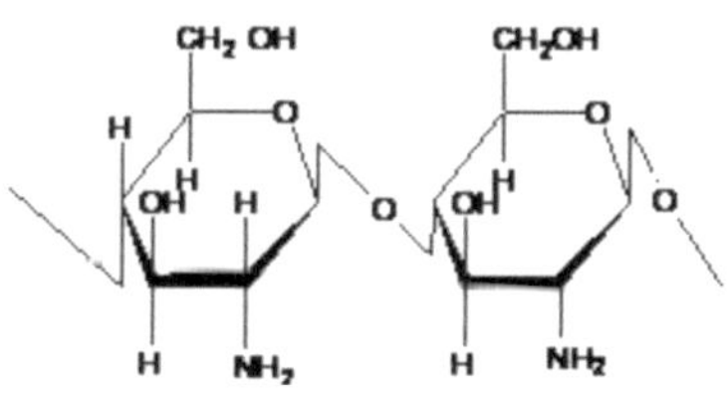

Figure 3. Molecular structure of cellulose.

The physico-chemical behaviors of chitin and chitosan are quite different even though they show many similarities in the structures. Due to the presence of acetamide group, chitin mostly shows an inert behavior, while chitosan is highly reactive for free active amine groups. Also, the behavior also attributes to the degree of crystallinity because chitosan is less crystalline compared to chitin. Moreover, there are very few solvents allowing the dissolution of chitin, whereas the chitosan dissolve in almost all aqueous acid solvents. The chitosan carries positive charges after interacting with acid for the presence of free amino groups and dissolves at pH ≤ 6.0. Over the pH 6.0, more than 50% of dissolved chitosan molecules lose its positive charges and become insoluble.

3. Chitosan-based fiber production

3.1. Spinning process

3.1.1. Wet spinning

3.1.1.1. Introduction

Chitin is an acetylated form of polysaccharide. The presence of acetyl groups increases the interchain forces and the percentage of crystallization which leads to a better dry and wet strength compared to semi-crystallize chitosan fibers. Besides, the chitosan fibers having higher moisture regaining property show poor performance in developing the desired strength.

The most popular method for chitin and chitosan fiber production is a wet spinning method. In this process (**Figure 4**), the fibers of chitosan are produced using viscous chitosan solution by the extrusion process in a coagulation bath. The chitosan dissolution occurs in 1–10% acetic acid below the pH 6, and a viscous dope is formed. It passes through a candle system filtration unit to remove the undesired impurities and a reservoir for degassing under vacuum around 30 mbar for 5 h to confirm the complete removal of the air bubble. Then, the extrusion process starts to obtain fibers where the extrusion chamber is comprised of a reservoir, a metering pump, and a spinneret. The chitosan dope is extruded less than 1.5 bar pressure into a coagulation bath. The coagulation bath contains an aqueous solution of the coagulant such as NaOH [7], KOH [8], cupric ammonia [9], alcohol/calcium chloride/acetate [10], NaOH–Na_2SO_4, NaOH–AcONa [11], NaOH-40% methanol, $CuSO_4$–NH_4OH, $CuSO_4$-concentrated ammonia [12], etc. The presence of tailors mentioned above chemicals acts as coagulation retardants. In this process, the take-up rollers, drawing system, drying rollers, and winding up are optimized according to the production rate and required properties of fibers. The excess amount of coagulant on fiber is washed off using distilled water or aqueous methanol and ethanol. This washing bath is also called predrying bath. Some additional physical and chemical processes are practiced to enhance the desired mechanical property in drying treatments.

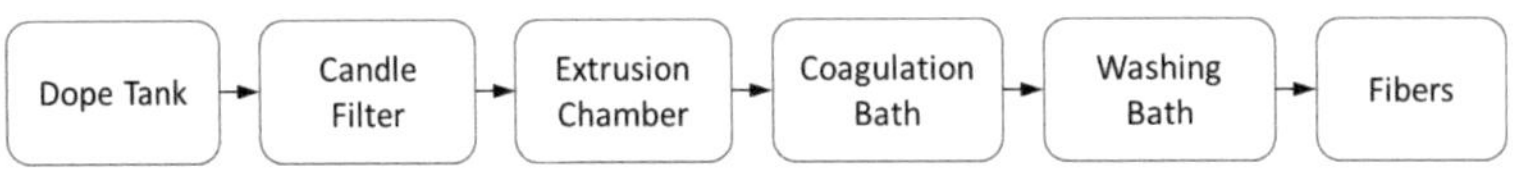

Figure 4. A typical block diagram of wet spinning.

3.1.1.2. Solvent system

A proper selection of solvent system and coagulation bath leads to produce desired fiber structures and properties. Many researchers have studied (**Table 2**) different types of solvent systems such as dichloroacetic acid-isopropyl ether-formic acid, trichloroacetic acid, dimethy-lacetamide-lithium chloride; a mixture of trichloroacetic acid-chloral hydrate-dichloromethane

http://dx.doi.org/10.5772/65259

(DCM) is used for dissolving chitin. Besides, the pH-sensitive behavior of chitosan is very useful in the processing of fiber. Acetic acid is efficient enough to dissolve chitosan, a simple solvent system, especially <pH 6. Moreover, coagulation bath containing NaOH is easier to prepare and strengthens the chitosan fibers within a short time.

Chitin dissolution (solvent system)	Chitosan dissolution (solvent system)
Trichloroaceticacid (40%)-chloral hydrate (40%)-methylene chloride (20%)	Dilute aqueous organic acid (acetic acid, formic acid)
LiCNS/Ca(CNS)$_2$/CaI$_2$/CaBr$_2$	Dimethylsulfoxide/*N*,*N*-dimethylformamide (DMF)
Lithium thiocyanate	p-Toluene sulfonic acid
Formic acid-dichloroacetic acid-isopropylether	Guanidine hydrochloric acid-urea
Sodium *N*-acylchitosan xanthate [*O*-(sodium thio) thiocarbonyl *N*-acylchitosan]-xanthate [*O*(sodium thio) thiocarbonyl cellulose]	Ionic liquid, 1-butyl-3-methylimidazolium bromide ([BMIM][Br])

Table 2. Solvent system for the dissolution of chitin and chitosan.

3.1.2. *Dry spinning process*

The chitosan used for dry spinning [13] should have the initial concentration over the critical concentration of chain entanglement which is equal or more than 0.5% (w/w). Acetic acid is frequently used for the dispersion of chitosan. Similarly, the above-mentioned process is followed up by the extrusion process. During the coagulation process, 125 ml of concentrated ammonia (20% w/w) in the gaseous state comes in contact with chitosan monofilament. The flow of gaseous ammonia is around 0.24 m^3/h for 7.5 s based on the relation between extruder and speed of the first roller around 3.4 m/min. The stretching is done by maintaining 14% of the maximum first roller speed that produces a straight fiber line during coagulation. It is important to maintain the speed exactly to avoid un-stretching. A second roller having higher speed is used to maintain a stretching ratio of 1:12 to the produced monofilament. In the drying process (**Figure 5**), the hot air around 110°C is passed through a 65 cm oven, while the air flow of 1.5 m^3/h is used and the retention time of monofilament in the oven is around 10 s. Then, the chitosan fiber is treated in the wet air circulating at 25°C and water content is 55% w/w for one week before final storage [13].

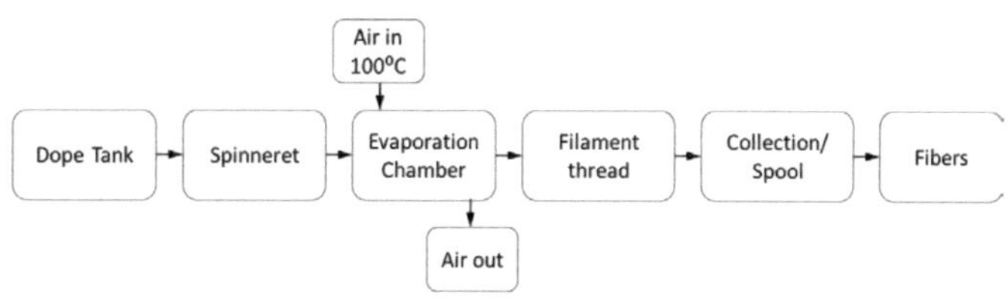

Figure 5. A typical block diagram of dry spinning.

3.1.3. Electrospun

In 1934, Antonin Formhals applied for a patent to manufacture textile yarn. The fabrication process is called electrostatic spinning or electrospinning. The central principle of this process is to apply an electrostatic potential to a polymer solution to overcome the surface tension which results in the formation of the nanofibers. An external high voltage source delivers the required electric potential in this process. The first step of the process (**Figure 6**) is to prepare the chitosan solution by placing in a pipette, which is charged by a positive electrode of the source. As a result, a repulsive force is developed that causes splitting of a charged jet from the pendant polymer droplet at the pipette tip. The single jet splits into multiple jets due to radial-charge repulsion which is driven electrically to the collector and neutralizes. The multifilament fluid jet dries, while the solvent evaporates as soon as the mixture comes out of the pipette tip. The solidified polymers are deposited in the collector.

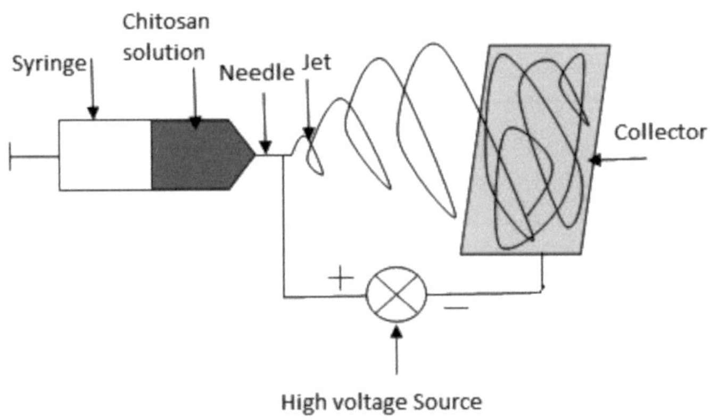

Figure 6. A typical diagram of electrospinning process.

In this process, solvent plays the main role in the spinnability of chitosan. Many physical and chemical properties of chitosan such as solution viscosity and conductivity affect the rheology of chitosan fibers. Many studies have been conducted for choosing suitable solvents such as acetic acid, dilute hydrochloric acid, formic acid [14], dichloroacetic acid, trifluoroacetic acid (TFA), etc., for the dissolution of chitosan. The understanding of solvent-chitosan interaction and the effect of solvent ionic strength, pH, and viscosity would help on the proper selection of the solvent.

The main problem of chitosan is its higher viscosity at higher concentration. For example, above 2% (w/w) of chitosan is highly viscous and loses its ability to flow. In contrast, low concentration of chitosan does not contain enough material to obtain solid and stable fibers. In the molecular level, every chitosan contains $-NH_2$ and $-OH$ groups which interact promptly at high viscous level. Therefore, the interchain interactions are higher among the chitosan chains at a higher concentration which results in a three-dimensional network formation and highly viscous gel formation [15]. The trifluoroacetic acid (TFA) is a proposed solvent that could dissolve around 8% of pure chitosan which is perfectly spinnable. A mixture of TFA and dichloromethane provides fibers in a range of 200–650 nm diameter [16].

3.2. Mechanical properties of fibers

The electrospinning process provides nanofibers for nonwoven. The diameter of fibers, uniform deposition at the nanolevel, thickness of the fibers, and bead formation are essential to validate a process. Concerning these morphological and structural properties, the tensile strength and tenacity are crucial parameters to quantify the quality of chitosan fibers. The chitosan fiber was obtained with 82 g/denier modulus of elasticity while 2 g/denier tenacity at a final dry jet stretching ratio of 3.7 [17]. Besides, chitosan fiber having the dry tenacity of 2.1 g/denier was obtained by the wet spinning process [8]. Ionic liquid, [Bmim]Cl (1-butyl-3-methylimidazolium chloride), is used as a solvent for the fabrication of the pure chitosan fibers by both wet and dry-wet processes. The dry fiber exhibits the strength and modulus at 1.5 and 71 cN/dtex, respectively, for the wet process, while 2.1 cN/dtex and above 3.5 cN/dtex are, respectively, found for the dry-wet process. The results are attributed to the orientation of the crystallites, which is increased by 40% due to the dry-jet-stretched ratio and inter-fibrillar amorphous zone restricting the inter-fibrillar sliding that causes plastic deformation [17]. Moreover, the polymorphism and crystallinity of chitosan membrane depend on the used solvent, the processing conditions, the form of dry or wet [18]. Synchrotron X-ray diffraction experiment confirms two types of allomorph existing in chitosan—tendon and annealed forms [17, 18]—depending on stretching ratio. In addition, the chitin fibers are also influenced by its polymorphism; α-chitin is profoundly hard, while β- and γ-chitin show superior quality on toughness and elasticity [19]. A very poor tensile strength is observed for the chitosan fibers, which are obtained after spinning process and exist in wet condition. Different processes are applied to improve mechanical strength such as epichlorohydrin in coagulation bath, glutaraldehyde in posttreatment, phosphate and phthalate ions containing the solution.

Physical, chemical, and mechanical characters of chitosan fibers and the ability to build in different forms facilitate chitosan nanofibers for the filtration process, for example, metal and dyes separation from water [20, 21], capturing of virus in air [22], etc. In addition, the technological developments in textile processing associated with chitin and chitosan would impact on reinforcing microfibers composite [23], dressing for wound skin [15, 24–26], carrying and delivering drugs [22, 27], tissue engineering [15, 28–30], vascular grafts [31], etc.

3.3. Chitosan blends for electrospinning process

Apart from the pure chitosan processing, many studies have reported on mixtures of chitosan with different polymers such as polyvinyl alcohol (PVA) [32], polyacrylamide [33, 34], polyethylene oxide (PEO), polyacrylonitrile (PAN) [35], caprolactone [14], and Zein (an agro-based protein) [36] to modify the chitosan solution property and facilitate the spinnability by easy flowing through the capillary tip. For example, the viscosity of chitosan solution decreases with the increasing amount of PEO. The main reasons for the change in viscosity is the reduction of intra- and intermolecular attraction among the chitosan chains and the formation of new bonding of PEO with the backbone of chitosan. The presence of PEO on chitosan structure interferes the intra- and intermolecular association, while new hydrogen bonds are formed between –OH group of chitosan and water. It results in a drastic reduction in the viscosity of chitosan solution [15]. Furthermore, PVA also facilitates interaction with chitosan in molecular levels that restrict the intra- and intermolecular interaction among chitosan chains and provide a better spinnable solution [16]. Moreover, inorganic nanoparticles such as sodium chloride, potassium chloride, or ions such as calcium and iron are used in doping chitosan in the electrospinning process [33]. The metal ions also help to break down the intra- and intermolecular hydrogen bonds and reduce the viscosity of chitosan solution that would enhance the chain entanglements. As a result, lower number of beads and reduced diameter of fibers are obtained from chitosan electrospinning process [33]. Some processes of chitosan blends and mechanical properties have been demonstrated below.

3.3.1. Nylon-6/chitosan blends fiber

Two solvents, 1,1,1,3,3,3-hexafluoroisopropanol (HFIP) and formic acid at a ratio of 90/10, have been used to dissolve nylon-6 and chitosan for the electrospinning process. The total weight of polymer has been maintained 6% by varying the concentrations of both polymers. The similar electrospinning (Section 3.1.3) process has been operated at room temperature [37]. The properties of the blend fibers depend on the blending weight ratio. The increased amount of chitosan increases the functional sites for capturing water molecules. A unique molecular interaction between nylon-6 and chitosan is formed due to developing a hydrogen bond which affects the mechanical property of the blend fibers. The tensile strength and crystallization behavior decrease due to the formation of an amorphous polymer during the electrospinning process. For example, 25% of chitosan in the blends shifts the melting point at 258.1°C, while pure nylon-6 exhibits the temperature at 268°C [37].

3.3.2. Chitosan-agarose blend fibers

Chitosan and agarose, one kind of red algae, are dissolved in a mixture of solvents, trifluoroacetic acid (TFA) and dichloromethane (DCM). The polymer has been maintained at 7%, while the solvent ratio between TFA and DCM is 7:3. In the electrospinning process, the operating electrostatic potential is 15 kV/12 cm and the solution flow rate is 0.5 ml/h. The blending contents of 30 and 50% agarose provide fibers with smooth and uniform surfaces. The main advantage of blending of chitosan and agarose is the production of nanosized fibers. The achieved fiber diameter is 520 ± 35 nm and 140 ± 9 nm for using 30 and 50% agarose, respectively. Besides, the pure chitosan fibers from electrospinning process exhibit around 1.76 ± 0.59 μm in diameter [38].

3.3.3. Poly lactic-co-glycolic acid (PLGA)-chitosan-PVA blends

The PVA and chitosan are dissolved in 2 wt.% acetic acid separately and blends at a ratio of 60:40. The mixture is used for the electrospinning process. Besides, PLGA of 6 wt.% is dissolved in a mixture of solvents, tetrahydrofuran (THF) and *N,N*-dimethylformamide (DMF). The electrospinning process is modified by adding an extra tip or syringe parallel to another tip for holding two polymer solutions separately. The collector collects the spinning fibers from both tips. In this process, chitosan/PVA and PLGA solutions are used separately in these tips and run the electrospinning process. The tensile strength values of three different electrospun nanofibers—PLGA, chitosan/PVA, and PLGA-chitosan/PVA—achieved by electrospun are 7.3, 4.3, and 2.6 MPa, respectively. The reason behind the result is no interaction present between PLGA and chitosan/PVA. Besides, PLGA-chitosan/PVA and chitosan/PVA nanofibers are crosslinked separately using 25% glutaraldehyde. The tensile strength of crosslinked PLGA-chitosan/PVA is found higher compared to uncrosslinked nanofibers. In contrast, the cross-linked chitosan/PVA nanofibers have lower tensile strength than that of uncrosslinked nanofibers. In the case of crosslinked PLGA-chitosan/PVA, some bonds are formed after crosslinking with glutaraldehyde [39].

4. Chitosan-based textile processing

4.1. Chitosan: a sizing and desizing agent for textile pretreatment

Sizing is the conventional process of fiber pretreatment for the resistance of warp breakage caused by highly stressed weaving process. In general, a sizing agent adheres to the surface of yarn and forms a coating. It provides the required strength and hydrophilicity for a successful weaving process. Polyvinyl alcohols and acrylics including all synthetic sizing agents need to be washed out at the end for a smooth finishing process, to avoid complications in the subsequent steps of fibers or fabric treatment. Chitosan is a better candidate in both aspects of biodegradability and efficiency, while synthetic agents are not. Chitosan is soluble in pH less than 6.0 since the pKa is 6.5 where more than 50% of chitosan molecules deprotonated and become insoluble.

Chitosan is hydrolyzed by concentrated hydrochloric acid 0.1 N [40], 0.2 N [41], 0.5–1 N [42] at high temperature (30–90°C) to achieve soluble chitosan which exhibits lower apparent viscosity. The temperature is associated with the activation energy, which catalyzed the hydrolyzing action of HCl. Besides, the content of nitrogen reduces from 7 to 6.5% since the hydrolysis of chitosan continues around 60 min. It attributes to chain scission at 1–4 glycosidic bonds of chitosan for the regular attack of concentrated HCl resulting in the lower molecular weight. In addition, the dissolution of short chain chitosan shows lower and stable apparent viscosity to 110 MPa s after 60 min probably due to the retention of crystallinity after degradation. The conformational change is attributed to the result of a transformation rod to coil transition [40]. In general, the sizability and desizability processes are expressed as size add-on and removal percentages, respectively. The add-on proportion of size mainly depends on the apparent viscosity and the solid content of size materials in the liquor. The tensile strength is found higher after drying the size coating by heat treatments, while the elongation at break decreases significantly. The effects attribute to the diffusion and deposition of the size materials in the fiber thread which increase the brittleness and rigidity. On the other hand, the removal efficiency of hydrolyzed chitosan is found 100% only with water at 90°C. Carboxymethylated chitosan, a modified form of chitosan, is readily soluble in water at neutral pH and could be used as sizing agent [42]. On the other hand, in combination with wax or starch by an extrusion process, a form of sizing agent is obtained [43]. The viscosity of chitosan solution increases with increasing chitosan concentration and limiting its performance as sizing agent arises from the high viscosity of chitosan at room temperature above 2% in the acidic medium. Carboxymethylated chitosan is achieved via hydrolysis using 0.5 N hydrochloric acid at a minimum of 90°C for 2 h. The $NaCO_3$ is used for the neutralization, and then, the filtration with washing is necessary for removing the sodium chloride salt. The final form is used as sizing agent [42]. Apart from the carboxymethylated chitosan, the addition of wax with the modified chitosan provides a reduction in surface tension which results in improved wetting property including adhesion behavior of modified sizing agent [43]. The viscosity is reduced by 0.5% of pure chitosan after the small proportion of wax addition. Moreover, the sizing film also reduces the friction between metal and fibers resulting in the better performance in the weaving process. Some modified chitosans such as *N*-[(2-hydroxy-3-trimethyl ammonium) propyl] chitosan chloride (HTACC) and *N*-[(4-dimethyl aminobenzyl)imino] chitosan (DBIC) have been developed for better performance of solubility and pretreatment of polyester fabric [44].

4.2. Chitosan: an auxiliary in textile dyeing process

The dyeing process is a major step in fabric processing, especially for consumer esthetic test. In this process, dye molecules diffuse and interact with functional sides of fibers or fabrics. Therefore, various types of fibers or fabrics made of cotton [45], polyester [46], rayon [46], nylon [47], wool [48, 49], silk [50], and composites [51] are processed using different functional dyes such as sulfur dye, vat dye [46], reactive dyes [45, 46, 52–54], etc. Therefore, the methodological variations are found based on dyes and fibers/fabrics such as Jig dye, Pad dye, and Jet dye [46] for the dyeing process. All methods require quite high energy, large amount of water and salts until the dye molecules distribute and diffuse uniformly. In addition, these processes release a high amount of complex effluents containing all chemicals together. Therefore, the loss of untreated dye and salt and highly efficient effluent treatment all have very high equivalent cost. Moreover, the drawbacks also impact on aquatic environment-related health. The use of chitosan or modified chitosan has beneficial effects on dyeing process which reduces the amount of untreated dye in the bath without using any salt. As a result, a simple effluent treatment would be enough to process released effluent from dye house.

In the dyeing process of cotton fibers or fabrics, reactive dyes are the mostly used due to its better-wet fastness, brilliance, hue variability, easy processability, and applicability [55]. In addition, reactive dye molecules form covalent bond through reaction with hydroxyl groups of cellulose via nucleophilic addition or substitution reaction. In general, reactive dyes and cotton fabrics or fibers hold anionic groups which affect the reaction due to charge repulsion in the dye bath. As a result, large quantities of salts are used for the screening of charges in dying process. A comprehensive and comparative optimization process studies have been conducted by Ramadan et al. [56] regarding dyeing treatment of chitosan-treated cotton fabrics. A cotton fabric is immersed in an aqueous solution of 2% chitosan and 50 mg/100 ml sodium periodate ($NaIO_4$). The liquor ratio is maintained at 1:50 for 60 min when the dye bath temperature is kept constant at 60°C. The treated cotton fiber is washed with 1% aqueous acetic acid solution and rinsed with water for several times, and the drying temperature was maintained at 60°C. Two steps are followed for the fabric treatment with chitosan to compare the dyeing and mechanical properties of the treated fabrics. The underlying mechanism of the process is to create aldehyde groups in the molecular structure of chitosan at C1 through the action of $NaIO_4$. In addition, C3 and C4 positions hold secondary hydroxyl groups; position C6 contains primary hydroxyl and C2 holds the amino groups. Moreover, the chain scission occurs at C3, C6, C2 positions and 1–4 glycosidic linkage under the attack of oxidizer [57]. The IR analysis confirms the presence of aldehyde and carboxyl groups in the molecular structure of treated cotton, which interacts with chitosan strongly. There is a possibility to degrade the fabrics due to oxidizing, but the presence of chitosan may reduce this drawback. The presence of chitosan also protects the fabric from degradation against the attack of the oxidizer and maintains similar mechanical properties (i.e., tensile strength, elongation at break) to the untreated bleached fabric [56]. In addition, highly soluble, uniform structure, low viscous, and better film-forming chitosan is achieved from the fragmented chitosan [58].

In the second step, the dyeing process is conducted using the dried cotton fabric in 1% reactive dyes while 5% sodium chloride in a liquor ratio of 1:50 at room temperature. The temperature of the dye bath is increased at 60°C and left it to be stable for 45 min. After the process, the fabric is rinsed with water for several times and treated with 1% wetting agent using a liquor ratio of 1:50 for half an hour at 60°C. Several washes are required using hot and normal water after the treatment and dried at room temperature. The enhanced color yield (K/S) is observed compared to the untreated fabric due to the Coulombic attractions between positively charged cotton surface modified by chitosan and anionic dye molecules (reactive and acid).

4.3. Chitosan: a binder in printing process

Fabric printing process requires a dye or a pigment, a binder and thickener. The main function of binder is to maintain a stable viscosity, enhances the droplet formation, provides adequate adhesion to the textile surface by forming a film, and binds the color or pigment molecules. In the industrial application, most of the binders are synthesized from styrene-butadiene, or vinyl acetate-acrylate copolymers or styrene-acrylate. Very few researchers have conducted experiments on using chitosan as binder and thickener due the weak performance compared to commercial binders. In the first step of preparation, 3% (w/w) chitosan solution is prepared using preferred amount of water and acetic acid. The pigment is added to the solution and stirred to achieve a 3% homogeneous pigment paste for printing. In the case of poor dispersion of pigments, the amount of acetic acid is increased. Additionally, a very high amount of chitosan (≥3%) in acetic acid generates very highly viscose solution. The achieved properties of the process exhibit lower efficiency and effectiveness of chitosan as a binder compared to the commercial Alcoprint pastes (BASF). The tendency of yellowing is higher after the treatment with increased amount of chitosan and the increase of the curing temperature. Moreover, same process parameters in different batches show variation in yellowing propensity. Furthermore, a share thinning property is exhibited by chitosan solution when the viscosity is measured. Besides, the viscosity is lower than commercial Alcoprint pastes (BASF) when the share rate is less than 50 per s. In contrast, almost similar viscosity profile is found at higher share rate, but yield point is lower than the commercial paste. It indicates that the required force or stress of the chitosan to start flowing is very low. Besides, the wet pick-up ratio of chitosan solution is 75%, while the commercial paste shows 85%. It results in 50% less color strength using chitosan solution after the determination of K/S values. It probably attributes to the aggregation of colorant in the chitosan solution where the dispersion is tough to maintain. In addition, the bending length and stiffness are higher for chitosan-used printed fabric compared to Alcoprint pastes, the commercial one. Eventually, the wash fastness and spreading behavior are not significantly different between chitosan and commercial tested binders [59].

4.4. Chitosan: a finishing agent for durable press for textile

The cotton fabric is hydrophilic due to a high amount of hydroxyl groups. Therefore, the cotton cellulose becomes flexible and exhibits lower crystallinity when it is swelled. A plastic deformation occurs due to the shifting of cellulose molecules during washing treatment. It results in shrinkage and wrinkles of fabric. For overcoming the problem, the crosslinking agent is used to increase the elasticity and prevent the displacement of cellulose which reverses the plastic deformation. Chitosan exhibits similar molecular structure to cellulose and would be applied to cotton finish. During this process, chitosan is fragmented by oxidizing agent such as hydrogen peroxide (H_2O_2). The fragmentation process is carried out using a mixture of hydrochloric acid and chitosan. After 30 min of stirring, hydrogen peroxide is added with an increased temperature of 60°C for 2 h. A vacuum filter is used for the separation of two phases. The residue is washed with water to neutralize, and the filtrate is mixed with ethanol to precipitate the chitosan for 24 h. All precipitates are dried and stored in airtight container. The main application of the fragmented chitosan is to develop durable press finishing process on fabric. In this process, the fragmented chitosan is added to distilled water and stirred until it dissolves completely. The finishing agent contains 8–10% (w/w) dimethyloldihydroxyl ethylene urea (DMDHEU), 0.8% $MgCl_2$, required amount of softener, and the fragmented chitosan solution. The fabric is then immersed in the finishing solution maintaining a pick-up ratio of 80%. The treated fabric is preheated at 80°C for 5 min and then cured at 150°C for 3 min [60].

The fragmentation of chitosan chain facilitates its diffusion into the fabric to form a network with fibers that results in increased recovery angle. The curing time and temperature have a significant effect on the result of final wrinkle property of fabric.

4.5. Chitosan: an antistatic finishing agent on textile

Static surface charge is developed on polymers due to very less amount of moisture content. Moreover, the electrical conductivity is also an important factor on developing static charge along the synthetic filament through a mechanism induced by an electric field. Therefore, antistatic chemical compounds influence the propensity of accumulating static charges by aborting the production of electrostatic charges or by enhancing the electrical conductivity or by both mechanisms. Chitosan can hold in high moisture content regardless the extent of humidity that opens the opportunity to apply chitosan as an antistatic finishing agent on textile surface. In general, the polymers show the resistance of specific electrical resistance in the order of 10^{16} Ω cm that is very close to the value of a good insulator. On the other hand, deionized water exhibits a magnitude that is 10^{8} times lower than that of polymers but mineral water show 10^{3} times higher electrical conductivity compared to pure water. Therefore, chitosan-treated fabrics enable to absorb a very significant amount of water from the atmosphere after exposure to the humid environment and result in the increase in electrical conductivity following the lower propensity to produce static charges. Chitosan having excellent moisture content property is an excellent antistatic agent [61]. Antistatic fabric is developed by fixing the chitosan on the fabric. In the fixing process, butane-1,2,3,4-tetracarboxylic acid (BTCA) is used in the presence of the catalyst, sodium hypophosphite (SHP). Chitosan (1% w/v) dissolution in acetic acid (1%) is the first step of the process. Liquor is produced by the addition of 6 g BTAC and 6 g SHP in the chitosan solution. A sample of fabric is then immerged into the liquor and padded to maintain wet pick-up ratio 100%. The treated fabric is required to dry at 80°C for 5 min, and the curing process is done at 140–170°C for 5–20 min according to the final finish of the fabric. At the end of the process, the temperature of the fabric decreases at room temperature and it is washed with acetic acid for removing extra chitosan. It is then washed again with deionized water and dried by air. The qualitative and quantitative analyses are done by potentiometer and compared with the untreated fabric. Another study has been carried out for the antistatic fictionalization of PET film and fabric by applying oligochitosan. The processes are almost similar except the drying and curing temperature and time. In the later process, drying temperature is maintained at 70°C for 1 h and the curing process is done at 110°C for 30 min [62]. A new *N*-substituted quaternary ammonium chitosan derivative facilitates for the utilization of antistatic functional textile. The chitosan derivative is synthesized through the *N*-alkylation of 4-formayl-*N*-methylpyridinium iodide and [3-(4-formylphenoxy)propyl]trimethylammonium iodide [63].

4.6. Chitosan: an antimicrobial agent for textile

Innovative ideas are in practice to develop a new form of chitosan for improving the efficacy of conventional chitosan applications in textile. Chitosan-treated fabrics and fibers are used for the production of antibacterial fabrics for working environment such as a hospital, biotechnology research lab, cosmetics, fundamental industries, and so on. The charged amino group of chitosan interacts with the cell wall of microbes that causes degradation of protein and intracellular constituents. It also affects the permeability of essential nutrients by altering the cell membrane and causes their death finally [64]. For example, chitosan exhibits direct interaction with the cell wall of *Aspergillus niger*. As a result, a controlled fungal growth and delays in the spore germination occur.

Despite having the potential as antibacterial property, it requires, at a high concentration, for desired efficacy without having any release property in bandage, sutures, etc. The innovation in synthesis, production, and application leads to form nanosized chitosan that could be beneficial for the presence of high amount of active sites available due to the large amount of surface area. In general, a single antibacterial agent has some limitations to perform against both gram positive and gram negative including a broad range of microbes. Therefore, the combination of chitosan and other antibacterial components such as silver has gained more and more attention to fight against large varieties of the microbe by focusing on easier the synthesis and highly efficient in the application [65]. For example, chitosan and silver particles both are used for the synthesis of chitosan-silver nanoparticles which provide excellent antibacterial property on textile application. The application of chitosan and its derivatives on fabric/fiber is followed by the covalent crosslinking into cellulose or wool substrates. The durable press finishing is one of the conventional processes to fix chemicals on fiber surface that also enhances the quality of textiles such as low-shrinkage capacities, anti-odor, wrinkle resistance, and antibacterial properties. For this purpose, a suitable crosslinking agent is required. It should be nontoxic and nonirritating during the use and storage; nevertheless, formaldehyde is one of the mostly used compounds and presents the limitation in those aspects due to its spontaneous toxic vaporization [64]. There are some other crosslinkers such as glutaricdialdehyde, butane tetracarboxylic acid, citric acid (CA), potassium permanganate, and sodium hypophosphite that form the covalent bond between chitosan and the cellulose fabrics by an ester bond.

A high intensity of UV lamp at a wavelength of 254 nm is applied to the raw fibers and exposed for 4 h before the crosslinking treatment. The content of chitosan is 40% less in the final treated fiber compared to the raw fibers without UV irradiation before the treatment with chitosan. It usually changes the surface polarity of fibers via photooxidation and forms active carboxylic groups by photodestruction of glycosidic bonds and rings. The fixation of chitosan with UV-irradiated cellulose fiber is more stable than the raw fibers according to the thermogravimetric analysis. In this process, a fixed amount of chitosan and NaH_2PO_4 is mixed in distilled water. Then, citric acid (CA) is added to the aqueous solution as an acidifier and crosslinking agent. The process could be carried out excluding NaH_2PO_4 based on the requirement of mechanical properties. The raw cellulose fibers are dipped in the solution and raised the temperature at 70°C in the pH range from 3.8 to 4.1. The pH is not recommended over 4.5 to perform the crosslinking reaction and avoids fibers deterioration. The treatment continues for 5 min, and fibers are removed for curing at 130°C up to 3 min. After the curing process, the samples are dried at room temperature under vacuum, before rinsing with distilled water and washing with acetic acid to remove the excess unreacted chitosan. The treated samples are rinsed again with distilled water and dried at room temperature under vacuum until to reach a stable mass.

The mechanism involved in the process follows two steps. In the first step, the crosslinking reaction between cellulose and polycarboxylic acid/CA takes place by the formation of carboxylic anhydride. In the second step, the rings open to react with the polysaccharide through the alcohol groups and form the ester bond. The use of maximum 2.3 w/w% NaH_2PO_4 accelerates the fixation through esterification due to the presence of the phosphate and the partial sodium salt of the polycarboxylic acid. The increased amount of NaH_2PO_4 affects the chitosan loading with fibers due to the neutralization of free carboxylic groups and inhibits esterification of the hydroxyl moieties present in chitosan. The unreacted amino moieties act as active antibacterial property.

Sustainability requires a process safer in terms of raw materials and finished products excluding any toxic by-product. In this aspect, reference [64] proposes UV radiation for

http://dx.doi.org/10.5772/65259

generation of reactive sites on chitosan that can fix many compounds through crosslinking without using any toxic chemical compounds.

Moreover, the antibacterial colloid is prepared by blending chitosan and AgCl–TiO_2 at a ratio of 1:5. A cotton fabric is dipped, and a padder is used to nip the treated fabric as though 70% pick-up ratio is achieved. The predrying temperature is maintained at 120°C and run for 2 min. Subsequently, a curing process is done at 150°C for 3 min. The tensile strength and bending length of chitosan-treated cotton fabric enhanced with the increasing of chitosan concentration. It affects the stiffness of the treated fabric by increasing the value. Moreover, chitosan-treated cotton fibers exhibit lower whiteness index (WI) with the increase in chitosan concentration from 1 to 5 g/L. The degree of WI is found six times lower than that the concentration of 5 g/L [66]. Modified chitosan is also used for the hydrophobic drug delivery [67].

4.7. Chitosan: a flame and fire retardant material for textile

Flame retardant textiles are highly recommended, especially where the working environment is highly concerned regarding fire or flame. The flame retardant property is considered when the flame is not permitted to propagate and extinguished at the same time of withdrawing the fire source. Phosphorus-based materials are interesting and well known as flame retardant due to low toxicity and absence of halogen compounds. These materials are quite efficient as a flame retardant for cellulose and its derivatives through dehydration and char production. In phosphorylation process, cellulose is treated with diammonium phosphate in the presence of urea.

Chitosan could be paired with phosphorous derivative to build an intumescent flame retardant compound. Chitosan is a nitrogen-containing polysaccharide, biodegradable, biocompatible, and environmentally nontoxic. It provides a char-forming property when it is used as intumescent additives due to its hydroxyl and amino groups on molecular skeleton. In addition, it functions as blowing agent and releases nitrogen as a result of molecular destruction. Therefore, a potential nitrogen-phosphorous bonding could be established as flame retardant materials. Chitosan-based flame retardant materials have been designed by chitosan-phytic acid [68], chitosan-sodium polyphosphate [69], chitosan phosphate (chitosan-orthophosphoric acid) [70], chitosan phosphate-nickel [71], chitosan melamine phosphate (chitosan-melamine-sodium hexametaphosphate) [72], chitosan-diammonium hydrogen phosphate [73], etc. It is expected that the presence of ammonium nitrogen in chitosan would provide the synergistic effect with phosphate groups against flame or fire [74]. The preparation of chitosan phosphate has been conducted using many processes. For example, 23 g chitosan has been added to a solution that contains urea (40 g), phosphoric acid (40 ml), and dimethylformamide (350 ml) at 100°C for 5 h. At the end of the process, the product is filtered and washed with isopropyl alcohol (50%) and dried at 60°C. In this process, a commercial resin, knittex FLC, has been provided by Ciba-Geigy (Switzerland). The resin allows the capturing of phosphate by crosslinking and by forming a network to fix the chitosan with cellulose. In addition, the increased amount of resin concentration (2–8%) leads to a reduction in tensile strength (140–114 kg) and elongation at break (21–15%). In contrast, the tensile strength and elongation at break increase for using a higher concentration of chitosan phosphate. It also guides to increase the amount of phosphate (0.095–0.314%) content in the coated fabric and shows higher residue contents by releasing very less volatile components compared to untreated fabric. A compensation effect works between the concentration of chitosan phosphate and resin. The cotton surface treated with 8% chitosan phosphate and 6% resin shows 36.557% residue, while the decomposition temperature is reduced by 25°C compared to the untreated fabric [70].

Furthermore, apart from a reaction mixture, layer-by-layer (LBL) deposition processes are also examined for developing a nanolayer formation [68] on substrate or fabric. In this process, the solution of chitosan is prepared using HCl and phytic acid salt (2 wt.%) in deionized water. The branched polyethyleneimine (1 wt.%) is used to increase the adhesion of cotton fabric as a primary layer. The fabric is dried and dipped sequentially in positively and negatively charged solutions. The process continues up to desired level of the bilayer to achieve effective layers for flame retardancy. After the dipping process, it is wrung to release excess solution and dried at 70°C for 2 h. The thinnest coating which is approximately 10 nm thick has been achieved at pH 4. The thickness contains 30 bilayers on the cotton fabric surface where one bilayer consists of one positive and negatively charged layers. More than 90% residues were left, while the flame propagation is completely stopped during the vertical burn test. In addition to chitosan phosphate, Shuang and his coworkers have focused on the metal ion binding ability of chitosan phosphate that results in a synergistic effect and enhanced flame retardant property by adding nickel ions [71]. The process involves 2 g chitosan and 30 ml methanesulfonic acid in magnetic stirrer putting on an ice bath in inert atmosphere to avoid moisture, by adding phosphorus pentoxide (10 g). The chitosan phosphate is achieved from the reaction and washed with acetone, methanol, and ether. The drying process is realized at 60°C using vacuum. Nickel (II) nitrate hexahydrate and the dried chitosan phosphate react in the ratio of 10:1 with each other at 60°C for 1 h in the presence of ethanol. The final product is washed with fresh ethanol to remove extra nickel (II) nitrate hexahydrate and dried at 60°C in a vacuum dryer. It is then blended with polyvinyl alcohol (PVA) that provides film or fibers or nanofibers [75] to synthesize flame- and chemical-resistant materials. The heat release rate of nickel chitosan phosphate (NiPCS) blended PVA decreases substantially compared to the raw PVA. The microscale combustion calorimeter test has exhibited the peak of heat release for PVA at 155 W/g, while NiPCS blended PVA having a peak at 40 W/g [71]. Besides, the total heat release rate is decreased by NiPCS from 18.2 to 10.4 kJ/g, which underlines the main action of NiPCS. Moreover, the increased amount of NiPCS enhances the char formation that resists the transfer of oxygen and heat. This process leads to delay the thermal decomposition of materials due to the improvement in thermal stability at high temperature.

4.8. Chitosan: a hydrophobic material for water repellent textile

Superhydrophobicity is observed in duck feather, wings of butterflies, the legs of water striders, etc. The specific energy of surface is a quantitative value to comprehend the understanding of how water droplets interact with a surface. It is highly relevant to the contact angle and roughness factors [76]. Nanoscale chitosan coating may be applied on the surface of cotton and polyester fibers to achieve rough surface. These rough surfaces are further treated with silicon [77] and fluoride [78] to reduce the surface energy. The fabrics are dipped for 1 min in a chitosan solution in 1% acetic acid before being squeezed. The wet fabric is neutralized by ammonia gas for 1 min and dried 80°C for 5 min [79]. The dipping-padding process with chitosan allows forming a thin film on the fabric surface. Later on, ammonia treatment changes the pH of the wet fabric, and insoluble-nanoscaled chitosan is precipitated on the fabric surface. The SEM study confirms that nanofiber-like shapes are deposited on the fiber surface, resulting in rough surface on microscale cotton fiber structure. Also, nanoflower-like precipitation is found for the same treatment with polyester fabrics. The deposition of chitosan nanosized particles forms proper roughness on cotton and polyester fibers. Besides, an emulsion of polysiloxane may be used for reducing the surface energy of the chitosan-treated fabric. The emulsion is prepared by hexadecyltrimethoxysilane (HDTMS, 85%), 3-glycidol-propyl-tri-methoxysilane (GPTMS, 97%), and sodium dodecyl sulfate (SDS). A mixture of 8 g HDTMS, 2 g GPTMS, and 1 g SDS is stirred at 75°C for 2 h maintaining pH about 2 using HCl. The operation temperature is reduced to room temperature, and the pH is maintained at 8.0 by ammonia. After the dipping, padding, washing, and drying, the curing process is done at 150°C for 1 min. Further washing is necessary if excess SDS is present on the fabric surface and need to cure again following the same process. In characterization, water contact angles (WCA) are found 152° and 148° for cotton and polyester, respectively. The cotton and polyester fabrics having WCA at 130° and 102°, respectively, after processing only with chitosan do not exhibit long duration hydrophobicity.

Another process based on fluorocarbon to reduce the surface energy has been developed for the preparation of superhydrophobic textile. In this process, anionic heptadecafluoro-1-octanesulfonate has been synthesized from a surfactant, $CF_3(CF_2)_7SO_3Li$. Chitosan nanoparticles are formed using 0.2% chitosan solution and mixed with soluble heptadecafluoro-1-octanesulfonate in acetic acid. The dispersion is stirred for 1 h and left it for 2 h before characterization. The size and surface charge of the chitosan NPs are controlled by the ratio of chitosan and $CF_3(CF_2)_7SO_3Li$ concentrations. For the treatment of fabric, the nanoparticles dispersion sprayed over the textile surface. The contact angle of the treated fabric is 157° ± 2.2°. It has been noticed that long-term water contact would degrade the particles due to the presence of hydroxyl groups in the chitosan complex. The hydroxyl groups are blocked by treating the fabric with chlorodimethyl-1H,1H,2H,2H-perfluorodecylsilane (90%) in 1% heptane solution. The contact angle has been improved after this treatment and found at 160° ± 2° [78].

4.9. Chitosan: a coating on textile for UV blocking

A coating of chitosan-graphene is designed for the process of developing a cotton fabric as UV blocker. A graphene nanosheet (1–3 nm) is added to chitosan solution by stirring vigorously around 5 h. A cotton fabric is submerged in the dispersion of chitosan-graphene for 2 h. In padding process, the fabric passes through two dips and two nips for achieving 80% wet pick up on average. The washing and drying treatments are maintained separately. The predrying process is continued at 70°C for 10 min and curing process at 110°C for 10 min. The ultraviolet protection factor (UPF) of the treated fabric using 1% graphene is 60 times greater than that of untreated cotton. The well-dispersed graphene particle in the chitosan matrix without any aggregation is considered for the high value of UPF [80].

5. Chitosan microcapsules for textile applications

The microencapsulation technique is widely used in pharmaceutical, cosmetic, food, biotechnology, textile industries to obtain control release of active components for a particular purpose. In general, the microencapsulation process is applied to fabrics for improving durability, desirable aroma, hygienic use, fire resistance, wound healing activity, and so on. A tiny amount of active components is surrounded by a continuous polymeric phase in the microencapsulation process. The active component is introduced in different terms such as core, internal phase, encapsulated, while the continuous polymeric phase is described as coating, wall, shell, external phase, and membrane. The continuous polymeric phase is one of the vital issues for application-based microencapsulation process. Chitosan is very efficient to encapsulate the core materials in microencapsulation process. The core materials are usually found in two different phases—solid and liquid—and chosen for specific application. The grapefruit seed oil, a liquid core material, is used to attain durable fragrance and antibacterial properties by microencapsulation process in the textile application [81]. Chitosan is one of the shell materials which are used for microencapsulation due to nontoxic, biodegradable, and naturally extracted. Depending on the process, different types of solvents, crosslinking agents, and surfactants could be used to modify the mechanical strength of chitosan shell for releasing active core materials.

5.1. Microencapsulation processes

There are many different microencapsulation techniques such as coacervation [82], interfacial polymerization or *in situ* polymerization, air suspension coating, emulsion hardening process, pan coating, spray drying [83, 84], centrifugal extraction [85], etc. (**Figure 7**).

The simple coacervation technique is usually followed by single colloidal solute while complex coacervation process involving more than one colloidal solute. The coacervation is one of the mostly used techniques for the development of smart textile and will be discussed. Coacervation can be defined as the separation of a macromolecular solution into two immiscible liquid phases: a dense coacervate phase and a dilute equilibrium phase. The general outline of this method consists in four consecutive steps carried out under stirring, i.e., (i) dispersion of the active substance in a solution of a surface-active hydrocolloid; (ii) precipitation of the hydrocolloid onto the dispersed droplets by lowering the solubility of the hydrocolloid (nonsolvent, pH change, temperature, or electrolyte); (iii) addition of a second hydrocolloid to induce the polymer-polymer complex in the case of complex coacervation; and (iv) stabilization and hardening of the microcapsules by crosslinking agent. Chitosan is used as shell material that is sensitive to the pH and ionic strength of the solution. The pKa value of chitosan is 6.5 and becomes insoluble above this pH since more than 50% chitosan molecules are deprotonated. The complex coacervation technique requires anionic materials that are used for the ionic interaction with protonated chitosan such as vanillin [83], carrageenan [82], alginate [86], gelatine [87, 88], gum Arabic [89], etc. Besides, the addition of crosslinking agent improves the mechanical property of chitosan microcapsules such as sodium tripolyphosphate [90–92], sodium hydroxide [93, 94], genipin [87, 88, 95] glutaraldehyde [96], tannic acid [97], etc. In the processing of microcapsule with chitosan, a soluble chitosan is prepared in solvent such as acetic acid [89], citric acid [83], etc. In an acid medium, chitosan is protonated and bears a positively charged group below pH 6.0. Besides, O/W emulsion is prepared separately using an anionic material which can interact with chitosan. The anionic material stabilizes the emulsion and resists the droplets from aggregation. A homogenizer of high-speed rpm provides the desired droplet size. After that, the chitosan solution is added into the emulsion and using around 800–1000 rpm neutralization of charge and formation of coacervate are achieved in the chitosan solution. The coacervation process can produce fine particles after the charge neutralization and deposition on the surface of emulsion droplets. A crosslinking agent is used to form a strong network by binding the deposited chitosan complex and form a shell around the droplets. The pH, temperature, and concentration ratio vary according to specific system. For example, a chitosan-gum arabic coacervation process requires pH 3.6 for the high yield of coacervation and homogenization speed at 11,000 rpm for less than 10 μm average capsule size [89].

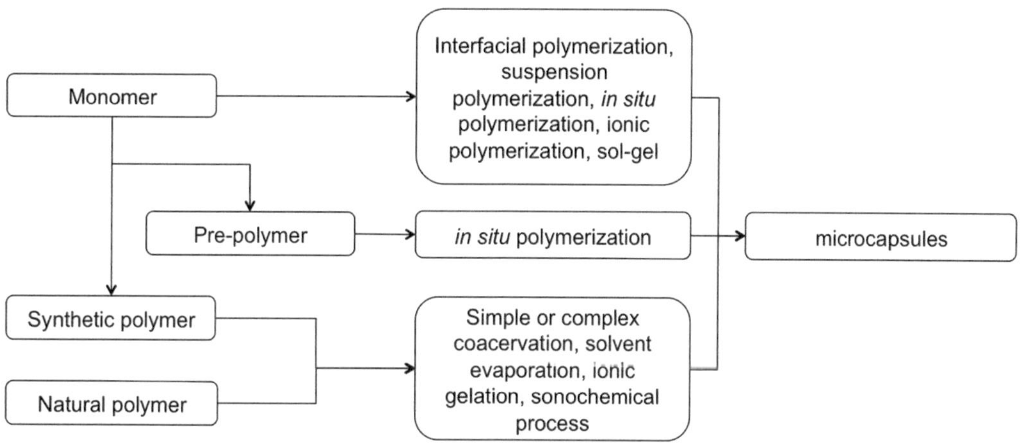

Figure 7. Different textile applications and techniques of chitosan-based microcapsules.

5.2. Textile processing with chitosan-based microcapsules

Citronella essential oil, a mosquito repellent agent, is extracted from *Cymbopogonnardus*, a crop which is mostly found in Argentina. The natural repellent is usually volatile which limits the direct applications on the textile surface. Hence, chitosan-encapsulated citronella essential oil is applied on textile for the fabrication of mosquito repellent textile [93]. Moreover, linseed oil [98], phase change materials [99], and lemon essential oil (*Citrus limon* L.) [94] are also encapsulated by using chitosan for the formation of shell material in textile applications (**Table 3**).

Apart from the microcapsule formulation, textile processing with the microcapsules is also important for achieving the real benefit of smart textiles. The mosquito repellent textile would be a proper example to review the textile processing with the chitosan-based microcapsules below.

Fei and Xin [100] have developed a DEET-containing textile made of cotton for the application of mosquito repellent and antibacterial activity. Chitosan and polyacrylate copolymers have been involved in the process for encapsulating the DEET *in situ*. Chitosan solution (0.5% w/v) is prepared using 0.6% (w/v) acetic acid in a flask which is equipped with a condenser, magnetic stirrer, and nitrogen flow facility. Butyl acrylate (BA) and *N,N*-diethyl-*m*-toluamide (DEET) are added to the chitosan solution, and the temperature is maintained at 80°C for 10 min with stirring. In general, DEET is easily dissolved in poly butyl acrylate, and the dissolution is uniformly homogeneous. Tert-butyl hydroperoxide is added to the mixture at 80°C for 4 h. A white emulsion will be achieved, and the emulsion is sprayed on a 4-mm cotton fabric. The fabric sample should be bleached before the treatment, and maximum 10 mg/cm^2 emulsion is maintained. Chinese standard GB/T 17322.10-1998, a standard test, is followed to check the mosquito repellency at 25°C and 65% humidity compared to untreated fabric sample. The percentage of mosquito repellency is determined based on the difference between control and treated fabric. For example, ED_{90} indicates 90% effective dose. In this process, 4% DEET is enough to achieve the ED_{90} value, while the content is 1.5 mg/cm^2 for 4 h protection. Apart from the formulation, *Vitex negundo* leaf extract is also used as a mosquito repellent and is encapsulated using chitosan as shell materials [101]. Sodium alginate (95 ml, 0.6% w/v) and 18 mM calcium chloride are added to the leaf extract. Chitosan is added to this solution by stirring for 30 min and left overnight at room temperature. It is then centrifuged at 1500 rpm for 15 min. The coacervate phase will be separated out, washed, and stored at 5°C for further analysis.

Application	Microencapsulation technique
Limonene oil microencapsulated textile for antibacterial activity	Coacervation process by NaOH dripping technique
Cosmato textile-chitosan/vanillin microcapsules	Spray drying technique
Microcapsulation of citronella oil	Coacervation process by NaOH dripping technique
Microcapsulation of linseed oil	Multilayer coacervation method
Phase change material encapsulation	Chitosan as binder

Table 3. Different techniques and applications of chitosan-based microencapsulation processes for textile.

The release rate of the core materials through the chitosan shell depends on the temperature, humidity, and time. On top of that, the release property of active component depends on chitosan concentration and temperature [93]. For example, the release of citronella oil is 20–25% when the operational temperature is maintained at 40–60°C. Besides, chitosan wall membrane concentration 0.2–0.5% is considered as a good microcapsule for a better release property. The release property is entirely inhibited when the operational temperature is 80°C. The mechanism underlying the release behavior is the contraction of chitosan membrane due to increase in temperature. As a result, the pore size reduced and the release of core materials stopped [93]. On the other hand, the chitosan-core material interaction also affects the release property. In this regard, the concentration of chitosan and crosslinking agent should be optimized properly.

6. Conclusion and future perspectives

Chitosan is biodegradable in human health, plant, and environment organisms. Concerning economic aspect, it is a by-product of fishing industry. Textile is one of the most usable consumer products in our daily life although the processes are mostly hazardous. The whole supply chain of textiles still requires more research on how to establish the sustainability regarding health safety and cost. The use of chitosan could open the door for developing a sustainable industrial practice in fiber production and fabric treatments for its nontoxic property and low cost. Moreover, the blended and modified chitosans that extend the availability of various high-end functional textiles with reasonable price will enhance the research and development on chitosan-based textile processing.

Author details

Jagadish Roy[1,2], Fabien Salaün[1,2*], Stéphane Giraud[1,2], Ada Ferri[3] and Jinping Guan[4]

*Address all correspondence to: fabien.salaun@ensait.fr

1 University Lille Nord de France, Lille, France

2 ENSAIT, GEMTEX, Roubaix, France

3 Department of Materials Science and Chemical Engineering, Polytechnic University of Turin, Turin, Italy

4 College of textile and Clothing Engineering, Soochow University, Suzhou, Jiangsu, China

References

[1] Niekraszewicz B, Niekraszewicz A. The structure of alginate, chitin and chitosan fibers. In: Handbook of Textile Fibre Structure Natural, Regenerated, Inorganic and Specialist Fibres, vol. 2. Woodhead Publishing Limited; 2009. p. 266–304.

[2] Bagheri-Khoulenjani S, Taghizadeh SM, Mirzadeh H. An investigation on the short-term biodegradability of chitosan with various molecular weights and degrees of deacetylation. Carbohydr Polym. 2009;78:773–8.

[3] Vårum KM, Myhr MM, Hjerde RJN, Smidsrød O. *In vitro* degradation rates of partially N-acetylated chitosans in human serum. Carbohydr Res. 1997;299:99–101.

[4] Vandevord PJ, Matthew HWT, Desilva SP, Mayton L, Wu B, Wooley PH. Evaluation of the biocompatibility of a chitosan scaffold in mice. J Biomed Mater Res. 2002;59(3):585–90.

[5] Sashiwa H, Aiba SI. Chemically modified chitin and chitosan as biomaterials. Prog Polym Sci. 2004;29:887–908.

[6] Mourya VK, Inamdar NN. Chitosan-modifications and applications: Opportunities galore. React Funct Polym. 2008;68:1013–51.

[7] Knaul JZ, Creber KAM. Coagulation rate studies of spinnable chitosan solutions. J Appl Polym Sci. 1997;66:117–27.

[8] East GC, Qin Y. Wet spinning of chitosan and the acetylation of chitosan fibers. J Appl Polym Sci. 1993;50:1773–9.

[9] Delta TD. Writing 1. p. 1–47.

[10] Tamura H, Tsuruta Y, Itoyama K, Worakitkanchanakul W. Preparation of chitosan filament applying new coagulation system. Carbohydr Polym. 2004;56:205–11.

[11] Hirano S, Nagamura K, Zhang M, Ki S, Geul B. Chitosan staple fibers and their chemical modification with some aldehydes. Carbohydr Polym. 1999;38:293–8.

[12] Tokura S, Nishimura S, Nishi N, Nakamura K, Hasegawa O, Sashiwa H, et al. Preparation and some properties of various deacetylated chitin fibers. Transaction. 1987;43:288–93.

[13] Notin L, Viton C, Lucas JM, Domard A. Pseudo-dry-spinning of chitosan. Acta Biomater. 2006;2:297–311.

[14] Schueren L Van Der, Steyaert I, Schoenmaker B De, Clerck K De. Polycaprolactone/chitosan blend nanofibres electrospun from an acetic acid/formic acid solvent system. Carbohydr Polym. 2012;88:1221–6.

[15] Bhattarai N, Edmondson D, Veiseh O, Matsen FA, Zhang M. Electrospun chitosan-based nanofibers and their cellular compatibility. Biomaterials. 2005;26:6176–84.

[16] Ohkawa K, Cha D, Kim H, Nishida A, Yamamoto H. Electrospinning of chitosan. Macromol Rapid Commun. 2004;25:1600–5.

[17] Notin L, Viton C, David L, Alcouffe P, Rochas C, Domard A. Morphology and mechanical properties of chitosan fibers obtained by gel-spinning: Influence of the dry-jet-stretching step and ageing. Acta Biomater. 2006;2:387–402.

[18] Ogawa K, Yui T, Miya M. Dependence on the preparation procedure of the polymorphism and crystallinity of chitosan membranes. Biosci Biotechnol Biochem. 1992;56:858–62.

[19] Agboh OC, Qin Y. Chitin and chitosan fibers. Polym Adv Technol. 1997;8:355–65.

[20] Crini G. Recent developments in polysaccharide-based materials used as adsorbents in wastewater treatment. Prog Polym Sci. 2005;30:38–70.

[21] Saha S, Sarkar P. Arsenic mitigation by chitosan-based porous magnesia-impregnated alumina: Performance evaluation in continuous packed bed column. Int J Environ Sci Technol. 2016;13:243–56.

[22] Li X, Wu P, Gao GF, Cheng S. Carbohydrate-functionalized chitosan fiber for influenza virus capture. Biomacromolecules. 2011;12:3962–9.

[23] Pinho ED, Martins A, Araújo JV, Reis RL, Neves NM. Size also matters in biodegradable composite microfiber reinforced by chitosan nanofibers. MRS Proc. 2014;1621:59–69.

[24] Knill CJ, Kennedy JF, Mistry J, Miraftab M, Smart G, Groocock MR, et al. Alginate fibres modified with unhydrolysed and hydrolysed chitosans for wound dressings. Carbohydr Polym. 2004;55:65–76.

[25] Miraftab M, Smart G, Kennedy JF, Knill CJ, Mistry J, Groocock MR. Novel chitosan-alginate fibres for advanced wound dressings. In: Medical Textiles and Biomaterials for Healthcare: Incorporating Proceedings of MEDTEX03 International Conference and Exhibition on Healthcare and Medical Textiles. 2005. p. 37–49.

[26] Jung B, Theato P. Chemical strategies for the synthesis of protein—polymer conjugates. Adv Polym Sci. 2012;244:19–44.

[27] Zamani M, Prabhakaran MP, Ramakrishna S. Advances in drug delivery via electrospun and electrosprayed nanomaterials. Int J Nanomedicine. 2013;8:2997–3017.

[28] Schauer CL, Lelkes PI et al. NIH public access. Biomaterials. 2013;33:9167–78.

[29] Sun K, Li ZH, Zakhem E, Bitar KN, Notin L, Viton C, et al. Morphology and mechanical properties of chitosan fibers obtained by gel-spinning: Influence of the dry-jet-stretching step and ageing. Acta Biomater. 2015;2(4):999–1011.

[30] Toskas G, Cherif C, Hund RD, Laourine E, Mahltig B, Fahmi A, et al. Chitosan(PEO)/silica hybrid nanofibers as a potential biomaterial for bone regeneration. Carbohydr Polym. 2013;94:713–22.

[31] Liu Y, Xiang K, Chen H, Li Y, Hu Q. Composite vascular repair grafts via micro-imprinting and electrospinning. AIP Adv. 2015;5(4):0–8.

[32] Barrère M, Ganachaud F, Bendejacq D, Dourges MA, Maitre C, Hémery P, et al. Controlled release properties of chitosan encapsulated volatile Citronella Oil microcapsules by

thermal treatments. Colloids Surf B Biointerfaces [Internet]. Third Ed. 2014;3:292–7. doi: 10.1016/j.carbpol.2012.04.013

[33] Su P, Wang C, Yang X, Chen X, Gao C, Feng XX, et al. Electrospinning of chitosan nanofibers: The favorable effect of metal ions. Carbohydr Polym. 2011;84:239–46.

[34] Hadipour-Goudarzi E, Montazer M, Latifi M, Aghaji AAG. Electrospinning of chitosan/sericin/PVA nanofibers incorporated with *in situ* synthesis of nano silver. Carbohydr Polym. 2014;113:231–9.

[35] Kim SS, Lee J. Antibacterial activity of polyacrylonitrile-chitosan electrospun nanofibers. Carbohydr Polym. 2014;102:231–7.

[36] Torres-giner S, José M, María J. Novel antimicrobial ultrathin structures of zein/chitosan blends obtained by electrospinning. Carbohydr Polym. 2009;77:261–6.

[37] Zhang H, Li S, Branford White CJ, Ning X, Nie H, Zhu L. Studies on electrospun nylon-6/chitosan complex nanofiber interactions. Electrochim Acta. 2009;54:5739–45.

[38] Teng SH, Wang P, Kim HE. Blend fibers of chitosan-agarose by electrospinning. Mater Lett. 2009;63:2510–2.

[39] Duan B, Yuan X, Zhu Y, Zhang Y, Li X, Zhang Y, Yao K. Polymer A nanofibrous composite membrane of PLGA—chitosan/PVA prepared by electrospinning. Eur Polym J. 2013;42:2013–22.

[40] Czechowska-Biskup R, Wojtasz-Pająk A. Solutions of hydrochloric acid as simple solvents of chitosan for viscosity-and light-scattering-based molecular weight. Polish Chitin Soc. 2007;P-11:87–94.

[41] Kasaai MR, Arul J, Charlet G. Fragmentation of chitosan by acids. Sci World J. 2013;2013:11.

[42] Hebeish A, Higazy A, El-Shafei A. New sizing agents and flocculants derived from chitosan. In: Starch/Staerke. 2006. p. 401–10.

[43] Stegmaier T, Wunderlich W, Hager T, Siddique AB, Sarsour J, Planck H. Chitosan—A sizing agent in fabric production—Development and ecological evaluation. CLEAN—Soil, Air, Water. 2008;36(3):279–86.

[44] Noppakundilograt S, Buranagul P, Graisuwan W, Koopipat C, Kiatkamjornwong S. Modified chitosan pretreatment of polyester fabric for printing by ink jet ink. Carbohydr Polym. 2010;82:1124–35.

[45] Wijesena RN, Tissera ND, De Silva KMN. Coloration of cotton fibers using nano chitosan. Carbohydr Polym. 2015;134:182–9.

[46] Akovali G. Advances in Polymer Coated Textiles, 1st edn. Advances in Polymer Coated Textiles. Shawbury, Shrewsbury, Shropshire, SY4 4NR, UK: Smithers Rapra Technology Ltd; 2012. 447 p.

[47] Sadeghi-kiakhani M, Safapour S. Improvement of dyeing and antimicrobial properties of nylon fabrics modified using chitosan-poly (propylene imine) dendreimer hybrid. J Ind Eng Chem. 2016;33:170–7.

[48] Dev VRG, Venugopal J, Sudha S, Deepika G, Ramakrishna S. Dyeing and antimicrobial characteristics of chitosan treated wool fabrics with henna dye. Carbohydr Polym. 2009;75:646–50.

[49] Periolatto M, Ferrero F, Vineis C, Rombaldoni F. Multifunctional finishing of wool fabrics by chitosan UV-grafting: An approach. Carbohydr Polym. 2013;98:624–9.

[50] Periolatto M, Ferrero F, Vineis C. Antimicrobial chitosan finish of cotton and silk fabrics by UV-curing with 2-hydroxy-2-methylphenylpropane-1-one. Carbohydr Polym. 2012;88:201–5.

[51] Tang X, Tian M, Qu L, Zhu S, Guo X, Han G, et al. A facile fabrication of multifunctional knit polyester fabric based on chitosan and polyaniline polymer nanocomposite. Appl Surf Sci. 2014;317:505–10.

[52] Ibrahim NA, Hashem M, El-Sayed WA, El-Husseiny S, El-Enany E. Enhancing antimicrobial properties of dyed and finished cotton/polyester fabrics. Carbohydr Polym. 2009;78:502–10.

[53] Mohamed FA, Ali NF, El-mohamedy RSR. Original research article the dye ability and antimicrobial activity of wool fibers dyed with reactive dyes and pre-treated with chitosan. Int J Curr Microbiol App Sci. 2015;4(11):587–96.

[54] Singha K, Maity S, Singha M. The salt-free dyeing on cotton: An approach to effluent free mechanism. Can chitosan be a Potential Option? Int J Text Sci. 2012;1:69–77.

[55] Bhuiyan MAR, Shaid A, Khan MA. Cationization of cotton fiber by chitosan and its dyeing with reactive dye without salt. Chem Mater Eng. 2014;2:96–100.

[56] Ramadan MA, Samy S, Hebeish AA. Eco-friendly pretreatment of cellulosic fabrics with chitosan and its influence on dyeing efficiency. In: Kumbasar EPA, editor. Natural Dyes, 1st ed. InTech; 2011.

[57] Ma Z, Wang W, Wu Y. Oxidative degradation of chitosan to the low molecular water-soluble chitosan over peroxotungstate as chemical scissors. PLoS One. 2014;9:2–8.

[58] Vold, Bjørn E Christensen IMN. Periodate oxidation of chitosans with different chemical compositions. Carbohydr Res. 2005;340:679–84.

[59] Bahmani SA, East GC, Holme I. The application of chitosan in pigment printing. Color Technol. 2000;116:94–9.

[60] Huang KS, Wu WJ, Chen JB, Lian HS. Application of low-molecular-weight chitosan in durable press finishing. Carbohydr Polym. 2008;73:254–60.

[61] Abdel-Halim ES, Abdel-Mohdy FA, Al-Deyab SS, El-Newehy MH. Chitosan and monochlorotriazinyl-B-cyclodextrin finishes improve antistatic properties of cotton/polyester blend and polyester fabrics. Carbohydr Polym. 2010;82:202–8.

[62] Kim JH, Lee SY. Surface modification of PET film and fabric by oligo-chitosan treatment. Fibers Polym. 2014;15:2489–94.

[63] Suzuki K, Oda D, Shinobu T, Saimoto H, Shigemasa Y. New selectively N-substituted quaternary ammonium chitosan derivatives. Polym J. 2000;32:334–8.

[64] Alonso D, Gimeno M, Olayo R, Vazquez-Torres H, Sepulveda-Sanchez JD, Shirai K. Cross-linking chitosan into UV-irradiated cellulose fibers for the preparation of antimicrobial-finished textiles. Carbohydr Polym. 2009;77:536–43.

[65] Ali SW, Rajendran S, Joshi M. Synthesis and characterization of chitosan and silver loaded chitosan nanoparticles for bioactive polyester. Carbohydr Polym. 2011;83:438–46.

[66] Arain RA, Khatri Z, Memon MH, Kim IS. Antibacterial property and characterization of cotton fabric treated with chitosan/AgCl-TiO_2 colloid. Carbohydr Polym. 2013;96:326–31.

[67] de Oliveira Pedro R, Schmitt CC, Neumann MG. Syntheses and characterization of amphiphilic quaternary ammonium chitosan derivatives. Carbohydr Polym. 2016;147:97–103.

[68] Laufer G, Kirkland C, Morgan AB, Grunlan JC. Intumescent multilayer nanocoating, made with renewable polyelectrolytes, for flame-retardant cotton. Biomacromolecules. 2012;13:2843–8.

[69] Charuchinda S, Srikulkit K. Co-application of sodium polyphosphate and chitosan to improve flame retardancy of cotton fabric. J Sci Res Chula Univ. 2005;30:97–107.

[70] Hebeish A. Chitosan phosphate induced better Thermal characteristics to cotton fabric. J Appl Polym Sci. 2006;103:2021–6.

[71] Hu S, Song L, Pan H, Hu Y. Thermal properties and combustion behaviors of chitosan based flame retardant combining phosphorus and nickel. Ind Eng Chem Res. 2012;51:3663–9.

[72] Leistner M, Abu-odeh AA, Rohmer SC, Grunlan JC. Water-based chitosan/melamine polyphosphate multilayer nanocoating that extinguishes fire on polyester-cotton fabric. Carbohydr Polym. 2015;130:227–32.

[73] El-Tahlawy K. Chitosan phosphate: A new way for production of chitosan phosphate for production of eco-friendly flame-retardant cotton textiles. J Text Inst. 2008;99:185–91.

[74] Kandola BK, Horrocks AR, Price D, Road ED, Sciences C, Science D, et al. Flame-retardant treatments of cellulose and their influence on the mechanism of cellulose pyrolysis. J Macromol Sci Part C. 1996;36:721–94.

[75] Medeiros ES, Mattoso LHC, Ito EN, Gregorski KS, Robertson GH, Offeman RD, et al. Electrospun nanofibers of poly(vinyl alcohol) reinforced with cellulose nanofibrils. J Biobased Mater Bioenergy. 2008;2:231–42.

[76] Wenzel RN. Resistance of solid surfaces to wetting by water. Ind Eng Chem. 1936;28:988–94.

[77] Rollings DE, Veinot JGC. Polysiloxane nanofibers via surface initiated polymerization of vapor phase reagents: A mechanism of formation and variable wettability of fiber-bearing substrates. Langmuir. 2008;24:13653–62.

[78] Ivanova NA, Rutberg GI, Philipchenko AB. Enhancing the superhydrophobic state stability of chitosan-based coatings for textiles. Macromol Chem Phys. 2013;214:1515–21.

[79] Liu Y, Chen X, Xin JH. Hydrophobic duck feathers and their simulation on textile substrates for water repellent treatment. Bioinspir Biomim. 2008;3:046007.

[80] Tian M, Tang X, Qu L, Zhu S, Guo X, Han G. Robust ultraviolet blocking cotton fabric modified with chitosan/graphene nanocomposites. Mater Lett. 2015;145:340–3.

[81] Alonso D, Gimeno M, Sepulveda-Sanchez JD, Shirai K. Chitosan-based microcapsules containing grapefruit seed extract grafted onto cellulose fibers by a non-toxic procedure. Carbohydr Res. 2010;345:854–9.

[82] Briones AV, Sato T. Encapsulation of glucose oxidase (GOD) in polyelectrolyte complexes of chitosan-carrageenan. React Funct Polym. 2010;70:19–27.

[83] Yang Z, Zeng Z, Xiao Z, Ji H. Preparation and controllable release of chitosan/vanillin microcapsules and their application to cotton fabric. Flavour Fragr J. 2014;29:114–20.

[84] Estevinho BN, Damas AM, Martins P, Rocha F. The influence of microencapsulation with a modified chitosan (Water Soluble) on β-galactosidase activity. Dry Technol. 2014;32:1575–86.

[85] Cheng SY, Yuen CWM, Kan CW, Cheuk KKL. Development of cosmetic textiles using microencapsulation technology. Res J Text Appar. 2008;12:41–51.

[86] Li Y, McClements DJ. Controlling lipid digestion by encapsulation of protein-stabilized lipid droplets within alginate-chitosan complex coacervates. Food Hydrocoll. 2011;25:1025–33.

[87] Hussain MR, Maji TK. Preparation of genipin cross-linked chitosan-gelatin microcapsules for encapsulation of Zanthoxylum limonella oil (ZLO) using salting-out method. J Microencapsul. 2008;25:414–20.

[88] Guo C, Zhou L, Lv J. Effects of expandable graphite and modified ammonium polyphosphate on the flame-retardant and mechanical properties of wood flour-polypropylene composites. Polym Polym Compos. 2013;21(7):449–56.

[89] Butstraen C, Salaün F. Preparation of microcapsules by complex coacervation of gum Arabic and chitosan. Carbohydr Polym. 2014;99:608–16.

[90] Su ZQ, Zhang HL, Wu SH, Tao Y, Zang LQ. Preparation and characterization of water-soluble chitosan nanoparticles as protein delivery system. J Nanomater. 2010;2010.

[91] Yoksan R, Jirawutthiwongchai J, Arpo K. Encapsulation of ascorbyl palmitate in chitosan nanoparticles by oil-in-water emulsion and ionic gelation processes. Colloids Surf B Biointerfaces. 2010;76:292–7.

[92] Hassani S, Laouini A, Fessi H, Charcosset C. Preparation of chitosan-TPP nanoparticles using microengineered membranes—Effect of parameters and encapsulation of tacrine. Colloids Surf A Physicochem Eng Asp. 2015;482:34–43.

[93] Hsieh WC, Chang CP, Gao YL. Controlled release properties of chitosan encapsulated volatile Citronella Oil microcapsules by thermal treatments. Colloids Surf B Biointerfaces. 2006;53:209–14.

[94] Souza JM, Caldas AL, Tohidi SD, Molina J, Souto AP, Fangueiro R, et al. Properties and controlled release of chitosan microencapsulated limonene oil. Braz J Pharmacogn. 2014;24(6):691–8.

[95] Yang Z, Peng Z, Li J, Li S, Kong L, Li P, et al. Development and evaluation of novel flavour microcapsules containing vanilla oil using complex coacervation approach. Food Chem. 2014;145:272–7.

[96] Ofokansi KC, Kenechukwu FC, Isah AB, Okigbo EL. Formulation and evaluation of glutaraldehyde-crosslinked chitosan microparticles for the delivery of ibuprofen. Trop J Pharm Res. 2013;12:19–25.

[97] Shutava TG, Lvov YM. Nano-engineered microcapsules of tannic acid and chitosan for protein encapsulation. J Nanosci Nanotechnol. 2006;6:1655–61.

[98] Chatterjee S, Salaün F, Campagne C. Development of multilayer microcapsules by a phase coacervation method based on ionic interactions for textile applications. Pharmaceutics. 2014;6:281–97.

[99] Cui R, Liu X, Yu W, He L. Preparation and characterization of microencapsulated n-octadecane as phase change materials experimental material. J Fiber Bioeng Informatics. 2012;5:51–8.

[100] Fei B, Xin JH. N, N-diethyl-m-toluamide-containing microcapsules for bio-cloth finishing. Am J Trop Med Hyg. 2007;77:52–7.

[101] Ramasamy R, Rajan R, Velmurugan R. Development of mosquito repellent fabrics using *Vitex negundo* loaded nanoparticles. Malaya J Biosci. 2014;1:19–23.

Chapter 13

Chitosan-Based Thermosensitive Materials

Waldo Argüelles-Monal,
Maricarmen Recillas-Mota and
Daniel Fernández-Quiroz

Additional information is available at the end of the chapter

http://dx.doi.org/10.5772/65713

Abstract

Thermosensitive polymers are materials capable of undergoing a reversible phase transition in aqueous media in response to a variation of the temperature. They have attracted high scientific interest for advanced applications in diverse areas, such as biotechnology, biomedical, environmental, food industry and other fields. At the same time, chitosan is a promising marine polysaccharide that has long been used in applications such as drug, peptide or gene delivery systems. Being the most abundant marine polysaccharide, chitin and chitosan do not exhibit thermoresponsive properties, but some of their derivatives do. In the present chapter, the efforts to produce chitosan-based thermosensitive materials are reviewed. Particularly, the properties and applications of chitosan-glycerophosphate thermogelling system are examined; the methods of synthesis of chitosan copolymers grafted with poly(*N*-isopropylacrylamide) or poly(*N*-vinylcaprolactam), their physicochemical properties and most of their prominent applications are discussed as well.

Keywords: chitosan, thermosensitive material, chitosan-glycerophosphate, chitosan-*g*-*N*-isopropylacrylamide, chitosan-*g*-poly(*N*-vinylcaprolactam)

1. Introduction

Smart polymers are usually defined as 'macromolecules capable of undergoing rapid, reversible phase transitions from a hydrophilic to a hydrophobic microstructure. These transitions are triggered by small shifts in the local environment, such as slight variations in temperature, pH, ionic strength, or the concentration of specific substances like sugars' [1]. Among them, thermosensitive water-soluble materials have attracted high scientific interest for advanced applications in diverse areas, such as biotechnology, biomedical, environmental, food industry and other fields.

Chitosan (Cs) is a linear polymer obtained by extensive deacetylation of chitin, the most abundant marine polysaccharide. It is mainly composed by two kinds of β(1→4)-linked structural units: 2-amino-2-deoxy-D-glucose and *N*-acetyl-2-amino-2-deoxy-D-glucose. Chitosan is a very interesting polymer for biomedical applications because of its biocompatibility, biodegradability and low toxicity. It can also be applied for the treatment of residual waters, in agriculture, food, cosmetics and textile industries, among others [2].

Because of the polyelectrolyte nature of chitosan, these thermoresponsive materials are also sensitive to changes of pH in the medium. Aqueous polymer solutions that could be transformed in situ into hydrogels by changes in environmental conditions, such as temperature and pH, have an advanced scientific interest due to their specific technological applications as sensors, actuators, controllable membrane for separations and modulators for delivery of drugs for use in medicine, biotechnology and other fields.

A wide variety of chitosan thermosensitive materials has been generated, like nanostructures, scaffolds, membranes, cryogels and paramagnetic beads, to cite some of them. In the present chapter, the efforts to produce chitosan-based thermosensitive materials, as well as their most relevant physicochemical properties and applications, will be considered.

2. Chitosan-glycerophosphate system

2.1. Physicochemical characteristics and mechanism of gelation

Chitosan is an ionic polysaccharide that does not yield physical gels by itself. It was up to 2000, when Chenite et al. published a couple of articles describing one of the rare true physical chitosan hydrogels [3, 4]. They described a chitosan formulation with glycerophosphate (GP), which is capable to render physical gels upon heating. In fact, when adding glycerophosphate to chitosan aqueous solutions, the polymer remains in solution at neutral pH and room temperature, while a heat-induced gelation can be triggered upon heating around the physiological temperature. Due to these interesting characteristics, this system has found a wide interest in the biomedical field including drug delivery and tissue engineering.

The rheological behaviour of a Cs-GP nearly neutral aqueous solution (pH 7.15) exhibits a strong rise of the storage modulus, G', upon heating, indicating that the liquid solution turned into a solid-like gel in the vicinity of 37 °C. In turn, during the cooling run, there is a decrease of G', revealing a tendency of the gel to return to the liquid state [3]. Cho et al. have classified the viscoelastic behaviour of this system in three well-distinctive regions: (1) a liquid-like behaviour manifested at low temperatures, (2) the thermal gelation process characterized by an increment in G' and G" moduli as the temperature rises and (3) a slow terminal gelation after the heat-induced cross-linking process [5].

The temperature of incipient gelation increases as the degree of deacetylation decreases, while the molecular weight showed no significant effect on the temperature of gelation [3]. If transparent hydrogels are needed to be prepared, it is necessary to previously reacetylate chitosan up to a degree of deacetylations 35–55%, under homogeneous conditions [6].

The morphology of Cs-GP hydrogel as observed by laser scanning confocal microscopy displays a heterogeneous microstructure, suggesting that the kinetic gelation mechanism of this system may be nucleation and growth. Power spectra reveal a fractal-like morphology in the gel [7].

The mechanism of gelation of chitosan in the presence of glycerophosphate involves several interactions such as screening of electrostatic repulsion, ionic cross-linking and hydrophobic and hydrogen-bonding interactions [5, 8].

The effects of experimental parameters on the characteristics of the gelation process have been documented. *Chitosan and glycerophosphate concentration*: the gelation temperature decreased with increasing both β-GP and polymer concentration, while the mechanical properties of the gel become enhanced due to an increase of intermolecular interactions and entanglements [8]. *Temperature*: it is interesting to note the behaviour of the pH and conductivity during heating. While any change in pH could not be appreciated, the conductivity displayed a monotonous increment as the temperature rises. This behaviour was associated to a reduction of chitosan solubility and an enhancement on the screening of electrostatic repulsion, therefore increasing hydrophobic interactions and improving conditions for gel formation [5]. According to the same authors, temperature most probably modifies hydrogen bond distribution and favours polymer-polymer interactions over those of polymer-solvent interactions. *Ionic strength*: an increment of the ionic strength gives favourable conditions for gel formation due to the higher screening of electrostatic repulsions, thus promoting the hydrophobic interactions in the polymer physical network [5].

In a continuation of these works, Cho et al. used oscillatory shear data to verify the scaling behaviour at the gel point. Their conclusions showed that the power-law index was dependent to some extent on chitosan concentration and temperature [9].

Moreover, an interesting research published by Supper et al. [10] demonstrated the specific role of the polyol-phosphate molecules on the thermo-physical gelation process. On the one hand, the phosphate moiety neutralizes the charges along the Cs chains up to the physiological pH, keeping them in the solution state at room temperature. On the other hand, the polyol unit is responsible for the thermal sensitivity of the Cs solution. Both the chemical structure and size of the polyol moiety play a significant role in controlling the gelation process [10]. In **Figure 1** this gelation mechanism is clearly depicted.

Similar conclusions have been achieved by Qiu et al. when they investigated the influence of urea and isobutanol on the thermogelling process of Cs-GP system. Urea appears to be unfavourable to the gelation process, because it disrupts hydrogen bonding and retards the formation of hydrophobic domains, but the addition of isobutanol speeded the sol-gel transition by strengthening the hydrophobic interactions. This opposite effect opens the possibilities of tuning the gelling process of Cs-GP system [11].

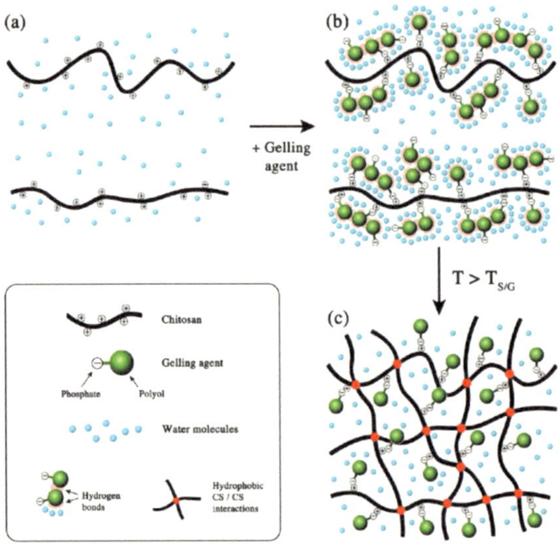

Figure 1. Representation of the gelation mechanism of Cs/polyol-phosphate solutions according to Supper et al. [10]. Reprinted with permission from Ref. [10]. Copyright 2013, American Chemical Society.

2.2. Applications

For biomedical purposes, the sterilization of the Cs-GP system and other biomedical characteristics are important to be considered. There have been documented some investigations about the influence of different sterilization procedures on the thermogelling ability of this system. Autoclaving and steam sterilization may significantly affect the molecular weight and viscosity of chitosan, but do not impair its ability to form gels upon heating [12, 13]. However, γ-radiation markedly changed its thermogelling properties [13]. Regarding the angiogenic potential of Cs-GP system, results indicate that this system does not contribute to enhance angiogenesis, while the presence of human bone marrow-derived mesenchymal stem cells resulted in an increased angiogenic response after 3 days of placement on the chick chorioallantoic membrane [14]. A lower acetylation of chitosan seems to be desirable for better biocompatibility. At the same time, even though biodegradation is slower, fewer fragments are generated, and this seems to lead to a minor pronounced immune response [15]. Newer studies in this sense confirm that Cs-GP is a biocompatible hydrogel, extracts of which can stimulate mesenchymal stem cell proliferation at certain concentrations. According to authors' conclusions, 'this material is therefore a promising vehicle for cell encapsulation and injectable tissue-engineering applications' [16].

Chitosan thermosensitive hydrogel has the advantage to form in situ a hydrogel at physiological temperature, avoiding the necessity of surgical implants. This fact underlies its applications in biomedical field including local drug delivery and tissue engineering [17]. Cs-GP system has a great potential as scaffold material in tissue engineering and regenerative medicine due to its good biocompatibility, minimal immune reaction, high antibacterial nature, good adhesion to cells and the possibility to be moulded in various geometries [18–26]. For similar reasons, the other key pharmaceutical application of this material is a smart-controlled release system [4, 18, 20, 27–38], because the hydrogel is able to keep the drug level within the therapeutic window during extended periods of time, thus avoiding frequent low doses and undesirable secondary effects in patients. Important and interesting patents of biomedical applications of this system have been revised recently [39].

Composite nanomaterials of Cs-GP hydrogel and silver nanoparticles (NPs) with potential applications in medicine due to their antibacterial activity have also been documented [40, 41].

3. Chitosan thermosensitive derivatives

Chitosan macromolecule is prone to chemical modifications, due to the high reactivity of their functional groups: primary and secondary hydroxyl groups at C-3 and C-6 positions, respectively, and the highly reactive amino group at C-2 position. The chemical structure of chitosan is represented in **Figure 2**. The most common purposes for modifying chitosan include improvement of its solubility at neutral or alkaline pH and to impart specific functional properties [42]. Among the different chemical approaches to modify chitosan, the grafting procedures open a huge range of possibilities to achieve versatile molecular designs for new advanced materials.

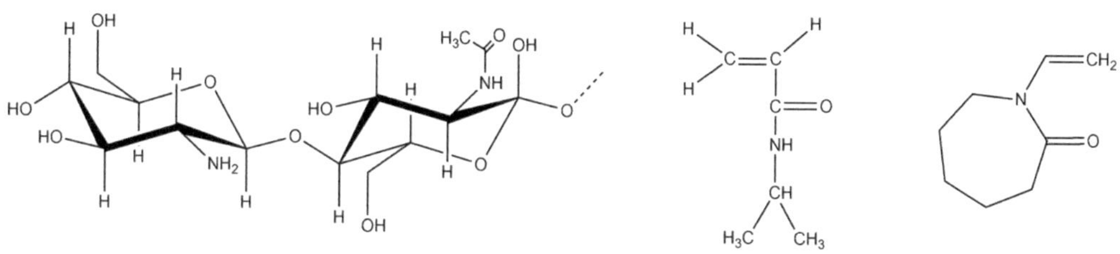

Figure 2. Structural units of chitosan, *N*-isopropylacrylamide and *N*-vinylcaprolactam.

A graft copolymer contains a long sequence of structural units (often referred to as the backbone polymer, in this case chitosan), with one or more branches (grafts) of long sequences of another monomer. Grafting copolymerization can be conducted in both heterogeneous and homogeneous conditions [43]. The graft copolymers can be principally synthesized by three strategies: *grafting through*, *grafting from* and *grafting onto*, in dependence of the method of preparation. The *grafting from* reaction is based on the in situ polymerization of the grafting monomers from a preformed macromolecular backbone that is chemically modified by the introduction of active sites. The *grafting onto* is an interesting technique that consists in a coupling reaction between end-functional groups of the graft chains onto pendant functional groups of the backbone chain [44]. This procedure has become an efficient tool for the preparation of graft copolymers with well-defined structure.

Poly(*N*-isopropylacrylamide), PNIPAm, and poly(*N*-vinylcaprolactam), PVCL, are two well-recognized water-soluble thermosensitive polymers (**Figure 2**). They both undergo hydrophilic to hydrophobic phase transition when temperature increases, exhibiting lower critical solution temperature (LCST) behaviour. Below the LCST, the polymer is soluble, but when temperature increases, they experience a reversible volume phase transition.

Chitosan has been grafted with PNIPAm and PVCL. Other chitosan derivatives such as chitosan-poly(ethylene glycol) copolymer and *N*-isobutyryl chitosan also display thermoreversible behaviour.

In this section, an overview about the methods to synthesize chitosan-based copolymers will be described, as well as their physicochemical properties and potential applications.

3.1. Chitosan-*graft*-poly(*N*-isopropylacrylamide)

3.1.1. Synthesis

The *grafting from* approaches to modify chitosan with NIPAm could be performed under simple homogeneous conditions by one-pot free radical polymerization. With this purpose cerium ammonium nitrate has been frequently employed as initiator [45], as well as a variety of thermal initiators such as azobisisobutyronitrile (AIBN) [46] and ammonium or potassium persulfate [47–50]. Cerium ion is an efficient redox agent capable of undergoing radical polymerization under soft conditions, in acidic aqueous media at low temperatures [51–54].

This copolymer has also been prepared by the *grafting onto* strategy. With this purpose, a two-step synthesis should be conducted. Firstly, carboxyl-terminated PNIPAm is obtained by free radical polymerization in the presence of 3-mercaptopropionic acid as chain-transfer agent. Then, the homopolymer is grafted onto chitosan chain using a condensing agent such as EDC/NHS [55, 56].

Nowadays, more organized polymer structures are required to attend specific biotechnological applications. The conventional radical polymerization employed to prepare PNIPAm-COOH homopolymers by the *grafting onto* strategies does not allow a control over the degree of polymerization of PNIPAm chains, and hence no predefined molecular architecture of chitosan-*g*-PNIPAm copolymer could be synthesized.

Atom transfer radical polymerization (ATRP) is an alternative in which the polymer length is controlled by the synthesis of well-defined graft chains. Bao et al. have obtained a Cs-*g*-PNIPAm copolymer by the following approach: azide-ended PNIPAm homopolymer was firstly prepared through ATRP. Simultaneously, alkynyl pendant Cs derivative is prepared by the amidation of Cs with 4-pentynoic acid in the presence of EDC/NHS. Finally, the Cs copolymer was completed by the click reaction of alkynyl Cs with PNIPAm-N_3 under mild click chemistry conditions. The click reaction proved to be an efficient coupling method for grafting Cs [57]. The same research group has also reported the synthesis of a comb-type Cs(-*g*-PDMAEMA)-*g*-PNIPAm terpolymer using a similar strategy of copolymerization [58].

Chen et al. have proposed a four-step route where the grafting reaction was directed towards the C-6 position; the method involves [59]:

- Protection of amino groups of chitosan by *N*-phthaloylation.
- Preparation of the bromoisobutyryl-terminated *N*-phthaloyl chitosan macroinitiator.
- Synthesis of the copolymer via ATRP from the chitosan macroinitiator with NIPAm.
- Removal of the *N*-phthaloyl groups regenerates amino groups.

Don et al. synthesized Cs-*g*-PNIPAm by means of a two-step route: in order to include vinyl carboxylic acid groups in the backbone, chitosan was first modified with maleic anhydride to produce MA-Cs. In the second step, NIPAm monomer was grafted onto MA-Cs via UV-initiated free radical polymerization [60]. Later, this group proposed a new grafting route in which chitosan amino groups were protected by *N*-phthaloylation, and then the vinyl functional group was introduced at the C-6 position by reaction with *m*-tetramethylxylene isocyanate, followed by deprotection of amino groups. Finally, PNIPAm was grafted to the vinyl Cs by UV-initiated free radical polymerization [61].

3.1.2. Properties

Due to the interesting dual-responsive behaviour of Cs-*g*-PNIPAm copolymers, the properties of these materials have been thoroughly studied using different techniques such as micro-differential scanning calorimetry (μDSC), dynamic light scattering (DLS), NMR, UV-Vis spectroscopy and rheological measurements, among others [46, 51, 52, 54, 58–60, 62–65].

During heating of Cs-*g*-PNIPAm solutions, μDSC traces show a sharp endothermic peak associated with the phase transition at LCST. Associated enthalpy values are proportional to NIPAm content in the copolymer. Comparable exothermic peaks were obtained during cooling, giving rise to fully reversible transition [52].

Hydrophilic and hydrophobic interactions are important factors governing thermosensitive properties of NIPAm polymers. At temperatures below LCST, water molecules form regular ice-like structures around hydrophobic methyl groups. An increase in temperature results in a breakdown of the hydrophobic hydration. As a result, hydrophobic interactions between methyl groups from different NIPAm-grafted blocks are promoted, giving rise to a polymer network. From a thermodynamic point of view, such a phase transition should generate a conformational entropy loss upon polymer association, which should be compensated by the translational-entropy gain of expelled water molecules. Therefore, there is a total entropy increment upon phase transition that overcomes the observed endothermic enthalpy, thus giving rise to a decrement in Gibbs free energy [52, 66].

Variation of viscoelastic G′ and G″ moduli during heating also confirms the existence of a sol-gel transition: as the phase transition takes place, there is a marked increase in storage modulus and a moderate decrement in loss modulus (**Figure 3a**). This phenomenon has been interpreted as the result of the formation of hydrophobic junctions at the expense of the net amount of sol fraction, giving rise to the formation of more elastic networks [52].

It is well known that the LCST value of PNIPAm may be changed conveniently if the hydrophobicity of the system is altered, either by changes in macromolecular composition or by changes in overall hydrophilicity of the surrounding environment [50, 66–68]. This response is also observed for Cs-*g*-PNIPAm copolymers. Wang et al. successfully regulated the LCST adding acrylamide as a comonomer during the synthesis. By this way, the LCST of Cs-*g*-poly(NIPAm-co-AAm) was increased from 33 to 38 °C when 5.5% of AAm was included [69]. An increment in LCST from 33 to 44 °C as the chitosan feed concentration with respect to NIPAm was raised up is also reported [67].

The influence of the environmental conditions on the LCST of Cs-*g*-PNIPAm is noticeable. For example, it was observed that the transition temperature of the copolymer shifted to lower temperatures with increasing concentration of alcohols [70]. Meanwhile, addition of salt to PNIPAm solutions is known to disrupt the regular ice-like structure of water molecules around NIPAm moieties resulting in a decrement of the transition temperature [71, 72]. Such an effect was appreciated for the copolymer: the addition of NaCl to hydrochloric stoichiometric solutions of the copolymer decreased the LCST and caused an increase in enthalpy change [52].

The fully reversible behaviour of the phase transition of this copolymer is an upmost property. On the one hand, it is noticeable from the fast and reversible variation of the viscoelastic moduli during heating-cooling cycles [52]. On the other hand, the same behaviour is evident from the continuous swelling-shrinking cycles, induced by stepwise periodic changes in temperature for a polyelectrolyte complex membrane of Cs-*g*-PNIPAm [63] (**Figure 3b**).

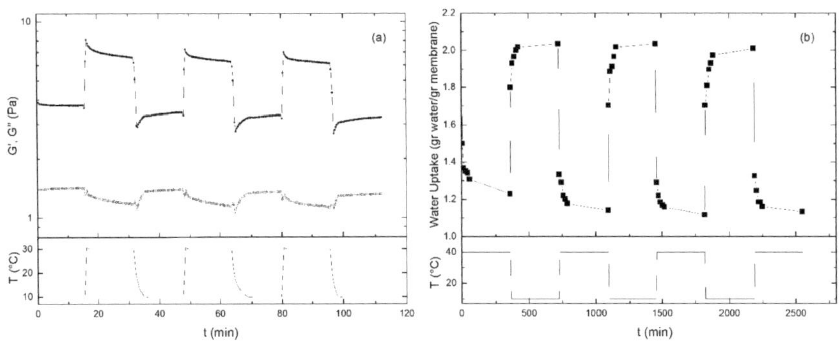

Figure 3. (a) Variations in mechanical moduli, G′ (,) and G″ (–), for 1% (w/w) solution of Cs-*g*-NIPAm in 10% aqueous acetic acid to stepwise periodic changes in temperature between 10 and 30 °C (ω = 1 rad s^{-1}; γ = 5%); reprinted with permission from Ref. [52]. Copyright 2009, American Chemical Society. (b) Variation of swelling in pure water (pH = 5.9) for a Cs-*g*-PNIPAm polyelectrolyte complex membrane to stepwise periodic changes in temperature between 10 and 40 °C; reprinted from Ref. [63]. Copyright 2011, with permission from Elsevier.

There are evidences that Cs-*g*-PNIPAm undergoes micellization processes above the phase transition temperature. On the one hand, Chen et al. noticed that aqueous solutions of Cs-*g*-PNIPAm (which amino groups were protected during the synthesis) showed pH-dependent behaviour, and at temperatures above the LCST, the copolymer self-assembled into micelles with chitosan core [59]. On the other hand, recent studies have demonstrated that during heating, Cs-*g*-oligo(NIPAm) copolymers self-assembled into aggregates due to the hydrophobization of NIPAm blocks. At 25 °C, oligo(NIPAm) chains are hydrophilic exhibiting expanded structures. In change, at 37 °C, a clear transition is observed, the side-chain segments become hydrophobic and the molecules began to fold [54]. In the same way, Cs(-*g*-PDMAEMA)-*g*-PNIPAm terpolymer experiments thermo- and pH-responsive micellization behaviour in aqueous solutions. Moreover, this terpolymer could form three-layer *onion-like* micelles at 25 °C when pH is above 7 [58].

3.1.3. Applications

Cs-*g*-(PNIPAm) is a potential thermosensitive in situ gel-forming material for ocular drug delivery that may enhance ocular absorption, efficacy, bioavailability and pharmacokinetic properties. Experimental results suggested that, at physiological pH, the copolymer hydrogel can interact with the mucus and cornea cell membrane, increasing the drug residential time [46].

Chitosan/PNIPAm hydrogels have been considered as a useful tool to enhance oral bioavailability of low-solubility drugs such as naproxen [73], paclitaxel [69], caffeine [74], etc.

Raskin et al. have proposed a mucoadhesive amphiphilic nanogel based on the micellization of CS-*g*-oligo(NIPAm) and stabilized through the ionotropic gelation of chitosan. These polymeric micelles are self-assembled and positively charged nanocarriers with potential for improved mucosal administration of hydrophobic drugs [54].

A formulation of curcumin-loaded biodegradable thermoresponsive Cs-*g*-PNIPAm nanoparticles was prepared by ionic cross-linking method. The in vitro drug release was prominent at temperatures above LCST. The curcumin-loaded nanoparticles (NPs) showed specific toxicity on cancer cells and increased apoptosis on PC3 cells [55].

Gui et al. have developed some inorganic/organic hybrid composite hydrogels using Cs-*g*-NIPAm as a shell that combine thermo-/pH sensitivity, fluorescence and biocompatibility giving an attractive option for biological and biomedical applications [64, 75]. In the first report, experimental data evidence that Adriamycin-loaded microspheres could effectively improve drug release and accumulation in targeted tumour cells or tissues [64]. On the second article, authors prepared doxorubicin-loaded nanospheres that exhibited remarkable fluorescence/thermo-/pH sensitivity with high anticancer activity [75].

Recently, interpenetrated cryogel scaffolds of PNIPAm and chitosan have been prepared via free radical polymerization in the presence of cross-linkers. These materials were evaluated as potential bioartificial liver devices. The cell-seeded cryogel proved their capacity to successfully purify plasma, supporting liver function in terms of both detoxification and synthesis of important metabolites [65].

3.2. Chitosan-*graft*-poly(*N*-vinylcaprolactam)

3.2.1. Synthesis

To the best of our knowledge, the first report of the preparation of chitosan-*graft*-PVCL (Cs-*g*-PVCL) copolymer was documented by Kudyshkin et al. [76]. They reported the synthesis of this thermosensitive material by radical graft polymerization. This reaction was performed under homogeneous conditions using a mixture of solvents at 60 °C. Unfortunately, this reaction is reported to be accompanied by a decrease in the molecular weight of chitosan, probably caused by the cleavage of the glycoside bonds by potassium persulfate [76, 77]. In the *grafting from* technique, the radical chain polymerization of a monomer is triggered by the thermal decomposition of the initiator. Simultaneously, the polymer radicals are formed by chain transfer between the propagating radical and polymer. It is well known that this method results in a mixture of homopolymer and graft copolymer, as well as ungrafted backbone polymer [43]. The *grafting from* approach allows the synthesis of graft copolymers by one-step reaction, but its principal drawback is the lack of control on the chain length of the grafted polymer.

A novel approach to obtain Cs-*g*-PVCL based on *grafting onto* copolymerization was documented by Prabaharan et al. This strategy consists of the reaction between functional groups from two different polymers. In this research, the first carboxyl-terminated PVCL homopolymer (PVCL-COOH) was synthesized by radical chain polymerization, using AIBN as initiator and 3-mercaptopropionic acid as chain-transfer agent. Then, PVCL-COOH chains were grafted onto chitosan backbone via amidation reaction using EDC/NHS as coupling agents [78].

After this pioneering work, the synthesis of the chitosan-*graft*-PVCL by the *grafting onto* approach has been documented by other authors [79–85]. The formation of amide bonds between PVCL-COOH and chitosan amino groups has been confirmed by FTIR [78, 80, 85], ^{1}H-NMR [78, 85] and Raman [85] spectroscopy techniques. EDC/NHS system is the activator agent usually used by many authors. The main drawback of this system is the formation of the secondary products, which are difficult to remove leading to low yields of functionalization [86, 87]. In this regard, DMTMM entails some benefits over the former system because it selectively promotes the formation of the amide bond in aqueous solution in a wide range of pH. The coupling reaction is thought to be initiated by addition of a carboxylate anion to DMTMM to give an activated ester, which undergoes attack by an amino to give the corresponding amide [88, 89].

The main advantage of the *grafting onto* procedure is the possibility to control the molecular architecture of the copolymer. In this sense, our group has reported the synthesis of PVCL-COOH samples with different molecular weights by means of controlled radical polymerization and their subsequent grafting onto the chitosan backbone using DMTMM as activator agent. This approach allows a control of the chain length of the grafted PVCL chains, as well as the spacing between grafted side chains onto the chitosan backbone [85].

Cs-*g*-PVCL copolymer has also been synthesized by gamma radiation [90]. This method allows obtaining functionalized materials without remaining residues. The reaction solution is irradiated with a ^{60}Co γ-source using doses between 10 and 50 kGy. The ionizing radiation is a powerful tool to achieve functionalization of polymers requiring no additional reactants [91]. However, it has been documented that γ-radiation leads to the scission of 1–4 glycosidic bond of chitosan, thus reducing the molecular weight [92, 93].

3.2.2. Properties

The Cs-*g*-PVCL system shows properties of both, chitosan and PVCL. As it is well known, chitosan is a pH-sensitive, non-toxic, biodegradable, biocompatible linear polyelectrolyte [2], while PVCL is a non-ionic, biocompatible, thermosensitive water-soluble polymer with a phase transition temperature in the physiological range [94]. This graft copolymer showed pH and temperature sensitiveness that could be very interesting for the development of smart-controlled release systems, as well as active scaffold for tissue engineering and regenerative medicine.

The grafting parameters, such as grafting percentage and grafting efficiency, are greatly influenced by the reaction conditions [95]. These parameters have been evaluated by ^{1}H-NMR [78], thermogravimetric analysis (TGA) [85] and gravimetrically [79, 85, 90].

Kholmuminov et al. observed a relatively rapid decrease on the effective viscosity of Cs-*g*-PVCL solutions at shear rates up to 100 s^{-1}. These authors suggested that this behaviour is related to the presence of flexible PVCL chain blocks that easily unwind and orient in a flow when increasing the shear rate [77].

The phase transition behaviour of Cs-*g*-PVCL in aqueous solutions has been investigated by turbidimetry [78–80, 85], DSC [79, 85] and DLS [96]. The LCST of this copolymer has been estimated between 32 and 42 °C, and it was close to the transition of the corresponding homopolymers.

It has been demonstrated that the properties of the Cs-*g*-PVCL are controlled by the molecular architecture of the copolymer: length of the grafted side chains and their spacing along the chitosan backbone [85].

On the one hand, the temperature of the phase transition of the copolymer depends on the spacing between grafted PVCL chains along chitosan backbone. Studies by μDSC evidenced that the longer the spacing between PVCL-grafted chains, the lower the LCST of the polymer, as well as the enthalpy associated with the phase transition (**Figure 4a**) [85]. It is obvious that as the spacing between grafted chains is smaller, the phase transition is more cooperative, thus facilitating the hydrophobic intercatenary interactions between PVCL chain segments during the dehydration process at the critical temperature. From a thermodynamic point of view, the phase transition produces a loss of conformational entropy due to the aggregation of PVCL chain segments. This phenomenon must be counterbalanced by the translational-entropy gain when water molecules are expelled out from the excluded volume of PVCL macromolecules during phase separation. Thus, the larger the hydrophobic portions are, the greater the entropy gain is, which explains why the cooperative hydrophobic interactions are favoured when the PVCL chains are closer in space [85].

On the other hand, the longer the grafted PVCL chain, the lower the phase transition temperature (**Figure 4b**) [85, 96]. This phenomenon has been attributed to the fact that increasing the PVCL chain length, the polymer-polymer interactions become more and more cooperative. As a result, longer hydrophobic segments appear, thus favouring polymer-polymer long-range interactions giving rise to phase separation phenomena at lower temperatures [85].

Since this system exhibits a dual temperature and pH sensitiveness, the ionic strength and pH of the medium have also an influence on the properties of the Cs-*g*-PVCL aqueous solutions, as evidenced from DLS and ζ-potential studies [96]. According to the DLS measurements, at temperatures above the phase separation, the size of the macromolecular aggregates is greater as the medium is more acidic, and a slight increment on the transition temperature is also observed (**Figure 4d**) [96]. Concerning the influence of the ionic strength on the copolymer solutions, a noticeable effect of the electrostatic screening on the hydrodynamic sizes of the copolymer coils, which causes that the phase transition takes place at lower temperatures, giving bigger macromolecular aggregates (**Figure 4e**) was found [96].

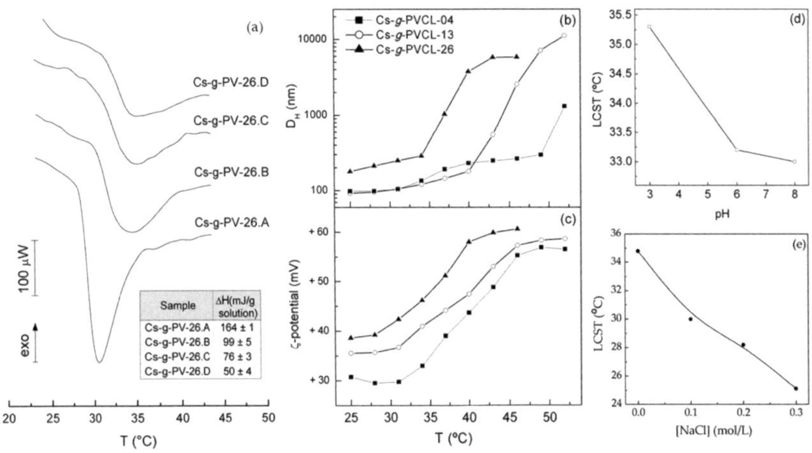

Figure 4. (a) Heating μDSC scans of 10 wt% aqueous solutions of Cs-*g*-PV-26.A, Cs-*g*-PV-26.B, Cs-*g*-PV-26.C and Cs-*g*-PV-26.D samples (heating rate, 0.6 °C min^{-1}; reference, water). Reprinted from Ref. [85]. Copyright 2015, with permission from Elsevier. Dependence of the (b) hydrodynamic diameter, D_H, and (c) ζ-potential of Cs-*g*-PVCL on temperature of aqueous solutions of Cs-*g*-PVCL for different number-average molecular weights of PVCL-grafted chains. Cs-*g*-PVCL samples with Mn 04, 13, and 26 kDa at 2 mg mL^{-1} (pH 6). Heating rate, 0.25 °C min^{-1}. Variation of LCST of the Cs-*g*-PVCL-26 solution (2 mg mL^{-1}) with (d) pH of the medium and (e) ionic strength. Heating rate, 0.25 °C min^{-1}. Reprinted from Ref. [96]. Copyright 2016, with permission from Springer.

Pérez-Calixto et al. have recently addressed the preparation of *N*-vinylcaprolactam and *N,N*-dimethylacrylamide (DMAAm) binary-grafted system onto cross-linked chitosan by γ-radiation. The incorporation of DMAAm hydrophilic comonomer increased the phase transition temperature from 34 to 37 °C, as well as the swelling degree due to intermolecular interactions with amino groups of chitosan molecule [97].

It is important to remark that Fernández-Quiroz et al. have reported that all the Cs-*g*-PVCL copolymers they synthesized were soluble in water at neutral pH, at room temperature. From a thermodynamic point of view, this fact suggests a significant improvement of the copolymer solvation via hydrogen bonding between PVCL-nitrogen unshared electron pair and water molecules. A better polymer-solvent interaction induces higher entropy of mixing giving rise to the dissolution of the macromolecule. This property could be of considerable interest for biomedical applications, where it is important to keep a pH near to the physiological one [85].

3.2.3. *Applications*

Thermoresponsive chitosan derivatives based on Cs-*g*-PVCL have been chiefly studied for biomedical applications as drug delivery and tissue engineering. Their biocompatibility, low toxicity, pH and temperature sensitiveness have drawn a great scientific interest.

In this sense, beads and nanomaterials based on Cs-*g*-PVCL copolymer have been prepared by ionotropic gelation using sodium tripolyphosphate (TPP) as ionic cross-linking agent [78, 80–84]. Studies carried out by Prabaharan et al. indicated that the copolymer has a swelling degree higher at pH 2.2 than at pH 7.4, decreasing with increased environmental temperature. These copolymer beads exhibited a slower release of ketoprofen, as compared with chitosan, revealing a different release profile in dependence on the pH and temperature. According to MTT assay, the copolymer showed no obvious cytotoxicity [78].

Rejinold et al. studied the behaviour of Cs-*g*-PVCL nanoparticles as 5-fluorouracil carrier for its delivery to cancer cells. The copolymer showed a temperature-induced phase transition in the range of 38–45 °C in aqueous solutions, and its NPs display nearly spherical shape with an average diameter of 150 nm, which increased up to 180–200 nm when loaded with the drug. According to the drug delivery, cytotoxicity, in vitro cell uptake and apoptosis studies, these NPs could be a promising candidate for cancer drug delivery [80].

Recently, several studies relative to Cs-*g*-PVCL nanoparticles loaded with curcumin (Cs-*g*-PVCL-CRC-NPs) have been investigated as a potential anticancer drug delivery carrier system [81–84]. These NPs exhibit similar morphology to those in Ref. [80]. The rate of drug release was dependent on pH and temperature of the medium and showed an increase in delivery at temperatures above LCST in acidic pH conditions. These NPs exhibited specific toxicity to cancer cells at above their LCST [84].

In other articles, this group also describes Cs-*g*-PVCL-CRC-NPs in combination with metallic nanoparticles to assess their feasibility as a potential system for radio frequency (RF)-assisted cancer therapy [81–83]. With this purpose, gold nanoparticles (Au-NPs) were incorporated in Cs-*g*-PVCL-CRC-NPs in order to induce RF-assisted heating. These NPs showed uniform spherical shape, with particle size around 170 nm and ζ-potential of +18 mV. These NPs proved to be beneficial for combined RF therapy for treating breast and colon cancers [81, 83].

In a recent study, iron oxide nanoparticles (Fe_3O_4 NPs) were also incorporated to Cs-*g*-PVCL-CRC-NPs. Fe_3O_4 NPs require lower-background RF heating than the Au-NPs. In this report, 80 W for 2 min was required to heat the system up to 42 °C, and curcumin was controlled and released in cancer cells. These nanoparticles showed cellular internalization on array of the cancer cells, which decrease in cell viability and increase in cellular apoptosis [82].

Indulekha et al. documented the behaviour of a thermoresponsive polymeric gel based on Cs-*g*-PVCL as an on-demand transdermal drug delivery carrier for pain management. In this article, the delivery behaviour was analyzed by loading acetamidophenol (a model hydrophilic drug) and etoricoxib (a model hydrophobic drug). This material showed a pulsatile drug release (ON-OFF mechanism), giving an enhanced release for both drugs at temperature above LCST and pH 5.5. Histopathological results proved that the gel is biocompatible [79].

3.3. Other chitosan derivatives

Selective modification of chitosan via *N*-acylation with carboxylic anhydrides in dilute acetic acid/methanol under mild conditions has been documented by Hirano et al. [98–100]. These gels are colourless, transparent, rigid and stable on heating. Among them, the gel formation during the *N*-isobutyrylation of chitosan has been investigated in our laboratory [101]. *N*-Isobutyryl chitosan has structural and steric similarities with NIPAm.

Even when it is known that *N*-acylation leads to irreversible gel formation, *N*-isobutyryl chitosan behaves as a thermally sensitive hydrogel material similar to other thermally synthetic systems such as PNIPA. Indeed, after the derivatization reaction, the gel was thoroughly washed with water at 60 °C in order to get rid of methanol, acetic acid and any excess of reactant. When this hydrogel was submitted to cyclic stepwise changes in temperature, there is a pulsatile reversible response of the viscoelastic moduli, which persisted after four cycles [101].

Lastly, it should be mentioned that other chitosan-*g*-poly(ethylene glycol) thermoreversible copolymers have been also recently proposed [102–104], which are potentially suitable in biomedicine, especially for drug release and tissue engineering applications.

4. Concluding remarks

In this chapter, a review about the methods of synthesis of chitosan thermosensitive derivatives is presented, as well as their featured properties and key potential applications.

When adding glycerophosphate to chitosan solutions at room temperature, the polymer remains in solution at neutral pH, while a gelation can be triggered upon heating around the physiological temperature. This thermo-gelation process involves several interactions such as screening of electrostatic repulsion, "ionic" cross-linking, hydrophobic effect and hydrogen-bonding interactions.

Some chitosan derivatives have also been extensively investigated. Among them, chitosan-*grafted*-poly(*N*-isopropylacrylamide) and poly(*N*-vinylcaprolactam) are two well-recognized chitosan-based thermosensitive materials (**Figure 5a** and **b**, respectively). Other derivatives such as chitosan-*g*-poly(ethylene glycol) and *N*-isobutyryl chitosan also display thermoreversible behaviour. These polymer materials exhibit a lower critical solution temperature behaviour. Below the LCST, polymer chains are soluble, but when temperature increases, they experience a hydrophilic-to-hydrophobic phase transition. In all cases, hydrophobic interactions play a key role, which are associated to a dehydration process at the critical temperature. Thus, the phase transition produces a loss of conformational entropy due to the aggregation of grafted chain segments in an extended polymer network. This phenomenon is counterbalanced by the translational-entropy gain when water molecules are expelled out from the excluded volume of the copolymers during phase separation.

Due to the polyelectrolyte nature of chitosan, all these materials are sensitive to changes in both temperature and pH. A great variety of thermosensitive materials have been produced, such as nanostructures, micelles, membranes, hydrogels, cryogels and paramagnetic beads, to cite some of them. These characteristics make them susceptible to be used in advanced and exciting technological applications such as sensors, actuators, controllable membrane for separations and in biomedical and biotechnological fields including drug delivery and tissue engineering (**Figure 5**).

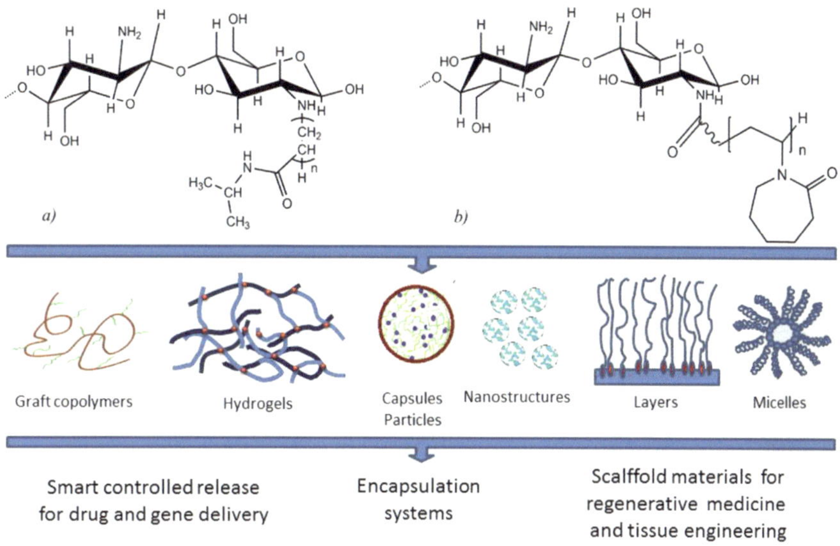

Figure 5. Chemical structure of (*a*) chitosan-*g*-PNIPAm, (*b*) chitosan-*g*-PVCL copolymers and some of their key applications.

Author details

Waldo Argüelles-Monal*, Maricarmen Recillas-Mota and Daniel Fernández-Quiroz

*Address all correspondence to: waldo@ciad.mx

Laboratory of Natural Polymers, Research Center for Food and Development, Guaymas, Sonora, Mexico

References

[1] Igor Galaev, Bo Mattiasson, editors. Smart Polymers: Applications in Biotechnology and Biomedicine. Second edition. Boca Raton, FL: CRC Press, Taylor & Francis Group; 2008.

[2] Peniche C, Argüelles-Monal W, Goycoolea FM. Chitin and chitosan: major sources, properties and applications. In: Gandini A, Belgacem M, editors. Monomers Polym. Compos. Renew. Resour., Amsterdam: Elsevier; 2008, p. 517–42.

[3] Chenite A, Chaput C, Wang D, Combes C, Buschmann MD, Hoemann CD, Leroux JC, Atkinson BL, Binette F, Selmani A. Novel injectable neutral solutions of chitosan form biodegradable gels in situ. Biomaterials 2000;21:2155–61.

[4] Ruel-Gariépy E, Chenite A, Chaput C, Guirguis S, Leroux J-C. Characterization of thermosensitive chitosan gels for the sustained delivery of drugs. Int J Pharm 2000;203:89–98.

[5] Cho J, Heuzey M-C, Bégin A, Carreau PJ. Physical gelation of chitosan in the presence of β-glycerophosphate: the effect of temperature. Biomacromolecules 2005;6:3267–75.

[6] Berger J, Reist M, Chenite A, Felt-Baeyens O, Mayer JM, Gurny R. Pseudo-thermosetting chitosan hydrogels for biomedical application. Int J Pharm 2005;288:17–25.

[7] Crompton KE, Prankerd RJ, Paganin DM, Scott TF, Horne MK, Finkelstein DI, Gross KA, Forsythe JS. Morphology and gelation of thermosensitive chitosan hydrogels. Biophys Chem 2005;117:47–53.

[8] Cho J, Heuzey M-C, Bégin A, Carreau PJ. Chitosan and glycerophosphate concentration dependence of solution behaviour and gel point using small amplitude oscillatory rheometry. Food Hydrocoll 2006;20:936–45.

[9] Cho J, Heuzey M-C. Dynamic scaling for gelation of a thermosensitive chitosan-β-glycerophosphate hydrogel. Colloid Polym Sci 2008;286:427–34.

[10] Supper S, Anton N, Seidel N, Riemenschnitter M, Schoch C, Vandamme T. Rheological study of chitosan/polyol-phosphate systems: influence of the polyol part on the thermo-induced gelation mechanism. Langmuir 2013;29:10229–37.

[11] Qiu X, Yang Y, Wang L, Lu S, Shao Z, Chen X. Synergistic interactions during thermosensitive chitosan-β-glycerophosphate hydrogel formation. RSC Adv 2011;1:282–9.

[12] Jarry C, Chaput C, Chenite A, Renaud M, Buschmann M, Leroux J. Effects of steam sterilization on thermogelling chitosan-based gels. J Biomed Mater Res 2001;58:127–35.

[13] Jarry C, Leroux J-C, Haeck J, Chaput C. Irradiating or autoclaving chitosan/polyol solutions: effect on thermogelling chitosan-β-glycerophosphate systems. Chem Pharm Bull (Tokyo) 2002;50:1335–40.

[14] Ahmadi R, Burns AJ, de Bruijn JD. Chitosan-based hydrogels do not induce angiogenesis. J Tissue Eng Regen Med 2010;4:309–15.

[15] Molinaro G, Leroux J-C, Damas J, Adam A. Biocompatibility of thermosensitive chitosan-based hydrogels: an in vivo experimental approach to injectable biomaterials. Biomaterials 2002;23:2717–22.

[16] Ahmadi R, de Bruijn JD. Biocompatibility and gelation of chitosan–glycerol phosphate hydrogels. J Biomed Mater Res A 2008;86A:824–32.

[17] Zhou HY, Jiang LJ, Cao PP, Li JB, Chen XG. Glycerophosphate-based chitosan thermosensitive hydrogels and their biomedical applications. Carbohydr Polym 2015;117:524–36.

[18] Hoemann CD, Sun J, Légaré A, McKee MD, Buschmann MD. Tissue engineering of cartilage using an injectable and adhesive chitosan-based cell-delivery vehicle. Osteoarthritis Cartilage 2005;13:318–29.

[19] Crompton KE, Goud JD, Bellamkonda RV, Gengenbach TR, Finkelstein DI, Horne MK, Forsythe JS. Polylysine-functionalised thermoresponsive chitosan hydrogel for neural tissue engineering. Biomaterials 2007;28:441–9.

[20] Cheng Y-H, Yang S-H, Su W-Y, Chen Y-C, Yang K-C, Cheng WT-K, Wu S-C, Lin F-H. Thermosensitive chitosan–gelatin–glycerol phosphate hydrogels as a cell carrier for nucleus pulposus regeneration: an in vitro study. Tissue Eng Part A 2009;16:695–703.

[21] Nair S, Remya NS, Remya S, Nair PD. A biodegradable in situ injectable hydrogel based on chitosan and oxidized hyaluronic acid for tissue engineering applications. Carbohydr Polym 2011;85:838–44.

[22] Dessì M, Borzacchiello A, Mohamed THA, Abdel-Fattah WI, Ambrosio L. Novel biomimetic thermosensitive β-tricalcium phosphate/chitosan-based hydrogels for bone tissue engineering. J Biomed Mater Res A 2013;101:2984–93.

[23] Niranjan R, Koushik C, Saravanan S, Moorthi A, Vairamani M, Selvamurugan N. A novel injectable temperature-sensitive zinc doped chitosan/β-glycerophosphate hydrogel for bone tissue engineering. Int J Biol Macromol 2013;54:24–9.

[24] Rahmati M, Samadikuchaksaraei A, Mozafari M. Insight into the interactive effects of β-glycerophosphate molecules on thermosensitive chitosan-based hydrogels. Bioinspir Biomim Nanobiomater 2016;5:67–73.

[25] Iliescu M, Hoemann CD, Shive MS, Chenite A, Buschmann MD. Ultrastructure of hybrid chitosan–glycerol phosphate blood clots by environmental scanning electron microscopy. Microsc Res Tech 2008;71:236–47.

[26] Cui J, Jiang B, Liang J, Sun C, Lan J, Sun X, Huang H, Sun K, Xu X. Preparation and characterization of chitosan/β-GP membranes for guided bone regeneration. J Wuhan Univ Technol Mater Sci Ed 2011;26:241–5.

[27] Ruel-Gariépy E, Leclair G, Hildgen P, Gupta A, Leroux J-C. Thermosensitive chitosan-based hydrogel containing liposomes for the delivery of hydrophilic molecules. J Control Release 2002;82:373–83.

[28] Ruel-Gariépy E, Shive M, Bichara A, Berrada M, Le Garrec D, Chenite A, Leroux J-C. A thermosensitive chitosan-based hydrogel for the local delivery of paclitaxel. Eur J Pharm Biopharm 2004;57:53–63.

[29] Zan J, Chen H, Jiang G, Lin Y, Ding F. Preparation and properties of crosslinked chitosan thermosensitive hydrogel for injectable drug delivery systems. J Appl Polym Sci 2006;101:1892–8.

[30] Cheng Y-H, Yang S-H, Lin F-H. Thermosensitive chitosan-gelatin-glycerol phosphate hydrogel as a controlled release system of ferulic acid for nucleus pulposus regeneration. Biomaterials 2011;32:6953–61.

[31] Lee S, Saito K, Lee H-R, Lee MJ, Shibasaki Y, Oishi Y, Kim B-S. Hyperbranched double hydrophilic block copolymer micelles of poly(ethylene oxide) and polyglycerol for ph-responsive drug delivery. Biomacromolecules 2012;13:1190–6.

[32] Khodaverdi E, Tafaghodi M, Ganji F, Abnoos K, Naghizadeh H. In vitro insulin release from thermosensitive chitosan hydrogel. AAPS PharmSciTech 2012;13:460–6.

[33] Liu W-F, Kang C-Z, Kong M, Li Y, Su J, Yi A, Cheng X-J, Chen X-G. Controlled release behaviors of chitosan/α, β-glycerophosphate thermo-sensitive hydrogels. Front Mater Sci 2012;6:250–8.

[34] Depani BP, Naik AA, Nair HA. Preparation and evaluation of chitosan based thermoreversible gels for intraperitoneal delivery of 5-fluorouracil (5-FU). Acta Pharm 2013;63:479–491.

[35] Kushwaha SKS, Rai AK, Singh S. Formulation of thermosensitive hydrogel containing cyclodextrin for controlled drug delivery of camptothecin. Trop J Pharm Res 2014;13:1007–12.

[36] Supper S, Anton N, Seidel N, Riemenschnitter M, Curdy C, Vandamme T. Thermosensitive chitosan/glycerophosphate-based hydrogel and its derivatives in pharmaceutical and biomedical applications. Expert Opin Drug Deliv 2014;11:249–67.

[37] Faikrua A, Wittaya-areekul S, Oonkhanond B, Viyoch J. A thermosensitive chitosan/corn starch/β-glycerol phosphate hydrogel containing TGF-β1 promotes differentiation of MSCs into chondrocyte-like cells. Tissue Eng Regen Med 2014;11:355–61.

[38] Wu J, Su Z-G, Ma G-H. A thermo- and pH-sensitive hydrogel composed of quaternized chitosan/glycerophosphate. Int J Pharm 2006;315:1–11.

[39] Tahrir FG, Ganji F, Ahooyi T. Injectable thermosensitive chitosan/glycerophosphate-based hydrogels for tissue engineering and drug delivery applications: a review. Recent Pat Drug Deliv Formul 2015;9:107–20.

[40] Tsai M-L, Chang H-W, Yu H-C, Lin Y S, Tsai Y-D. Effect of chitosan characteristics and solution conditions on gelation temperatures of chitosan/2-glycerophosphate/nanosilver hydrogels. Carbohydr Polym 2011;84:1337–43.

[41] Chang H-W, Lin Y-S, Tsai Y-D, Tsai M-L. Effects of chitosan characteristics on the physicochemical properties, antibacterial activity, and cytotoxicity of chitosan/2-glycerophosphate/nanosilver hydrogels. J Appl Polym Sci 2013;127:169–76.

[42] Lizardi-Mendoza J, Argüelles Monal WM, Goycoolea Valencia FM. Chemical characteristics and functional properties of chitosan. In: Silvia Bautista-Baños, Gianfranco Romanazzi, Antonio Jiménez-Aparicio, editors. Chitosan Preserv. Agric. Commod., San Diego: Academic Press; 2016, p. 3–31.

[43] Odian G. Principles of Polymerization. Hoboken, New Jersey: John Wiley & Sons; 2004.

[44] Zhang M, Müller AHE. Cylindrical polymer brushes. J Polym Sci Part Polym Chem 2005;43:3461–81.

[45] Jenkins DW, Hudson SM. Review of vinyl graft copolymerization featuring recent advances toward controlled radical-based reactions and illustrated with chitin/chitosan trunk polymers. Chem Rev 2001;101:3245–74.

[46] Cao Y, Zhang C, Shen W, Cheng Z, Yu L, Ping Q. Poly(N-isopropylacrylamide)–chitosan as thermosensitive in situ gel-forming system for ocular drug delivery. J Control Release 2007;120:186–94.

[47] Wang L-Q, Tu K, Li Y, Fu J, Yu F. Micellization behavior of temperature-responsive poly (N-isopropylacrylamide) grafted dextran copolymers. J Mater Sci Lett 2002;21:1453–5.

[48] Lee C-F, Wen C-J, Chiu W-Y. Synthesis of poly(chitosan-N-isopropylacrylamide) complex particles with the method of soapless dispersion polymerization. J Polym Sci Part Polym Chem 2003;41:2053–63.

[49] Nie P, He X, Chen L. Temperature-sensitive chitosan membranes as a substrate for cell adhesion and cell sheet detachment. Polym Adv Technol 2012;23:447–53.

[50] Liu S, Zhang J, Cui X, Guo Y, Zhang X, Hongyan W. Synthesis of chitosan-based nanohydrogels for loading and release of 5-fluorouracil. Colloids Surf Physicochem Eng Asp 2016;490:91–7.

[51] Kim SY, Cho SM, Lee YM, Kim SJ. Thermo- and pH-responsive behaviors of graft copolymer and blend based on chitosan and N-isopropylacrylamide. J Appl Polym Sci 2000;78:1381–91.

[52] Recillas M, Silva LL, Peniche C, Goycoolea FM, Rinaudo M, Argüelles-Monal WM. Thermoresponsive behavior of chitosan-g-N-isopropylacrylamide copolymer solutions. Biomacromolecules 2009;10:1633–41.

[53] Mu B, Liu P, Li X, Du P, Dong Y, Wang Y. Fabrication of flocculation-resistant ph/ionic strength/temperature multiresponsive hollow microspheres and their controlled release. Mol Pharm 2012;9:91–101.

[54] Raskin MM, Schlachet I, Sosnik A. Mucoadhesive nanogels by ionotropic crosslinking of chitosan-g-oligo(NiPAam) polymeric micelles as novel drug nanocarriers. Nanomed 2016;11:217–33.

[55] Sanoj Rejinold N, Sreerekha PR, Chennazhi KP, Nair SV, Jayakumar R. Biocompatible, biodegradable and thermo-sensitive chitosan-g-poly (N-isopropylacrylamide) nanocarrier for curcumin drug delivery. Int J Biol Macromol 2011;49:161–72.

[56] Antoniraj MG, Kumar CS, Kandasamy R. Synthesis and characterization of poly (N-isopropylacrylamide)-g-carboxymethyl chitosan copolymer-based doxorubicin-loaded polymeric nanoparticles for thermoresponsive drug release. Colloid Polym Sci 2015;294:527–35.

[57] Bao H, Li L, Leong WC, Gan LH. Thermo-responsive association of chitosan-graft-poly(N-isopropylacrylamide) in aqueous solutions. J Phys Chem B 2010;114:10666–73.

[58] Bao H, Li L, Gan LH, Ping Y, Li J, Ravi P. Thermo- and pH-responsive association behavior of dual hydrophilic graft chitosan terpolymer synthesized via ATRP and click chemistry. Macromolecules 2010;43:5679–87.

[59] Chen C, Liu M, Gao C, Lü S, Chen J, Yu X, Ding E, Yu C, Guo J, Cui G. A convenient way to synthesize comb-shaped chitosan-graft-poly (N-isopropylacrylamide) copolymer. Carbohydr Polym 2013;92:621–8.

[60] Don T-M, Chen H-R. Synthesis and characterization of AB-crosslinked graft copolymers based on maleilated chitosan and N-isopropylacrylamide. Carbohydr Polym 2005;61:334–47.

[61] Don T-M, Chou S-C, Cheng L-P, Tai H-Y. Cellular compatibility of copolymer hydrogels based on site-selectively-modified chitosan with poly(N-isopropyl acrylamide). J Appl Polym Sci 2011;120:1–12.

[62] Bae JW, Go DH, Park KD, Lee SJ. Thermosensitive chitosan as an injectable carrier for local drug delivery. Macromol Res 2006;14:461–5.

[63] Recillas M, Silva LL, Peniche C, Goycoolea FM, Rinaudo M, Román JS, Argüelles-Monal WM. Thermo- and pH-responsive polyelectrolyte complex membranes from chitosan-g-N-isopropylacrylamide and pectin. Carbohydr Polym 2011;86:1336–43.

[64] Gui R, Wang Y, Sun J. Encapsulating magnetic and fluorescent mesoporous silica into thermosensitive chitosan microspheres for cell imaging and controlled drug release in vitro. Colloids Surf B Biointerfaces 2014;113:1–9.

[65] Jain E, Damania A, Shakya AK, Kumar A, Sarin SK, Kumar A. Fabrication of macroporous cryogels as potential hepatocyte carriers for bioartificial liver support. Colloids Surf B Biointerfaces 2015;136:761–71.

[66] Feil H, Bae YH, Feijen J, Kim SW. Effect of comonomer hydrophilicity and ionization on the lower critical solution temperature of N-isopropylacrylamide copolymers. Macromolecules 1993;26:2496–500.

[67] Jaiswal MK, Banerjee R, Pradhan P, Bahadur D. Thermal behavior of magnetically modalized poly(N-isopropylacrylamide)-chitosan based nanohydrogel. Colloids Surf B Biointerfaces 2010;81:185–94.

[68] Li G, Song S, Zhang T, Qi M, Liu J. pH-sensitive polyelectrolyte complex micelles assembled from CS-g-PNIPAM and ALG-g-P(NIPAM-co-NVP) for drug delivery. Int J Biol Macromol 2013;62:203–10.

[69] Wang Y, Xu H, Wang J, Ge L, Zhu J. Development of a thermally responsive nanogel based on chitosan-poly(N-isopropylacrylamide-co-acrylamide) for paclitaxel delivery. J Pharm Sci 2014;103:2012–21.

[70] Liu W, Huang Y, Liu H, Hu Y. Composite structure of temperature sensitive chitosan microgel and anomalous behavior in alcohol solutions. J Colloid Interface Sci 2007;313:117–21.

[71] Yang Y, Zeng F, Tong Z, Liu X, Wu S. Phase separation in poly(N-isopropyl acrylamide)/water solutions. II. Salt effects on cloud-point curves and gelation. J Polym Sci Part B Polym Phys 2001;39:901–7.

[72] Eeckman F, Amighi K, Moës AJ. Effect of some physiological and non-physiological compounds on the phase transition temperature of thermoresponsive polymers intended for oral controlled-drug delivery. Int J Pharm 2001;222:259–70.

[73] Li G, Guo L, Chang X, Yang M. Thermo-sensitive chitosan based semi-IPN hydrogels for high loading and sustained release of anionic drugs. Int J Biol Macromol 2012;50:899–904.

[74] Lee C-F, Wen C-J, Lin C-L, Chiu W-Y. Morphology and temperature responsiveness-swelling relationship of poly(N-isopropylamide-chitosan) copolymers and their application to drug release. J Polym Sci Part Polym Chem 2004;42:3029–37.

[75] Gui R, Wang Y, Sun J. Embedding fluorescent mesoporous silica nanoparticles into biocompatible nanogels for tumor cell imaging and thermo/pH-sensitive in vitro drug release. Colloids Surf B Biointerfaces 2014;116:518–25.

[76] Kudyshkin VO, Milusheva RY, Futoryanskaya AM, Yunusov MY, Rashidova SS. Synthesis of graft copolymers of N-vinylcaprolactam on chitosan. Russ J Appl Chem 2007;80:1750–2.

[77] Kholmuminov AA, Kudyshkin VO, Futoryanskaya AM, Avazova OB, Milusheva RY, Rashidova SS. Rheological properties of solutions of chitosan and its graft copolymer with N-vinylcaprolactam. Polym Sci Ser A 2010;52:939–41.

[78] Prabaharan M, Grailer JJ, Steeber DA, Gong S. Stimuli-responsive chitosan-graft-poly(N-vinylcaprolactam) as a promising material for controlled hydrophobic drug delivery. Macromol Biosci 2008;8:843–51.

[79] Indulekha S, Arunkumar P, Bahadur D, Srivastava R. Thermoresponsive polymeric gel as an on-demand transdermal drug delivery system for pain management. Mater Sci Eng C 2016;62:113–22.

[80] Sanoj Rejinold N, Chennazhi KP, Nair SV, Tamura H, Jayakumar R. Biodegradable and thermo-sensitive chitosan-g-poly(N-vinylcaprolactam) nanoparticles as a 5-fluorouracil carrier. Carbohydr Polym 2011;83:776–86.

[81] Sanoj Rejinold N, Thomas RG, Muthiah M, Chennazhi KP, Manzoor K, Park I-K, Jeong YY, Jayakumar R. Anti-cancer, pharmacokinetics and tumor localization studies of pH-, RF- and thermo-responsive nanoparticles. Int J Biol Macromol 2015;74:249–62.

[82] Sanoj Rejinold N, Thomas RG, Muthiah M, Lee HJ, Jeong YY, Park I-K, Jayakumar R. Breast tumor targetable Fe_3O_4 embedded thermo-responsive nanoparticles for radiofrequency assisted drug delivery. J Biomed Nanotechnol 2016;12:43–55.

[83] Sanoj Rejinold N, Thomas RG, Muthiah M, Chennazhi KP, Park I-K, Jeong YY, Manzoor K, Jayakumar R. Radio frequency triggered curcumin delivery from thermo and pH responsive nanoparticles containing gold nanoparticles and its in vivo localization studies in an orthotopic breast tumor model. RSC Adv 2014;4:39408–27.

[84] Sanoj Rejinold N, Muthunarayanan M, Divyarani VV, Sreerekha PR, Chennazhi KP, Nair SV, Tamura H, Jayakumar R. Curcumin-loaded biocompatible thermoresponsive polymeric nanoparticles for cancer drug delivery. J Colloid Interface Sci 2011;360:39–51.

[85] Fernández-Quiroz D, González-Gómez Á, Lizardi-Mendoza J, Vázquez-Lasa B, Goycoolea FM, San Román J, Argüelles-Monal WM. Effect of the molecular architecture on the thermosensitive properties of chitosan-g-poly(N-vinylcaprolactam). Carbohydr Polym 2015;134:92–101.

[86] Reyes-Ortega F, Parra-Ruiz FJ, Averick SE, Rodríguez G, Aguilar MR, Matyjaszewski K, San Román J. Smart heparin-based bioconjugates synthesized by a combination of ATRP and click chemistry. Polym Chem 2013;4:2800–14.

[87] Farkaš P, Bystrický S. Efficient activation of carboxyl polysaccharides for the preparation of conjugates. Carbohydr Polym 2007;68:187–90.

[88] Kunishima M, Kawachi C, Monta J, Terao K, Iwasaki F, Tani S. 4-(4,6-Dimethoxy-1,3,5-triazin-2-yl) 4 methyl morpholinium chloride: an efficient condensing agent leading to the formation of amides and esters. Tetrahedron 1999;55:13159–70.

[89] Kunishima M, Kawachi C, Hioki K, Terao K, Tani S. Formation of carboxamides by direct condensation of carboxylic acids and amines in alcohols using a new alcohol- and water-soluble condensing agent: DMT-MM. Tetrahedron 2001;57:1551–8.

[90] Montes JÁ, Ortega A, Burillo G. Dual-stimuli responsive copolymers based on N-vinylcaprolactam/chitosan. J Radioanal Nucl Chem 2014;303:2143–50.

[91] Wang J-P, Chen Y-Z, Ge X-W, Yu H-Q. Gamma radiation-induced grafting of a cationic monomer onto chitosan as a flocculant. Chemosphere 2007;66:1752–7.

[92] Choi W-S, Ahn K-J, Lee D-W, Byun M-W, Park H-J. Preparation of chitosan oligomers by irradiation. Polym Degrad Stab 2002;78:533–8.

[93] Hien NQ, Phu DV, Duy NN, Lan NTK. Degradation of chitosan in solution by gamma irradiation in the presence of hydrogen peroxide. Carbohydr Polym 2012;87:935–8.

[94] Imaz A, Forcada J. N-vinylcaprolactam-based microgels for biomedical applications. J Polym Sci Part Polym Chem 2010;48:1173–81.

[95] Jayakumar R, Prabaharan M, Reis RL, Mano JF. Graft copolymerized chitosan—present status and applications. Carbohydr Polym 2005;62:142–58.

[96] Fernández-Quiroz D, González-Gómez Á, Lizardi-Mendoza J, Vázquez-Lasa B, Goycoolea FM, Román JS, Argüelles-Monal WM. Conformational study on the thermal transition of chitosan-g-poly(N-vinylcaprolactam) in aqueous solution. Colloid Polym Sci 2016;294:555–63.

[97] Pérez-Calixto MP, Ortega A, Garcia-Uriostegui L, Burillo G. Synthesis and characterization of N-vinylcaprolactam/N,N-dimethylacrylamide grafted onto chitosan networks by gamma radiation. Radiat Phys Chem 2016;119:228–35.

[98] Hirano S, Ohe Y. Chitosan gels: a novel molecular aggregation of chitosan in acidic solutions on a facile acylation. Agric Biol Chem 1975;39:1337–8.

[99] Hirano S, Yamaguchi R. N-acetylchitosan gel: a polyhydrate of chitin. Biopolymers 1976;15:1685–91.

[100] Hirano S, Ohe Y, Ono H. Selective N-acylation of chitosan. Carbohydr Res 1976;47:315–20.

[101] Félix L, Hernández J, Argüelles-Monal WM, Goycoolea FM. Kinetics of gelation and thermal sensitivity of N-isobutyryl chitosan hydrogels. Biomacromolecules 2005;6:2408–15.

[102] Liang Y, Deng L, Chen C, Zhang J, Zhou R, Li X, Hu R, Dong A. Preparation and properties of thermoreversible hydrogels based on methoxy poly(ethylene glycol)-grafted chitosan nanoparticles for drug delivery systems. Carbohydr Polym 2011;83:1828–33.

[103] Bhattarai N, Matsen FA, Zhang M. PEG-grafted chitosan as an injectable thermoreversible hydrogel. Macromol Biosci 2005;5:107–11.

[104] Bhattarai N, Ramay HR, Gunn J, Matsen FA, Zhang M. PEG-grafted chitosan as an injectable thermosensitive hydrogel for sustained protein release. J Control Release 2005;103:609–24.

Chapter 14

Chitosan and Phthaloylated Chitosan in Electrochemical Devices

Zurina Osman and Abdul Kariem Arof

Additional information is available at the end of the chapter

http://dx.doi.org/10.5772/65656

Abstract

Chitin and chitosan are widely found in nature. Chitin can be obtained from fungi and in the lower animals. Chitosan can be derived from chitin. The process of chitosan derivation from chitin is called deacetylation. Chitosan is non-toxic, odourless, biocompatible in animal tissues and enzymatically biodegradable. It has found many applications in the fields of cosmetics, wound healing, dietetics and waste-water treatment. Chitosan holds many promising potentials, but its inability to dissolve in many of the common solvents has restricted its application. Hence, chitosan has been modified, and there are now many derivatives of chitosan. In the current chapter, we discuss chitosan and only the phthaloyl chitosan derivative. Their applications in several electrochemical devices are also discussed.

Keywords: chitosan, phthaloyl chitosan, conductivity, batteries, solar cells, supercapacitors, fuel cells

1. General characteristics

Nature contains many types of polysaccharides. One of the important polysaccharides is chitin. Chitin can be obtained from crabs, shrimps and other sea creatures. Chitin can also be derived from insects and algae. Chitosan is the product that can dissolve in dilute acids after chitin has been deacetylated for a sufficient amount of time [1, 2]. Chitosan is a linear polysaccharide.

In nature, chitosan can also be found in some fungi, diatoms, sponges, worms and molluscs [3]. Both chitin and chitosan are versatile and promising biomaterials.

The dilute acids that can dissolve chitosan include dilute acetic, hydrochloric, formic and butyric acids [4]. Chitosan is biodegradable, biocompatible, odourless and non-toxic. The linear polysaccharide also has the ability to form thin films, is resistant to chemical attack, has electrolytic properties, easy to cross-link or modify and has antibacterial property [5–7].

Chitosan has found uses in waste-water processing [8, 9], medicine for wound healing and dietetics [10, 11], agriculture [12], reverse osmosis, food industry [13–15] and also in cosmetics [16–18]. In this article, emphasis is on chitosan and chitosan-derived materials as hosts for ionic conduction.

The utilization of chitosan in energy devices depends on its chemical and mechanical properties. Chitosan films are good liquid absorbers. Thus, chitosan can achieve good ionic conductivity when immersed in liquid electrolytes. This is important for energy storage devices. However, when swollen, chitosan becomes less stable.

A chitin monomer has an amide functional group, $HNCOCH_3$, the infrared (IR) transmission peak is located at 1650 cm^{-1}. This has been partially replaced by the amine functional group, when chitin was deacetylated. The IR peak for the $-NH_2$ group is located at 1590 cm^{-1} in the chitosan spectrum. There is only one $-NH_2$ functional group in the chitosan monomer. The nitrogen atom in the amine group is an electron donor. So is the oxygen atom in the hydroxyl (–OH) functional group. There are two –OH groups in each monomer [19]. The free amine and hydroxyl functional groups allow chitosan to be altered chemically for specific applications. The $-NH_2$ group provides attachment for coordination or chelation of cations from organic/inorganic salts. Coordination of, for example, $LiCF_3SO_3$ [20] with N atom of the reactive functional group enables the formation of complexes that can function as an electrolyte material. Chitosan has a carboxyl group at 1650 cm^{-1} and NH_3^+ peak at 1514 cm^{-1}. According to Ritthidej [21], the NH_3^+ peak can disappear from the spectrum on storage of the film.

Chitosan has good film or membrane-forming ability. This is an advantage since most of the polymer electrolytes are prepared in film form via the solution cast technique [20, 22]. In forming a polymer electrolyte film, the cation of the salt must coordinate electrostatically with the electron donor atom of the polymer. The chitosan-based electrolyte films are homogeneous and have high mechanical strength [23, 24].

According to Sakurai and co-workers [24, 25], chitosan is not a totally amorphous polymer. From wide angle X-ray diffraction (WAXD), it is observed to be partially crystalline. These authors, however, did not investigate the effect of adding salt or plasticizer to chitosan. Alves et al. [26] reported that pure chitosan has a broad band centred at $2\theta \approx 20.6°$ and X-ray diffraction (XRD) patterns change with salt addition. This band implies the amorphous structure of the chitosan-based electrolytes indicating that the polymer chains are essentially disordered. The XRD patterns also exhibited changes depending on the salt concentration. According to Kurita et al. [27], a fully deacetylated chitosan is even more amorphous than that deacetylated up to 95%.

Muzzarelli has reported that chitosan decomposes in air at elevated temperatures [28]. The glass transition temperature, T_g, of the chitosan biopolymer was successfully observed by Sakurai and co-workers [25]. Using differential scanning calorimetry (DSC), they reported T_g at 203°C. According to Dong et al. [29], chitosan has a T_g between 140 and 150°C. The researchers also found that the number of alkali treatments or the degree of deacetylation of chitosan does not influence the T_g.

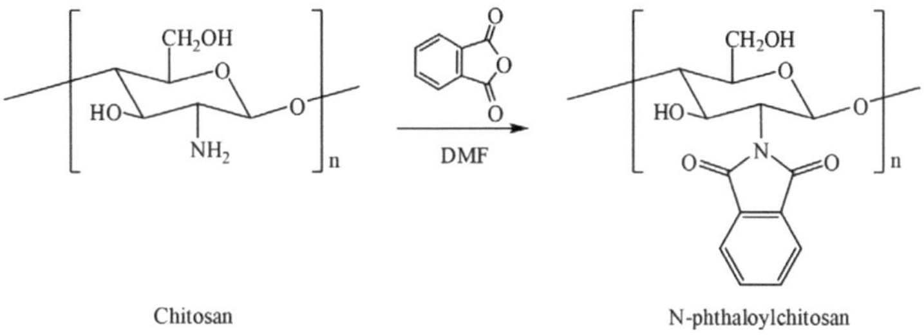

Figure 1. Reaction of N-phthaloylation.

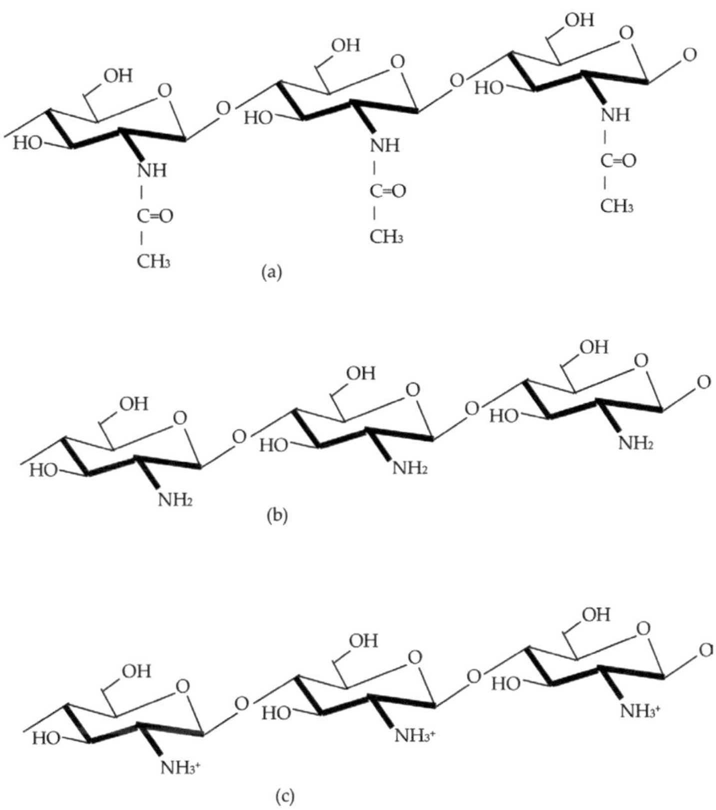

Figure 2. Chemical structure of (a) chitin, (b) chitosan and (c) phthaloyl chitosan.

As mentioned above, in swollen form, chitosan loses its mechanical strength. To overcome this, chitosan has been blended, cast in multilayers or added with inorganic/organic supports. Even then, its utilization is still restricted because of its poor solubility in many of the common solvents such as alkali, organic solvents and water. Hence, to make chitosan soluble in other solvents, it needs to be modified. This can be done by several processes such as acylation and phthaloylation. Phthaloylation is discussed here. Phthaloyl chitosan dissolves in dimethylsulfoxide (DMSO), dimethylacetamide (DMAc), dimethylformamide (DMF) and pyridine [30].

Unlike chitosan, the IR spectrum of phthaloyl chitosan exhibits characteristic peaks at 719, 1708 and 1772 cm^{-1}. Proton NMR (Nuclear Magnetic Resonance) exhibits peaks at 3.0 and 7.5 ppm [30]. Aziz et al. [31] reported that the diffractogram of phthaloyl chitosan showed only one broad peak at around $2\theta = 18°$. Phthaloyl chitosan can be considered as chitosan stripped off its hydrogen atoms, except for the hydrogen in hydroxyl group. It is therefore less rigid than chitosan since inter- and intra-hydrogen bonding have been removed. In the synthesis of phthaloyl chitosan, chitosan and phthalic anhydride were refluxed for 6 h between 100 and 120°C in dimethylformamide (DMF) and nitrogen environment. The temperature was then reduced to 60°C, and the mixture was left overnight. To precipitate out the product, the solution was poured into iced water. The product, N-phthaloyl chitosan (PhCh), was then washed with ethanol and dried in vacuum at 60°C. The synthesis process is shown in **Figure 1**. The chemical structures of chitin, chitosan and phthaloyl chitosan are shown in **Figure 2**.

2. Ionic conductivity

An important characteristic that any electrolyte should have is high ionic conductivity especially at room temperature. In polymer electrolytes, ionic conductivity is attributed to the motion of cations and anions of the doping salt. However, in polymers, both crystalline and amorphous phases exist and ionic conductivity can only occur in the amorphous phase. Hence, if the electrolyte is more amorphous, then more ions can be conducted through the enlarged amorphous region and the conductivity enhanced. The increased amorphousness of the electrolyte can be achieved through several ways. One such way is through doping the polymer with salt. Depending on the salt concentration, some portion of the crystalline phase of the polymer can be disrupted. Increasing the salt concentration can increase amorphousness, but to a certain extent. It has been observed that beyond a particular salt content and for a fixed polymer mass, the salt cannot be totally dissolved in the polymer, and the X-ray diffractogram showed peaks due to the salt [32].

The amorphousness of the electrolyte can also be increased by adding additives such as plasticizers and inert fillers. The plasticizers should have a high dielectric constant and low viscosity. The high dielectric constant helps to dissociate the salt into ions and probably also to dissociate contact ions. The low viscosity is required to promote ion mobility, which will be reduced if the viscosity of the plasticizer is high. The two additives also provide additional pathways for ionic conduction and this is manifested in the smaller activation energy or pseudo-activation energy of the chitosan (Ch) and PhCh ionic conduction hosts. **Table 1** lists representative of the conductivity value of some plasticized Ch- and PhCh-based electrolytes.

According to Croce et al. [34], plasticization can ensure conductivity of polymer electrolytes to achieve appreciable values at ambient temperature. However, plasticization deteriorates the

http://dx.doi.org/10.5772/65656

mechanical strength of the ion-conducting medium. Plasticization can also increase the electrolyte's reactivity towards lithium metal anode. The authors have also shown that inert metal oxides such as Al_2O_3, SiO_2 and TiO_2 can act as solid plasticizers or fillers to enhance ionic conductivity without sacrificing dimensionality of the electrolyte. The fillers must be inert to the contents of the polymer electrolyte. Xiao et al. [35] have investigated the influence of multiwall carbon nanotubes (MWCNTs) on chitosan/cellulose. Conductivity was enhanced on addition of MWCNT. To our knowledge, works concerning filler addition to Ch- and PhCh-salt complexes are scarce. Thus, this could be another area of research that could be carried out to complement work on electrolytes based on Ch and PhCh.

Sample	σ_{RT}/S m^{-1}	ε	η/mPa s	References
Ch- $LiCF_3SO_3$	1.70×10^{-3}	–	–	[32]
Ch-LiTSFI	2.00×10^{-3}	–	–	[33]
Ch-EC-$LiCF_3SO_3$	4.00×10^{-3}	89 (EC)	1.90 (EC)	[22]
Ch-SN-LiTSFI	6.00×10^{-2}	55 (SN)	1.69 (SN)	[33]
PhCh-TPAI-LiI	0.61	–	–	[30]
PhCh-NH_4SCN	2.42×10^{-3}	–	–	[31]
PhCh-EC-DMF-TPAI-I_2	0.55	37 (DMF)	0.80 (DMF)	[30]

EC = ethylene carbonate, DMF = dimethylformamide, SN = succinonitrile, $LiCF_3SO_3$ = lithium triflate, LiTSFI = lithium bis (trifluoromethylsulfonyl) imide, TPAI = tetrapropylammonium iodide.

Table 1. Room temperature conductivity (σ_{RT}) of Ch- and PhCh-based electrolytes, dielectric constant (ε) and viscosity (η) of the plasticizers.

Aziz et al. have prepared proton-conducting electrolytes based on phthaloylated chitosan that was incorporated with NH_4SCN as the doping salt [31]. The highest room temperature conductivity was 2.42×10^{-3} S m^{-1} for the 30 wt.% salt-containing sample.

Muthumeenal et al. have measured the conductivity of polyethersulfone/phthaloyl chitosan blend electrolyte dissolved in concentrated sulphuric acid. The highest room temperature conductivity achieved was 92×10^{-2} S m^{-1} [36]. The high H^+ conductivity can be attributed to the sulfonic acid groups and the presence of the polar groups in PhCh.

Azzahari et al. determined the conductivity of PhCh-iodide-mixed salt (LiI, CsI and BMII) gel polymer electrolyte systems using the method of electrochemical impedance spectroscopy [37]. They have also predicted the conductivity of the systems using response surface methodology and artificial neural network. The authors showed that artificial neural network gave a better prediction than response surface methodology.

Ionic transport in polymer electrolytes can be explained using the Lewis acid-base reaction theory [38]. In this theory, bases donate electron pairs and acids accept them. The oxide and hydroxide groups on the filler provide additional sites (apart from the heteroatoms in the polymer chain) for interaction with the cations of the salt and indirectly form conduction paths for the transport of ions via Lewis acid-base interactions between the O/OH groups on the surface of the filler and the ionic species. Dissanayake et al. [39] and Jayathilaka et al. [40] have demonstrated this concept through the conductivity enhancement of the electrolytes with the addition of fillers.

3. Transference number measurement (TNM)

Transference number is a parameter indicating as to how much an ionic species contribute to the overall conductivity of the electrolyte. Knowing the transference number, one can tell whether the major contributor towards the conductivity is the cation or anion. It can be expected and proven that chitosan-based polymer electrolytes are ionic conductors. Osman et al. [22] reported that the conductivity of a chitosan electrolyte comprising $LiCF_3SO_3$ as the ion source and ethylene carbonate (EC) as the plasticizer was in the order of 10^{-3} S m^{-1}. The ionic transference number obtained was 0.9. Aziz et al. [31] also measured the TN of PhCh-NH_4SCN.

Although these electrolytes are ionic conductors, it is vital to know, for some applications, whether the major contributor to the overall conductivity is the cation or anion. Work on this aspect for Ch- and PhCh-based electrolytes is scarce although there are some reports on the subject [41]. For battery application, it is appreciated that these electrolytes be major cationic conductors and for dye-sensitized solar cells (DSCCs) with I^-/I^{3-} redox mediators, I^- transport is more important than the cation transport.

4. Application in electrochemical devices

Ch and PhCh have been used to host ionic conductivity in batteries, supercapacitors, dye-sensitized photovoltaics and fuel cells. Critical review on Ch as integrative biomaterial for microdevices has been done by Koev et al. [42]. Here, we discuss application of the Ch and PhCh materials in the electrochemical devices.

4.1. Chitosan in batteries

Some battery characteristics fabricated with Ch-based electrolytes are listed in **Table 2**.

Jia et al. [46] developed a Ch electrolyte that was incorporated with a biocompatible choline nitrate $[Chl][NO_3]$ ionic liquid that possessed negligible vapour pressure, low viscosity and flammability as well as high ionic conductivity and electrochemical stability. They have demonstrated a compact bio-battery system with the use of this thin-film gel chitosan-choline nitrate electrolyte. A magnesium alloy anode and a polypyrrole-para(toluene sulfonic acid) cathode were used. Some characteristics of the cell are listed in **Table 3**. It was reported that the gel electrolyte was also mechanically robust.

Electrolyte	σ (S m^{-1})	V_{oc} (V)	Q	P	References
Ch:NH_4NO_3:EC* 18:12:70 (wt.%)	0.99	1.56	17 mA h	8.70 mW cm^{-2}	[43]
PVA:Ch:NH_4NO_3:EC* 10.8:7.2:12:70 (wt.%)	0.16	1.64	38 mA h	9.47 mW cm^{-2}	[44]
Ch:choline nitrate#	10^{-2} to 1 depends on choline nitrate content	1.80	1.76 mA h	3.9 mW cm^{-3}	[45]
Ch-SN-LiTSFI	6.00×10^{-2}	4.2	142 m Ah g^{-1}	468 mWhg^{-1}	[33]

*Zn+$ZnSO_4.7H_2O$//MnO_2
#Mg//Ppy: cell area assumed 1.1 cm^2.

Table 2. Battery characteristics, electrolyte conductivity, open circuit voltage (V_{oc}), capacity (Q) and power density (P).

Material	Conductivity (S m^{-1})	Selectivity index (S s m^{-3})	Methanol permeability (m^2 s^{-1})	References
PCh	2	1.25×10^{10}	1.62×10^{-11}	[57]
Ch/Cs_2-PTA-5 wt.%	0.6	1.1×10^{10}	5.6×10^{-11}	[49]
Ch flakes	–	–	3.1×10^{-10} (20°C)	[58]
Ch/15 wt.% mordenite/30 wt.% sorbitol/60	–	–	4.9×10^{-11}	[59]
Ch-40 wt.% HO_3SY	2.07	2.3×10^{10}	9.04×10^{-11}	[60]
Ch-zeolite	1.75		1×10^{-10}	[61]
Ch/PMA	1.5	5.6×10^{10}	2.7×10^{-11}	[62]
GS-Ch/PVP	2.4		7.3×10^{-12}	[63]
CGS-12/Nafion 112	7.5	~2×10^{11} (25°C)	~4×10^{-11}	[64]
Ch/SHNT	1.88 (25°C)	2×10^{10}	9.2×10^{-11}	[65]
Ch/sodium alginate (3:1)	4.2	–	4.6×10^{-12}	[66]

Table 3. Material characteristics for DMFC.

Apart for application as host for ionic conductors, chitosan was also used as a binder material for battery cathodes. Prasanna et al. [47] have investigated the potential of Ch as a binder material for the fabrication of cathode in lithium ion batteries. The cathode-active material was $LiFePO_4$. The cathode fabricated with chitosan binder showed a high ionic conductivity compared to the $LiFePO_4$ cathode with poly(vinylidene difluoride) binder. The lithium ion cell fabricated with the cathode using chitosan binder also exhibited a discharge capacity higher than the cell with poly(vinylidene difluoride) binder by ~32 mAh g^{-1}. The capacity retention after 30 cycles for the cell with chitosan binder was ~98.4% and that of the cell with PVDF binder was ~85%.

4.2. Chitosan in electrical double-layer capacitors (EDLCs)

Arof et al. [48] have prepared an electrolyte for electrical double-layer capacitor (EDLC) study. The electrolyte comprised a chitosan/iota (i)-carrageenan-blended polymer with H_3PO_4 as the proton source and plasticized with poly(ethylene glycol) (PEG). The highest conducting sample has conductivity of 6.29×10^{-2} S m^{-1} at room temperature. This electrolyte contained equal amounts of chitosan and i-carrageenan (37.5% by weight), 18.75 wt.% H_3PO_4 and 6.25 wt.% PEG. This electrolyte was used as separator cum electrolyte in an EDLC. The discharge showed stability for 30 cycles.

Apart from being used to host ionic conduction, chitosan has been used to produce activated carbon (AC) [49]. The AC has high specific surface area of $\approx$ 3500 m^2 g^{-1}. This is higher than the surface area of AC derived from durian shell [45]. The EDLC fabricated with the chitosan-based AC exhibited a capacitance of 338 F g^{-1} at 2 mV s^{-1} scan rate. The charge-discharge curves showed an inverted 'V' shape indicating excellent EDLC performance.

Izabela Stepniak et al. [7] have prepared a Ch/chitin-based membrane for use in an EDLC. To enable the membrane to host Li^+ ion conduction, the membrane was soaked in lithium acetate (LiOAc) solution. The researchers also fabricated an EDLC with Ch-LiOAc electrolyte for comparison. The specific discharge capacitances of the EDLCs with films of Ch/chitin and Ch were 96 and 87 F g^{-1}, respectively. On comparing the first and 10,000th charge/discharge characteristics for the EDLC with the Ch/chitin membrane, it was observed that the device showed excellent capacity retention. The inverted 'V' shape of the charge/discharge curves and the almost 'rectangular' shape of the cyclic voltammogram of the EDLC with Ch/chitin membrane confirm the excellent symbiotic nature among the materials in the cell. From this work, it can be inferred that Ch has potential for use as a host material in EDLC electrolyte. Its performance can be improved by blending with chitin extracted from Ianthella basta sponge.

4.3. Chitosan in polymer electrolyte fuel cells (PEFCs)

Application of chitosan electrolyte in fuel cells is a demanding task. Chitosan is receiving a lot of attention as materials for bioelectrolytes and electrodes [45]. Membrane is the core component of polymer electrolyte fuel cells (PEFCs). The search for low-cost, efficient and stable polymer electrolyte has led researchers to study the chitosan biopolymer electrolyte as alternative candidate for possible production of cheaper fuel cells. Vaghari et al. [51] have written an informative review on the use of chitosan-based electrolytes for fuel cells.

Glutaraldehyde cross-linked N-[(2-hydroxy-3-trimethyl-ammonium) propyl] chitosan chloride was blended with crosslinked quaternized PVA [18]. Quaternized PVA has a low mechanical strength and blending with the chitosan derivative improved their performance.

Majid and Arof [52] have applied chitosan-based electrolytes in PEFCs. The electrolyte systems comprise Ch, H_3PO_4 and Al_2SiO_5. The open-circuit voltage of the PEFCs was 0.9 V and the room temperature current density was greater than 200 A m^{-2}. This again showed that chitosan can be used to host ionic conductivity for fuel cell application.

Wan et al. [53] studied the ionic conductivity of chitosan electrolytes with different molecular weights and degree of deacetylation. They have proposed that chitosan can be used as electrolyte material for alkaline fuel cells.

Wang et al. [54] have incorporated quaternized chitosan with polystyrene. Tensile strength improved and the composite also showed better tolerance to bases. However, ionic conductivity decreased with polystyrene content.

A good direct methanol fuel cell (DMFC) should be fabricated with a polymer electrolyte membrane that only allows a low methanol permeability. Nafion membrane allows a high methanol crossover. Chitosan with desirable proton conductivity has been used in DMFCs [55]. The chitosan electrolyte was added with salt and plasticized for conductivity enhancement [56]. However, excessive water uptake can make the membranes fragile and less robust for fuel cell application. **Table 3** lists some materials based on chitosan for DMFC.

In **Table 3**, PCh-G1h is a blend of poly(vinyl alcohol) and chitosan with PVA/Ch weight ratio of 90/10 cross-linked in glutaraldehyde for 1 h [57]. Cs2-PTA is Cs2HPW12O40. Ch/Cs2-PTA-5 wt.% is chitosan doped with 5 wt.% caesium phototungstate salt [49]. Ch flakes [58] were prepared from chitin. The power density of the DMFC with the chitosan flakes is 27.78 W m^{-2}. The Ch/15 wt.% mordenite/30 wt.% sorbitol/60 was prepared at 60°C [59]. The hybrid membrane in [60] consisted of Ch and $-SO_3H$-modified zeolite. Data shown are for 2.0 mol L^{-1} methanol concentration. Methanol permeability was observed to increase with the decrease in zeolite size for the zeolite-filled chitosan membranes as reported in [61]. Ch/PMA [62] was prepared by adding phosphomolybdic acid chitosan, both in solution form. The methanol permeability is about one order of magnitude less than that of Nafion 117. Ch was blended in poly(vinyl pyrrolidone) or PVP (Ch:PVP of 4:1). The polymer blend was cross-linked with glutaraldehyde and sulphuric acid to form GS-Ch/PVP [63]. ChGS-12 is a structurally modified chitosan membrane that was developed by the authors [64]. In ChGS-12/Nafion 112 membrane, two layers of ChGS-12 were coated on Nafion 112 in order to synergize the low methanol permeability of ChGS-12 with the higher proton conductivity of Nafion 112. ChGS-12 is chitosan cross-linked with glutaraldehyde and sulfosuccinic acid. The maximum output power density of the DMFC was 662.5 W m^{-2}. SHNT [65] are halloysite nanotubes bearing sulphonate polyelectrolyte brushes. SHNT was incorporated in Ch matrix. The proton conductivity of CS/SHNT increases with SHNT content.

4.4. Chitosan in dye-sensitized solar cells (DSSCs)

Some examples of dye-sensitized solar cells using chitosan-based electrolytes are listed in **Table 4**.

EMImSCN is 1-ethyl 3-methylimidazolium thiocyanate ionic liquid. It has low viscosity. DSSC with 1-butyl-3-methylimidazolium iodide (BMII) used anthocyanin dye as the sensitizer.

As in studies on lithium ion batteries, chitosan was also used as a binder in TiO_2 photoelectrode of the DSSCs. The chitosan-based TiO_2 paste was prepared by mixing nano-TiO_2 particles in a chitosan colloidal solution [71]. The dye used was N719 or $Ru(dcbpy)_2(NCS)_2$ and the concentration was 0.5 mmol L^{-1} ethanol. The electrolyte comprised 1,2-dimethyl-3-propylimidazolium iodide with the I^-/I^{3-} redox mediator. The DSSC also contained 2.0% by weight of chitosan in the photoanode and exhibited the highest photon conversion efficiency of 4.16%. The Voc, Jsc and FF were 0.69 V, 10.15 mA cm^{-2} and 0.59, respectively. The amount of chitosan in the photoanode influences the efficiency.

Electrolyte	Dye	J_{sc} (mA cm^{-2})	V_{oc} (V)	η (%)	References
chitosan:NaI/I_2	–	1.05	0.35	0.13	[67]
chitosan:NaI/I_2 +150 wt.% EMImSCN	–	2.62	0.53	0.73	[67]
11 wt.% chitosan-9 wt.% NH_4I-80 wt.% BMII	ABR	0.90	0.36	0.15	[68]
11 wt.% chitosan-9 wt.% NH_4I-80 wt.% BMII	ARC*1	1.59	0.45	0.29	[68]
11 wt.% chitosan-9 wt.% NH_4I-80 wt.% BMII	ARC*2	2.09	0.62	0.38	[68]
11 wt.% (chitosan:PEO, wt. ratio 30:70)-9 wt.% NH_4I-80 wt. % BMII	ARC*2	2.52	0.40	0.39	[68]
11 wt.% phthaloyl chitosan-9 wt.% NH4I-80 wt.% BMII	ARC*2	3.47	0.36	0.43	[68]
11 wt.% (phthaloyl chitosan:PEO, wt. ratio 30:70)-9 wt.% NH_4I-80 wt. % BMII	ARC*2	3.50	0.34	0.46	[68]
12.02 wt.% phthaloyl chitosan-36.06 wt.% EC-36.06 wt.% DMF-14.42 wt.% TPAI-1.44 wt.% I_2	N3	12.72	0.60	5.00	[30]
PhCh:EC:PC:TPAI: LiI:I_2 15.82:31.65:31:65: 18.99:0.00:1.90	N719	7.38	0.72	3.50	[69]
PhCh:EC:PC:TPAI:LiI:I_2 15.82:31.65:31.65:15.82:3.16:1.90	N719	6.33	0.80	3.61	[69]
PhCh:EC:PC:TPAI:LiI:I_2 15.82:31.65:31.65:12.66:6.33:1.90	N719	7.25	0.77	3.71	[69]
PhCh:EC:PC:TPAI:LiI:I_2 15.82:31.65:31.65:9.49:9.49:1.90	N719	3.64	0.75	2.04	[69]
PhCh:EC:PC:TPAI:LiI:I_2 15.82:31.65:31.65:6.33:12.66:1.90	N719	3.64	0.70	1.77	[69]
disulphide/thiolate	MP	4.72	0.57	1.47	[70]
disulphide/thiolate	MP	3.88	0.58	0.60	[70]
disulphide/thiolate	MP	8.70	0.60	2.63	[70]
I_2/NaI	MP	5.40	0.62	1.75	[70]
I_2/NaI	MP	4.33	0.58	0.88	[70]

ABR: anthocyanin from black rice; ARC: anthocyanin from red cabbage; MP: mangosteen peel.
*1: in hydrochloric acid
*2: in tartaric acid

Table 4. Some characteristics of DSSCs with chitosan-based electrolytes.

Maiaugree et al. [70] studied DSSCs using an organic disulphide/thiolate solution as the electrolyte. The advantages of this electrolyte are its high transmittance and low corrosiveness. These authors have compared counter-electrodes of Pt, PEDOT-PSS and mangosteen peel carbon. DSSCs were also fabricated using these CEs, but liquid I2/NaI electrolyte.

5. Summary

As a summary, chitosan and its derivatives have potential for use as electrolytes or binders or as a source for activated carbon. We have given some examples of such uses in this article. Chitosan is a material useful for green technology.

Author details

Zurina Osman* and Abdul Kariem Arof

*Address all correspondence to: zurinaosman@um.edu.my

Centre for Ionics University of Malaya, Physics Department, University of Malaya, Kuala Lumpur, Malaysia

References

[1] Muzzarelli RAA. Natural Chelating Polymers. Pergamon Press: New York, USA; 1973.

[2] Bolker HI. Natural and Synthetic Polymers: An Introduction. Marcel Dekker Incorporated: New York, USA; 1974.

[3] Kurita K. Controlled functionalization of the polysaccharide chitin. Prog Polym Sci 2001;26:1921–71. doi:10.1016/S0079-6700(01)00007-7.

[4] Demarger-Andre S, Domard A. Chitosan carboxylic acid salts in solution and in the solid state. Carbohydr Polym 1994;23:211–9. doi:10.1016/0144-8617(94)90104-X.

[5] Peng Z, Li Z, Shen Y, Zhang F, Peng X. Fabrication of gelatin/chitosan microspheres with different morphologies and study on biological properties. Polym Plast Technol Eng 2012;51:739–43. doi:10.1080/03602559.2012.663037.

[6] Liu Y, An M, Qiu HX, Wang L. The properties of chitosan-gelatin scaffolds by once or twice vacuum freeze-drying methods. Polym Plast Technol Eng 2013;52:1154–9. doi: 10.1080/03602559.2013.798818.

[7] Stepniak I, Galinski M, Nowacki K, Wysokowski M, Jakubowska P, Bazhenov V V., et al. A novel chitosan/sponge chitin origin material as a membrane for supercapacitors – preparation and characterization. RSC Adv 2016;6:4007–13. doi:10.1039/C5RA22047E.

[8] Northcott KA, Snape I, Scales PJ, Stevens GW. Dewatering behaviour of water treatment sludges associated with contaminated site remediation in Antarctica. Chem Eng Sci 2005;60:6835–43. doi:10.1016/j.ces.2005.05.049.

[9] Crini G. Recent developments in polysaccharide-based materials used as adsorbents in wastewater treatment. Prog Polym Sci 2005;30:38–70. doi:10.1016/j.progpolymsci.2004.11.002.

[10] Berger J, Reist M, Mayer JM, Felt O, Gurny R. Structure and interactions in chitosan hydrogels formed by complexation or aggregation for biomedical applications. Eur J Pharm Biopharm 2004;57:35–52. doi:10.1016/S0939-6411(03)00160-7.

[11] Ng LT, Swami S. IPNs based on chitosan with NVP and NVP/HEMA synthesised through photoinitiator-free photopolymerisation technique for biomedical applications. Carbohydr Polym 2005;60:523–8. doi:10.1016/j.carbpol.2005.03.009.

[12] Chung YC, Wang HL, Chen YM, Li SL. Effect of abiotic factors on the antibacterial activity of chitosan against waterborne pathogens. Bioresour Technol 2003;88:179–84. doi:10.1016/S0960-8524(03)00002-6.

[13] Suntornsuk W, Pochanavanich P, Suntornsuk L. Fungal chitosan production on food processing by-products. Process Biochem 2002;37:727–9. doi:10.1016/S0032-9592(01)00265-5.

[14] Ham-Pichavant F, Sèbe G, Pardon P, Coma V. Fat resistance properties of chitosan-based paper packaging for food applications. Carbohydr Polym 2005;61:259–65. doi:10.1016/j.carbpol.2005.01.020.

[15] Devlieghere F, Vermeulen A, Debevere J. Chitosan: Antimicrobial activity, interactions with food components and applicability as a coating on fruit and vegetables. Food Microbiol 2004;21:703–14. doi:10.1016/j.fm.2004.02.008.

[16] Sun L, Du Y, Yang J, Shi X, Li J, Wang X, et al. Conversion of crystal structure of the chitin to facilitate preparation of a 6-carboxychitin with moisture absorption-retention abilities. Carbohydr Polym 2006;66:168–75. doi:10.1016/j.carbpol.2006.02.036.

[17] Rinaudo M. Chitin and chitosan: properties and applications. Prog Polym Sci 2006;31:603–32. doi:10.1016/j.progpolymsci.2006.06.001.

[18] Xiong Y, Liu QL, Zhang QG, Zhu AM. Synthesis and characterization of cross-linked quaternized poly(vinyl alcohol)/chitosan composite anion exchange membranes for fuel cells. J Power Sources 2008;183:447–53. doi:10.1016/j.jpowsour.2008.06.004.

[19] Agrawal P, Strijkers GJ, Nicolay K. Chitosan-based systems for molecular imaging. Adv Drug Deliv Rev 2010;62:42–58. doi:10.1016/j.addr.2009.09.007.

[20] Osman Z, Arof AK. FTIR studies of chitosan acetate based polymer electrolytes. Electrochim Acta 2003;48:993–9. doi:10.1016/S0013-4686(02)00812-5.

[21] Ritthidej GC, Phaechamud T, Koizumi T. Moist heat treatment on physicochemical change of chitosan salt films. Int J Pharm 2002;232:11–22. doi:10.1016/S0378-5173(01)00894-8.

[22] Osman Z, Ibrahim ZA, Arof AK. Conductivity enhancement due to ion dissociation in plasticized chitosan based polymer electrolytes. Carbohydr Polym 2001;44:167–73. doi:10.1016/S0144-8617(00)00236-8.

[23] Muzzareli RAA, Ferrero A, Pizzoli M. Light-scattering, X-ray diffraction, elemental analysis and infrared spectro- photometry characterization of chitosan, a chelating polymer. Talanta 1972;19:1222–6. doi:http://dx.doi.org/10.1016/0039-9140(72)80068-7.

[24] Sakurai K, Maegawa T. Chitin and Chitosan. Second Asia Pacific Symp. Bangkok, Thail., 1996, p. 224–7.

[25] Sakurai K, Maegawa T, Takahashi T. Glass transition temperature of chitosan and miscibility of\rchitosan/poly(N-vinyl pyrrolidone) blends. Polymer 2000;41:7051–6.

[26] Alves R, Donoso JP, Magon CJ, Silva IDA, Pawlicka A , Silva MM. Solid polymer electrolytes based on chitosan and europium triflate. J Non Cryst Solids 2015;432:1–6. doi:10.1016/j.jnoncrysol.2015.10.024.

[27] Kurita K, Ikeda H, Yoshida Y, Shimojoh M, Harata M. Chemoselective protection of the amino groups of chitosan by controlled phthaloylation: facile preparation of a precursor useful for chemical modifications. Biomacromolecules 2002;3:1–4. doi:10.1021/bm0101163.

[28] Muzzarelli RAA. Chitin. Pergamon Press: Oxford, UK; 1977.

[29] Dong Y, Ruan Y, Wang H, Zhao Y, Bi D. Studies on glass transition temperature of chitosan with four techniques. J Appl Polym Sci 2004;93:1553–8. doi:10.1002/app.20630.

[30] Yusuf SNF, Azzahari -AD, Yahya R, Majid SR, Careem MA, Arof AK. From crab shell to solar cell: a gel polymer electrolyte based on N-phthaloylchitosan and its application in dye-sensitized solar cells. RSC Adv 2016;6:27714–24. doi:10.1039/C6RA04188D.

[31] Aziz NA, Majid SR, Arof AK. Synthesis and characterizations of phthaloyl chitosan-based polymer electrolytes. J Non Cryst Solids 2012;358:1581–90. doi:10.1016/j.jnoncrysol.2012.04.019.

[32] Arof AK, Osman Z, Morni N, Kamarulzaman N, Ibrahim ZA, Muhamad MR. Chitosan-based electrolyte for secondary lithium cells. J Mater Sci 2001;36:791–3.

[33] Taib NU, Idris NH. Plastic crystal-solid biopolymer electrolytes for rechargeable lithium batteries. J Memb Sci 2014;468:149–54. doi:10.1016/j.memsci.2014.06.001.

[34] Croce F, Appetecchi GB, Persi L, Scrosati B. Nanocomposite polymer electrolytes for lithium batteries. Nature 1998;394:456–8. doi:10.1038/28818.

[35] Xiao P, Deng ZQ, Manthiram A, Henkelman G. Calculations of oxygen stability in lithium-rich layered cathodes. J Phys Chem C 2012;116:23201–4. doi:10.1021/jp3058788.

[36] Muthumeenal A, Neelakandan S, Kanagaraj P, Nagendran A. Synthesis and properties of novel proton exchange membranes based on sulfonated polyethersulfone and N-phthaloyl chitosan blends for DMFC applications. Renew Energy 2016;86:922–9. doi: 10.1016/j.renene.2015.09.018.

[37] Azzahari A, Yusuf S, Selvanathan V, Yahya R. Artificial neural network and response surface methodology modeling in ionic conductivity predictions of phthaloylchitosan-based gel polymer electrolyte. Polymers 2016;8:22. doi:10.3390/polym8020022.

[38] Karuppasamy K, Antony R, Thanikaikarasan S, Balakumar S, Shajan XS. Combined effect of nanochitosan and succinonitrile on structural, mechanical, thermal, and electrochemical properties of plasticized nanocomposite polymer electrolytes (PNCPE) for lithium batteries. Ionics 2013;19:747–55. doi:10.1007/s11581-012-0806-9.

[39] Dissanayake MAKL, Bandara LRA K, Bokalawala RSP, Jayathilaka PARD, Ileperuma OA, Somasundaram S. A novel gel polymer electrolyte based on polyacrylonitrile (PAN) and its application in a solar cell. Mater Res Bull 2002;37:867–74. doi:10.1016/S0025-5408(02)00712-2.

[40] Jayathilaka PARD, Dissanayake MAKL, Albinsson I, Mellander BE. Effect of nano-porous Al_2O_3 on thermal, dielectric and transport properties of the (PEO)9LiTFSI polymer electrolyte system. Electrochim Acta 2002;47:3257–68. doi:10.1016/S0013-4686(02)00243-8.

[41] Chai MN, Isa MIN. Novel proton conducting solid bio-polymer electrolytes based on carboxymethyl cellulose doped with oleic acid and plasticized with glycerol. Sci Rep 2016;6:27328. doi:10.1038/srep27328.

[42] Koev ST, Dykstra PH, Luo X, Rubloff GW, Bentley WE, Payne GF, et al. Chitosan: an integrative biomaterial for lab-on-a-chip devices. Lab Chip 2010;10:3026–42. doi: 10.1039/c0lc00047g.

[43] Ng LS, Mohamad AA. Protonic battery based on a plasticized chitosan-NH4NO3 solid polymer electrolyte. J Power Sources 2006;163:382–5. doi:10.1016/j.jpowsour.2006.09.042.

[44] Kadir MFZ, Majid SR, Arof AK. Plasticized chitosan–PVA blend polymer electrolyte based proton battery. Electrochim Acta 2010;55:1475–82. doi:10.1016/j.electacta.2009.05.011.

[45] Jia X, Wang C, Zhao C. Biocompatible ionic liquid-biopolymer electrolyte enabled thin an. ACS Appl Mater Interfaces 2013;6:21110–7.

[46] Ma J, Sahai Y. Chitosan biopolymer for fuel cell applications. Carbohydr Polym 2013;92:955–75. doi:10.1016/j.carbpol.2012.10.015.

[47] Prasanna K, Subburaj T, Jo YN, Lee WJ, Lee CW. Environment-friendly cathodes using biopolymer chitosan with enhanced electrochemical behavior for use in lithium ion batteries. ACS Appl Mater Interfaces 2015;7:7884–90. doi:10.1021/am5084094.

[48] Arof AK, Shuhaimi NEA, Alias NA, Kufian MZ, Majid SR. Application of chitosan/iota-carrageenan polymer electrolytes in electrical double layer capacitor (EDLC). J Solid State Electrochem 2010;14:2145–52. doi:10.1007/s10008-010-1050-8.

[49] Hu Y, Wang H, Yang L, Liu X, Zhang B, Liu Y, et al. Preparation of chitosan-based activated carbon and its electrochemical performance for EDLC. J Electrochem Soc 2013;160:H321–6. doi:10.1149/2.062306jes.

[50] Tey JP, Careem MA, Yarmo MA, Arof AK. Durian shell-based activated carbon electrode for EDLCs. Ionics 2016:1–8. doi:10.1007/s11581-016-1640-2.

[51] Vaghari H, Jafarizadeh-Malmiri H, Berenjian A, Anarjan N. Recent advances in application of chitosan in fuel cells. Sustain Chem Process 2013;1:16. doi: 10.1186/2043-7129-1-16.

[52] Majid SR, Arof AK. FTIR studies of chitosan-orthophosphoric polymer electrolyte 2008;484:483–92. doi:10.1080/15421400801904286.

[53] Wan Y, Creber KAM, Peppley B, Bui VT. Ionic conductivity of chitosan membranes. Polymer 2003;44:1057–65. doi:10.1016/S0032-3861(02)00881-9.

[54] Wang J, He R, Che Q. Anion exchange membranes based on semi-interpenetrating polymer network of quaternized chitosan and polystyrene. J Colloid Interface Sci 2011;361:219–25. doi:10.1016/j.jcis.2011.05.039.

[55] Ramirez-Salgado J. Study of basic biopolymer as proton membrane for fuel cell systems. Electrochim Acta 2007;52:3766–78. doi:10.1016/j.electacta.2006.10.051.

[56] Mukoma P, Jooste BR, Vosloo HCM. Synthesis and characterization of cross-linked chitosan membranes for application as alternative proton exchange membrane materials in fuel cells. J Power Sources 2004;136:16–23. doi:10.1016/j.jpowsour.2004.05.027.

[57] Ming Yang J, Chih Chiu H. Preparation and characterization of polyvinyl alcohol/chitosan blended membrane for alkaline direct methanol fuel cells. J Memb Sci 2012;419–420:65–71. doi:10.1016/j.memsci.2012.06.051.

[58] Osifo PO, Masala A. Characterization of direct methanol fuel cell (DMFC) applications with H2SO4 modified chitosan membrane. J Power Sources 2010;195:4915–22. doi: 10.1016/j.jpowsour.2009.12.093.

[59] Yuan W, Wu H, Zheng B, Zheng X, Jiang Z, Hao X, et al. Sorbitol-plasticized chitosan/zeolite hybrid membrane for direct methanol fuel cell. J Power Sources 2007;172:604–12. doi:10.1016/j.jpowsour.2007.05.040.

[60] Wu H, Zheng B, Zheng X, Wang J, Yuan W, Jiang Z. Surface-modified Y zeolite-filled chitosan membrane for direct methanol fuel cell. J Power Sources 2007;173:842–52. doi: 10.1016/j.jpowsour.2007.08.020.

[61] Wang J, Zheng X, Wu H, Zheng B, Jiang Z, Hao X, et al. Effect of zeolites on chitosan/zeolite hybrid membranes for direct methanol fuel cell. J Power Sources 2008;178:9–19. doi:10.1016/j.jpowsour.2007.12.063.

[62] Cui Z, Xing W, Liu C, Liao J, Zhang H. Chitosan/heteropolyacid composite membranes for direct methanol fuel cell. J Power Sources 2009;188:24–9. doi:10.1016/j.jpowsour.2008.11.108.

[63] Smitha B, Sridhar S, Khan AA. Chitosan-poly(vinyl pyrrolidone) blends as membranes for direct methanol fuel cell applications. J Power Sources 2006;159:846–54. doi:10.1016/j.jpowsour.2005.12.032.

[64] Hasani-Sadrabadi MM, Dashtimoghadam E, Majedi FS, Hojjati Emami S, Moaddel H. A high-performance chitosan-based double layer proton exchange membrane with reduced methanol crossover. Int J Hydrogen Energy 2011;36:6105–11. doi:10.1016/j.ijhydene.2011.01.010.

[65] Bai H, Zhang H, He Y, Liu J, Zhang B, Wang J. Enhanced proton conduction of chitosan membrane enabled by halloysite nanotubes bearing sulfonate polyelectrolyte brushes. J Memb Sci 2014;454:220–32. doi:10.1016/j.memsci.2013.12.005.

[66] Smitha B, Sridhar S, Khan AA. Chitosan-sodium alginate polyion complexes as fuel cell membranes. Eur Polym J 2005;41:1859–66. doi:10.1016/j.eurpolymj.2005.02.018.

[67] Singh PK, Bhattacharya B, Nagarale RK, Kim K-W, Rhee H-W. Synthesis, characterization and application of biopolymer-ionic liquid composite membranes. Synth Met 2010;160:139–42. doi:10.1016/j.synthmet.2009.10.021.

[68] Buraidah MH, Teo LP, Yusuf SNF, Noor MM, Kufian MZ, Careem MA, et al. TiO_2/chitosan-$NH_4I(+I_2)$-BMII-based dye-sensitized solar cells with anthocyanin dyes extracted from black rice and red cabbage. Int J Photoenergy 2011;2011:11 p. doi: 10.1155/2011/273683.

[69] Yusuf SNF, Aziz MF, Hassan HC, Bandara TMWJ, Mellander B, Careem MA, et al. Phthaloylchitosan-based gel polymer electrolytes for efficient dye-sensitized solar cells. J Chem 2014;2014:8 p.

[70] Maiaugree W, Lowpa S, Towannang M, Rutphonsan P, Tangtrakarn A, Pimanpang S, et al. A dye sensitized solar cell using natural counter electrode and natural dye derived from mangosteen peel waste. Sci Rep 2015;5:15230. doi:10.1038/srep15230.

[71] Jin EM, Park K, Park J, Lee J, Yim S, Zhao XG, et al. Preparation and characterization of chitosan binder-based TiO_2 electrode for dye-sensitized solar cells. Int J Photoenerey 2013;2013:1–8.